# Lecture Notes in Mathematics 2133

More information about this series at http://www.springer.com/series/304

Farrukh Mukhamedov • Nasir Ganikhodjaev

# Quantum Quadratic Operators and Processes

 Springer

Farrukh Mukhamedov
Dept. of Comput. & Theor. Sciences
International Islamic University Malaysia
Kuantan, Malaysia

Nasir Ganikhodjaev
Dept. of Comput. & Theor. Sciences
International Islamic University Malaysia
Kuantan, Malaysia

ISSN 0075-8434                    ISSN 1617-9692    (electronic)
Lecture Notes in Mathematics
ISBN 978-3-319-22836-5           ISBN 978-3-319-22837-2    (eBook)
DOI 10.1007/978-3-319-22837-2

Library of Congress Control Number: 2015952068

Mathematics Subject Classification (2010): 37A50; 47D07; 37A30; 37A55; 46L53; 81P16; 0G07,60G99, 81S25; 60J28, 81R15, 35Q92, 37N25

Springer Cham Heidelberg New York Dordrecht London

Printed on acid-free paper

Springer International Publishing AG Switzerland is part of Springer Science+Business Media (www.springer.com)

*To our families*

# Preface

Nonlinear mappings appear throughout mathematics, and their range of applications is immense, including the theory of differential equations, the theory of probability, the theory of dynamical systems, mathematical biology, and statistical physics. Most of the simplest nonlinear operators are quadratic. Even in a one-dimensional setting, the behavior of such operators reveals their complicated structure. If one considers multidimensional analogues of quadratic operators, then the situation becomes more complicated, i.e., the investigation of the dynamical behavior of such operators is very difficult.

The history of quadratic stochastic operators and their dynamics can be traced back to Bernstein's work [18]. The continuous time dynamics of this type of operator was considered by Lotka [134] and Volterra [252]. Quadratic stochastic operators are an important source of analysis in the study of dynamical properties and for modeling in various fields such as mathematical economics, evolutionary biology, population and disease dynamics, and the dynamics of economic and social systems.

Unfortunately, up to now, there have been no books devoted to the dynamics of quadratic stochastic operators. This omission in the literature gave us the motivation to write a systematic book about such operators.

The general objectives of this book are: (i) to give the first systematic presentation of both analytical and probabilistic techniques used in the study of the dynamics of quadratic stochastic operators and corresponding processes; (ii) to establish a connection between the dynamics of quadratic stochastic operators with the theory of Markov processes; and (iii) to give a systematic introduction to noncommutative or quantum analogues of quadratic stochastic operators and processes.

The book addresses the most fundamental questions in the theory of quadratic stochastic operators: dynamics, constructions, regularity, and the connection with stochastic processes. This connection means that the dynamics of such operators can be treated as certain Markov or quadratic processes. This interpretation allows us to use the methods of stochastic processes for a better understanding of the limiting behavior of the dynamics of quadratic operators.

Below we provide an overview of the main topics discussed in this book and explain why they have been selected.

The starting point of our book is to introduce a quadratic stochastic operator $V :$ $S(X, F) \to S(X, F)$ defined on the set of all probability measures $S(X, F)$ on $(X, F)$ and to present some motivations to study such operators. The next step is to define and study stochastic processes that are related to the quadratic stochastic operators in the same way as Markov processes are related to linear transformations. After this, it is natural to develop analytic methods for such processes. The last step is to generalize the theory of quadratic stochastic operators and processes to different algebraic structures, including von Neumann algebras. Such quadratic operators are called quantum quadratic stochastic operators (q.q.s.o.s). In this direction, we study the asymptotic properties of dynamical systems generated by q.q.s.o.s. Moreover, we also investigate Markov and quantum quadratic stochastic processes associated with q.q.s.o.s.

An essential feature of our exposition is the first systematic presentation of both the classical and quantum theory of quadratic stochastic operators and processes. We combine analytical and probabilistic tools to get a better insight into the dynamics of both classical and quantum quadratic operators. Moreover, we use several methods from the theory of noncommutative probability, matrix analysis, etc.

Now we discuss the structure of the book in more detail. The book is divided into eight chapters; at the end of each chapter, we give some comments and references related to the chapter.

The first chapter is an introduction where we collect some models, which can be described by quadratic stochastic operators.

Chapter 2 is devoted to quadratic stochastic operators (q.s.o.s) defined on a finite-dimensional simplex. In this chapter, we essentially deal with asymptotical stability (or regularity) condition for such operators. Moreover, we show how the dynamics of q.s.o.s are related to some Markov processes. Some relations between the regularity of a q.s.o. and the corresponding Markov process are investigated.

In Chap. 3, we introduce quadratic stochastic processes (q.s.p.s) and give examples of such processes. Note that these quadratic processes naturally arise in the study of certain models with interactions, where interactions are described by quadratic stochastic operators. Furthermore, this chapter contains a construction of nontrivial examples of q.s.p.s. Given a q.s.p., one can associate two kinds of processes, which are called *marginal processes*. One of them is a Markov process. We prove that marginal processes uniquely define q.s.p.s. The weak ergodicity of q.s.p.s is also studied in terms of the marginal processes.

In Chap. 4, we develop analytical methods for q.s.p.s. We follow the lines of Kolmogorov's [121] paper. Namely, we will derive partial differential equations with delaying argument, for quadratic processes of types A and B, respectively.

In the previous chapters, we are considering classical (i.e., commutative) quadratic operators. These operators are defined over commutative algebras. However, such operators do not cover the case of quantum systems. Therefore, in Chap. 5 we introduce a noncommutative analogue of a q.s.o., which is called a *quantum quadratic stochastic operator* (q.q.s.o.). We show that the set of q.q.s.o.s is weakly compact. By means of q.q.s.o.s, one can define a nonlinear operator, which is called a *quadratic operator*. We also study the asymptotical stability of the dynamics of quadratic operators.

Chapter 6 is devoted to quantum quadratic stochastic operators (q.q.s.o.s) acting on the algebra of $2 \times 2$ matrices $\mathbb{M}_2(\mathbb{C})$. Positive, trace-preserving maps arise naturally in quantum information theory (see, e.g., [199]) and in other situations where one wishes to restrict attention to a quantum system that should properly be considered a subsystem of a larger system which it interacts with. Therefore, we first describe quadratic operators with a Haar state (invariant with respect to the trace). Then q.q.s.o.s with the Kadison–Schwarz property are characterized. By means of such a description, we provide an example of a positive q.q.s.o., which is not a Kadison–Schwarz operator. On the other hand, this characterization is related to a separability condition, which plays an important role in quantum information [17]. We also examine the stability of the dynamics of quadratic operators associated with q.q.s.o.s given on $\mathbb{M}_2(\mathbb{C})$.

In Chap. 7, we investigate a class of q.q.s.o.s defined on the commutative algebra $\ell^\infty$. We define the notion of a Volterra quadratic operator and study its properties. It is proved that such operators have infinitely many fixed points and the set of Volterra operators forms a convex compact set. In addition, its extreme points are described. Furthermore, we study certain limit behaviors of such operators and give some more examples of Volterra operators for which their trajectories do not converge. Finally, we define a compatible sequence of finite-dimensional Volterra operators and prove that any power of this sequence converges in the weak topology. Note that in the finite-dimensional setting such operators have been studied by many authors (see, for example, [74, 252]).

In Chap. 8, we define a quantum (noncommutative) analogue of quadratic stochastic processes. In our case, such a process is defined on a von Neumann algebra. In this chapter, we essentially study the ergodic principle for these processes. From a physical point of view, this principle means that for sufficiently large values of time a system described by the process does not depend on the initial state of the system.

This book is not intended to contain a complete discussion of the theory of quadratic operators, but primarily relates to the asymptotic stability of such operators and associated processes. Moreover, it reflects the interests of the authors in key aspects of this theory. There are many omitted topics that naturally fit into the purview of quadratic operators. However, we have tried to collect the existing references on quadratic stochastic operators. Some of these are discussed in the separate sections entitled "Comments and References."

This book is suitable as a textbook for an advanced undergraduate/graduate level course or summer school in quantum dynamical systems. It can also be used as a reference book by researchers looking for interesting problems to work on, or useful techniques and discussions of particular problems. It also includes the latest developments in the fields of quadratic dynamical systems, Markov processes, and quantum stochastic processes. Researchers at all levels are likely to find the book inspiring and useful.

Kuantan, Malaysia                                                Nasir Ganikhodjaev
Kuantan, Malaysia                                             Farrukh Mukhamedov
May 2015

# Acknowledgements

We want to express our warm thanks to our students (M.S. Saburov, A.F. Embong, N.Z.A. Hamza) and the mathematicians who read drafts and offered useful suggestions and corrections, including L. Accardi, W. Bartoszek, R. Ganikhodzhaev, F. Fidaleo, R.A. Minlos, M. Pulka, U.A. Rozikov, and Ya.G. Sinai.

We would also like to manifest our heartfelt gratitude towards our parents (Nabikhodja Ganikhodjaev, Jorahon Ziyautdinova, Maksut Mukhamedov, Munira Mukhamedova). We would like to show our love towards our motivating wives (Mahsuma Usmanova, Shirin Mukhamedova) for their patience during the process of this time-consuming book. Also, our thanks goes to Farzona Mukhamedova for her help in rectifying the flaws in our work. The authors also acknowledge the MOE grants FRGS14-116-0357 and FRGS14-135-0376.

Finally, the authors are also grateful to the referees for their useful suggestions, which allowed us to improve the presentation.

# Contents

# Chapter 1
# Introduction

Nonlinear (in particular, quadratic) mappings appear in various branches of mathematics and their applications: the theory of differential equations, probability theory, the theory of dynamical systems, mathematical economics, mathematical biology, statistical physics, etc. Usually dynamical systems are studied for discrete time and the corresponding differential equations are reduced to difference (recurrence) equations. Conversely, the recurrence equation

$$u_n = S(u_{n-1})$$

implying the equality

$$u_{n+1} - u_n = S(u_n) - u_n,$$

is replaced by

$$\frac{du}{dt} = S(u) - u. \tag{1.1}$$

The solution of Eq. (1.1) does not always vary with the behavior of the corresponding dynamical system.

The quadratic differential system

$$\dot{x}_i = \sum_{j,k} a^i_{jk} x_j x_k, \quad i = 1, 2, \dots, n \tag{1.2}$$

governs mathematical models for a large interacting population of $n$ constituents, where the numbers $x_i$ represent the fraction of constituents of type $i$, $i = 1, 2, \dots, n$

© Springer International Publishing Switzerland 2015
F. Mukhamedov, N. Ganikhodjaev, *Quantum Quadratic Operators and Processes*,
Lecture Notes in Mathematics 2133, DOI 10.1007/978-3-319-22837-2_1

and satisfy the conservation law $\sum_i x_i = 1$ and the tensors $\{a_{lm}^i\}$ of order $n$ having $n^3$ real constants which satisfy

$$a_{jk}^i = a_{kj}^i, \quad \sum_i a_{jk}^i = 0, \quad a_{jk}^i \geq 0, \quad \text{for all } j \neq i, \ k \neq i.$$

In the following two mathematical models (see [112–114]) we introduce and study quadratic stochastic operators and quadratic stochastic processes, respectively.

**Volterra's Model** Perhaps the best known work on quadratic models is Volterra's treatise on the biological struggle for life [252]. The homogeneous differential equations for these models are distinguished by the form

$$\dot{x}_i = x_i \sum_j c_{ij} x_j, \quad i = 1, 2, \ldots, n$$

where $x_i$ denotes the fraction of species of type $i$ at time $t$ and the $c_{ij}$ are biological constants satisfying $c_{ij} = -c_{ji}$.

**Boltzmann's Model** The following finite analogue of the Boltzmann model (see [24]) for gas dynamics provided the motivation to study quadratic models. In this model molecules of a dilute uniform gas have $n$ possible velocity characteristics which may be changed only through binary collision with other molecules. Consider a uniform gas composed of spherical molecules all of the same radius and mass. Let the velocity space of the molecules, which we assume to be $\mathbb{R}^3$, be partitioned into $n$ regions $R_1, R_2, \ldots, R_n$, i.e. $R_i \bigcap R_j = \emptyset$ for $i \neq j$, and $\bigcup_i R_i = \mathbb{R}^3$.

For each $i = 1, 2, \ldots, n$ let us define

$$x_i(t) = \begin{cases} \text{fraction of molecules which have velocities} \\ \text{lying in region } R_i \text{ at time } t. \end{cases}$$

Making appropriate assumptions, one can assert that

$$\begin{cases} \text{number of } \ell - m \\ \text{collisions / unit time} \end{cases} = \mu_{\ell m} x_\ell x_m$$

for some $\mu_{\ell m} > 0$. Let $p_{\ell m}^i$ be the fraction of $R_\ell$ molecules, that is, the molecules whose velocities belong to $R_\ell$ that collide with $R_m$. Evidently, $(p_{\ell m}^1, p_{\ell m}^2, \ldots, p_{\ell m}^n)$ is a probability vector for each pair $\ell, m$, and hence belongs to the $(n-1)$-dimensional simplex. Moreover,

$$\begin{cases} \text{number of } R_\ell \text{ molecules scattered into region } R_i \\ \text{due to } \ell - m \text{ collisions per unit time} \end{cases} = p_{\ell m}^i \mu_{\ell m} x_\ell x_m$$

from which evidently one finds

$$
\left\{ \begin{array}{l} \text{net change in population of region} \\ \text{due to binary collisions per unit time} \end{array} \right. = \sum_{\ell,m} p^i_{\ell m} \mu_{\ell m} x_\ell x_m - \sum_{\ell,m} p^\ell_{im} \mu_{im} x_\ell x_m
$$

$$
= \sum_{\ell,m} a^i_{\ell m} \mu_{\ell m} x_\ell x_m, \quad i = 1, 2, \ldots, n,
$$

for $a^i_{\ell m} = \frac{1}{2}\big((p^i_{\ell m} + p^i_{m\ell}) - (\delta_{i\ell} + \delta_{im})\big)\mu_{\ell m}$, where $\delta_{ij}$ is the Kronecker delta. Now if we assume that changes in $x_i$ depend only on binary collisions and not on external forces, the effects of the walls of the container, etc., then the fluctuation in population of $x_i$ can be studied by the system (1.2).

**Quadratic stochastic operators** The history of quadratic stochastic operators can be traced back to the work of Bernstein [18]. Quadratic dynamical systems are considered an important source of analysis in the study of dynamical properties and for modeling in various fields, such as population dynamics [34, 43, 44, 99, 100, 117, 133], physics [204, 249], economics [250] and mathematics [99, 117, 137, 139]. Some of the most important findings in the theory of quadratic stochastic operators emerged when Markov processes were employed to describe some physical and biological systems. One such system is given by quadratic stochastic operators related to population genetics [18]. A quadratic stochastic operator (in short q.s.o.) is usually used to describe the time evolution of species in biology, which surfaces as follows: let us consider a population consisting of $m$ species, and $x^0 = (x^0_1, \ldots, x^0_m)$ be a probability distribution of species in the initial generation, and $P_{ij,k}$ be the probability that individual in the $i^{th}$ and $j^{th}$ species interbreed to produce an individual $k$. Then, the probability distribution $x' = (x'_1, \ldots, x'_m)$ of the species in the first generation can be found by the total probability, i.e.

$$
x'_k = \sum_{i,j=1}^m P_{ij,k} x^0_i x^0_j, \quad k = 1, \ldots, m.
$$

This means that the correspondence $x^0 \to x'$ defines a map $V$ called the *evolution operator*. The population evolves by starting from an arbitrary state $x^0$, then passing to the state $x'' = V(V(x^0))$, and so on. Therefore, the states of the population are described by the following dynamical system $x^0, x' = V(x^0), x'' = V(V(x^0)), x''' = V^3(x^0), \ldots$ More precisely, the q.s.o. describes the distribution of the next generation if the current distribution of the generation is given. The fascinating applications of q.s.o.s to population genetics are given in [137].

Note that the heredity coefficients $\{P_{ij,k}\}$ define a binary operation "$\circ$" on $\mathbb{R}^m$ as follows:

$$
\mathbf{x} \circ \mathbf{y} = \left( \sum_{i,j=1}^m P_{ij,k} x_i y_j \right)^m_{k=1},
$$

where $\mathbf{x} = (x_1, \ldots, x_m)$, $\mathbf{y} = (y_1, \ldots, y_m)$. The pair $(\mathbb{R}^m, \circ)$ is called a *genetic algebra*. Several algebraic properties of genetic algebras are investigated in [137, 139, 188]. However, we won't be discussing genetic algebras in this book.

Quadratic Stochastic Processes. The subject of our research is a biological process, i.e. the evolution of a biological system. It should be noted that the ideas and considerations presented here may be used not only in biology, but also in other scientific subjects, in particular, in physics (see Boltzmann's model), chemistry, natural philosophy, etc. The set of distinct types for the given system will be denoted by $E$, and let us suppose that $E$ consists of at least two elements. It is necessary to give a distribution on the set $E$ to know the portion of each type in the given system. Suppose that to an arbitrary pair of types $x, y \in E$ there corresponds (probably randomly) a type $z \in E$, and this correspondence is called an *interaction of elements $x$ and $y$*. For a biological development, the incubation period, i.e. the time for realization of this interaction, is necessary. Let us suppose that this period is equal to 1. This circumstance is one of the most important assumptions in further considerations.

If at the moment $t_0$ the distribution of types is given by a probability measure $m_{t_0}$, then after the interaction of a pair of types $x$ and $y$ at the moment $t_0$, we obtain another distribution on $E$ at the moment $t > t_0 + 1$. It naturally depends on $m_{t_0}, x, y$ and the interactions of $x$ and $y$. Thus we say that the biological process is defined if we are given an initial distribution $m_{t_0}$ and the law of interaction for any pair of elements in $E$.

The Markov process is defined by a family of transition probabilities $P(s, x, t, A)$ which satisfy the Kolmogorov–Chapman equations. The theory of Markov processes and its generalizations to different algebraic structures, including von Neumann algebras, are well developed [39, 107, 129, 202].

The field of interacting particle systems is a large and growing section of probability theory that is devoted to the rigorous analysis of certain types of models which arise in biology, chemistry, economics, statistical physics, and other fields. In this book we consider models that are described by a family of functions $P(s, x, y, t, A)$ defined as follows: if particles $x$ and $y$ interact at time $s$, then with probability $P(s, x, y, t, A)$ one of the elements of the set $A$ will be realized at time $t$. Note that the Markov process defined by a family of transition functions $\{P(s, x, t, A)\}$ is a special case of the process defined by a family of functions $P(s, x, y, t, A)$ with $P(s, x, y, t, A) = [P(s, x, t, A) + P(s, y, t, A)]/2$.

The starting point of our book is to introduce a quadratic stochastic operator $V : S(X, F) \to S(X, F)$ defined on the set of all probability measures $S(X, F)$ on $(X, F)$ and to present some motivations to study such operators. The next step is to define and to study stochastic processes that are related to nonlinear transformations, namely, quadratic stochastic operators, in the same way as Markov processes are related to linear transformations. After this, it is natural to develop analytic methods for such processes. The last step is to generalize the theory of quadratic stochastic processes to different algebraic structures, including von Neumann algebras. In this direction, we study the asymptotic properties of dynamical systems generated by quadratic stochastic operators. We also investigate their connections with related Markov and quadratic stochastic processes. Furthermore, quantum analogues of these kinds of operators and processes will also be discussed.

.

# Chapter 2
# Quadratic Stochastic Operators and Their Dynamics

In this chapter we introduce quadratic stochastic operators defined on a simplex. We study the asymptotic stability of dynamical systems generated by quadratic stochastic operators. Moreover, we provide a stability criterion in terms of an associated nonhomogeneous discrete Markov process.

## 2.1 Quadratic Stochastic Operators

In what follows, we consider $\mathfrak{X} := \mathbb{R}^d$ equipped with the norm $\|\mathbf{x}\|_1 = \sum_{j=1}^{d} |x_j|$ ($\mathbf{x} = (x_j)$). By $S^{d-1}$ we denote the set of all probability vectors (distributions) i.e.

$$S^{d-1} = \{\mathbf{x} \in \mathbb{R}^d : x_j \geq 0, \sum_{j=1}^{d} x_j = 1\}.$$

Clearly, $S^{d-1}$ is a norm closed convex subset of $\mathfrak{X}$. In particular, $S^{d-1}$ is a simplex, i.e. any $\mathbf{x} \in S^{d-1}$ is written as follows $\mathbf{x} = \sum_{j=1}^{d} x_j \mathbf{e}_j$, where $\{\mathbf{e}_j\}_{j=1}^{d}$ is the standard basis of $\mathfrak{X}$. The support of $\mathbf{x} \in \mathfrak{X}$ is defined to be the set $\mathrm{supp}(\mathbf{x}) = \{j : x_j \neq 0\}$. For any subset $A \subseteq \{1, 2, \ldots, d\}$ we mean $\mathbf{x}(A) = \sum_{j \in A} x_j$.

The transformation $V : S^{d-1} \to S^{d-1}$ is called a *quadratic stochastic operator* (*q.s.o.*) if

$$V : (V\mathbf{x})_k = \sum_{i,j=1}^{d} p_{ij,k} x_i x_j, \quad (k = 1, \cdots, d), \tag{2.1}$$

© Springer International Publishing Switzerland 2015
F. Mukhamedov, N. Ganikhodjaev, *Quantum Quadratic Operators and Processes*,
Lecture Notes in Mathematics 2133, DOI 10.1007/978-3-319-22837-2_2

where

$$p_{ij,k} \geq 0; \quad p_{ij,k} = p_{ji,k}; \quad \text{and} \quad \sum_{k=1}^{d} p_{ij,k} = 1 \tag{2.2}$$

for arbitrary $i, j, k = 1, \cdots, d$. Note that the condition $p_{ij,k} = p_{ji,k}$ is not onerous, otherwise one can determine a new heredity coefficient

$$q_{ij,k} = \frac{p_{ij,k} + p_{ji,k}}{2}$$

preserving the operator $V$.

In population genetics, special attention is paid to a q.s.o. Roughly speaking, $V(\mathbf{p})$ represents a distribution of genotypes in the next generation if $\mathbf{p} \in S^{d-1}$ is a distribution of genotypes in the parents' generation. In this simplified model, the iterates $V^n(\mathbf{p})$, where $n = 0, 1, \dots$ , describe the evolution of the distribution of genotypes in a population. Hence, the quadratic stochastic operator $V$ is defined by a cubic matrix $(p_{ij,k})_{i,j,k=1}^{d}$.

In what follows, by a *fixed point* of $V$ we mean a vector $\mathbf{x} \in S^{d-1}$ such that $V\mathbf{x} = \mathbf{x}$.

*Remark 2.1.1* Note that the simplex $S^{d-1}$ can be considered as the set of all probability measures on the measurable space $(E, \mathscr{F})$, where $E = \{1, \cdots, d\}$ and the $\sigma$-algebra $\mathscr{F}$ is the power set of $E$, i.e. the set of all subsets of $E$. Hence, a quadratic stochastic operator transforms a measure on $E$ to another measure. Now we can generalize the notion of a quadratic stochastic operator for a set $E$ with arbitrary cardinality. Let $(E, \Im)$ be a measurable space and $\mathfrak{M}$ be the collection of all probability measures on $(E, \Im)$. The quadratic stochastic operator on $(E, \Im)$ is a mapping of the set $\mathfrak{M}$ into itself of the form: for $m \in \mathfrak{M}$ and arbitrary $B \in \Im$

$$(Vm)(B) = \int_E \int_E P(x, y, B) dm(x) dm(y), \tag{2.3}$$

where the functions $\{P(x, y, B) : x, y \in E, B \in \Im\}$ satisfy the following conditions:

(i) for fixed $x, y \in E$ one has $P(x, y, \cdot) \in \mathfrak{M}$;
(ii) for each fixed $B \in \Im$, $P(x, y, B)$ is regarded as a measurable function on $(E \times E, \Im \otimes \Im)$. Moreover, $P(x, y, B) = P(y, x, B)$ for any $x, y \in E$ and $B \in \Im$. Here $\Im \otimes \Im$ denotes the tensor product of two $\sigma$-algebras.

Note that from condition (i) it immediately follows that $P(x, y, B) \geq 0$ and $P(x, y, E) = 1$ for any $x, y \in E$.

Now by $L_b(E, \mathfrak{I})$ we denote the set of all bounded measurable functions on $(E, \mathfrak{I})$. Then, by means of $P(x, y, A)$, one can define a mapping $\tilde{P} : L_b(E, \mathfrak{I}) \to L_b(E \times E, \mathfrak{I} \otimes \mathfrak{I})$ by

$$\tilde{P}(f)(x, y) = \int_E f(u) P(x, y, du), \ f \in L_b(E, \mathfrak{I}).$$

Clearly, this operator is positive, i.e. $\tilde{P}f \geq 0$ whenever $f \geq 0$, and $\tilde{P}(I) = I$, where $I$ is the unit function, i.e. $I(x) = 1$ for all $x$. Note that this kind of approach will be used in Chap. 4 to define quantum analogues of quadratic stochastic operators.

One of the main problems in mathematical biology consists of the study of the asymptotical behavior of the trajectories of q.s.o.s. The difficulty of the problem depends on the given matrix $\mathbf{P}$. In this section we shall consider several particular cases of $\mathbf{P}$ for which the above mentioned problem is (particularly) solved.

We notice that any quadratic stochastic operator $V$ defines a bilinear mapping $\mathbf{Q} : \mathfrak{X} \times \mathfrak{X} \to \mathfrak{X}$ if we set $\mathbf{Q}(\mathbf{x}, \mathbf{y})_k = \sum_{i,j=1} p_{ij,k} x_i y_j$ for any $k \geq 1$. Clearly, $V(\mathbf{p}) = \mathbf{Q}(\mathbf{p}, \mathbf{p})$, and $\mathbf{Q}$ is monotone (i.e. $\mathbf{Q}(\mathbf{x}, \mathbf{y}) \geq \mathbf{Q}(\mathbf{u}, \mathbf{w})$ whenever $\mathbf{x} \geq \mathbf{u} \geq 0$ and $\mathbf{y} \geq \mathbf{w} \geq 0$) and it is bounded as $\sup_{\|\mathbf{x}\|_1, \|\mathbf{y}\|_1 \leq 1} \|\mathbf{Q}(\mathbf{x}, \mathbf{y})\|_1 = 1$. From conditions (2.2) we easily obtain that for any $\mathbf{u}, \mathbf{w} \in \mathfrak{X}$

$$\|\mathbf{Q}(\mathbf{u}, \mathbf{w})\|_1 = \sum_{k=1}^{d} \left| \sum_{i,j=1}^{d} u_i w_j p_{ij,k} \right| \leq \sum_{k=1}^{d} \sum_{i,j=1}^{d} |u_i| |w_j| p_{ij,k} = \|\mathbf{u}\|_1 \|\mathbf{w}\|_1.$$

It follows from the properties above that $\mathbf{Q}(S^{d-1} \times S^{d-1}) \subseteq S^{d-1}$.

*Remark 2.1.2* It is worth noting that any homogeneous Markov dynamics on $S^{d-1}$ defined by a stochastic matrix $P = (p_{j,k})_{j,k \geq 1}$ may be viewed as a quadratic mapping $V : S^{d-1} \to S^{d-1}$. In fact, let us define $p_{ij,k} = \frac{1}{2}(p_{i,k} + p_{j,k})$ for all $i, j, k \geq 1$. For any $\mathbf{x} \in S^{d-1}$ and $k$ we get

$$V(\mathbf{x})_k = \sum_{i,j=1}^{d} x_i x_j p_{ij,k} = \frac{1}{2} \sum_{i,j=1}^{d} x_i x_j p_{i,k} + \frac{1}{2} \sum_{i,j=1}^{d} x_i x_j p_{j,k}$$

$$= \frac{1}{2} (\mathbf{x} \circ P)_k + \frac{1}{2} (\mathbf{x} \circ P)_k = (\mathbf{x} \circ P)_k.$$

Clearly, for any $\mathbf{u}, \mathbf{v}, \tilde{\mathbf{u}}, \tilde{\mathbf{v}} \in \mathfrak{X}$

$$\|\mathbf{Q}(\mathbf{u}, \mathbf{v}) - \mathbf{Q}(\tilde{\mathbf{u}}, \tilde{\mathbf{v}})\|_1 \leq \|\mathbf{Q}(\mathbf{u}, \mathbf{v}) - \mathbf{Q}(\tilde{\mathbf{u}}, \mathbf{v})\|_1 + \|\mathbf{Q}(\tilde{\mathbf{u}}, \mathbf{v}) - \mathbf{Q}(\tilde{\mathbf{u}}, \tilde{\mathbf{v}})\|_1$$

$$\leq \|\mathbf{u} - \tilde{\mathbf{u}}\|_1 \|\mathbf{v}\|_1 + \|\mathbf{v} - \tilde{\mathbf{v}}\|_1 \|\tilde{\mathbf{u}}\|_1. \tag{2.4}$$

If all the vectors $\mathbf{u}, \mathbf{v}, \tilde{\mathbf{u}}, \tilde{\mathbf{v}}$ are from the unit ball, then

$$\|Q(\mathbf{u}, \mathbf{v}) - Q(\tilde{\mathbf{u}}, \tilde{\mathbf{v}})\|_1 \leq \|\mathbf{u} - \tilde{\mathbf{u}}\|_1 + \|\mathbf{v} - \tilde{\mathbf{v}}\|_1 .$$

In particular, for all $\mathbf{u}, \mathbf{v} \in S^{d-1}$

$$\|V(\mathbf{u}) - V(\mathbf{v})\|_1 = \|Q(\mathbf{u}, \mathbf{u}) - Q(\mathbf{v}, \mathbf{v})\|_1 \leq 2 \|\mathbf{u} - \mathbf{v}\|_1 .$$

It follows that $Q$ is continuous on $\mathfrak{X} \times \mathfrak{X}$ and uniformly continuous if applied to the vectors from the unit ball in $\mathfrak{X}$. In particular, $V$ is uniformly continuous on the unit ball in $\mathfrak{X}$.

Given an initial distribution $\mathbf{p} \in S^{d-1}$, we denote its trajectory, i.e. the sequence $(V^n(\mathbf{p}))_{n \geq 0}$, by $\gamma(\mathbf{p})$. The family of all possible trajectories $\{\gamma(\mathbf{p}) : \mathbf{p} \in S^{d-1}\}$ is denoted by $\Gamma(V)$.

**Definition 2.1.1** A quadratic stochastic operator $V$ is called *asymptotically stable* (or *regular*) if there exists a vector $\mathbf{p}^* \in S^{d-1}$ such that for all $\mathbf{x} \in S^{d-1}$ one has

$$\lim_{n \to \infty} \|V^n(\mathbf{x}) - \mathbf{p}^*\|_1 = 0.$$

One of the important examples of a regular q.s.o. is a contractive operator. Therefore, we are now interested in the case when the q.s.o. is a contraction. Note that this question was first studied in [117] (see also [138]).

To study it, we need some auxiliary results. First recall that for the simplex $S^{d-1}$ the tangent space is the hyperplane

$$\Delta_0^d = \{\mathbf{x} \in \mathbb{R}^d : s(\mathbf{x}) = 0\},$$

where

$$s(\mathbf{x}) = x_1 + x_2 + \cdots + x_d. \tag{2.5}$$

In what follows, we assume that any matrix $A = (a_{jk})_{j,k=1}^n$ acts on $\mathbb{R}^d$ (here, as before, $\mathbb{R}^d$ is considered with the norm $\|\cdot\|_1$) as follows

$$(A\mathbf{x})_k = \sum_{j=1}^d a_{jk} x_j, \quad \mathbf{x} = (x_j) \in \mathbb{R}^d.$$

Consider a q.s.o. $V$ given by (2.1). At a point $\mathbf{x} \in S^{d-1}$ the tangent map $d_\mathbf{x} V$ is the restriction to $\Delta_0^d$ of the linear map with Jacobian matrix $\left( \dfrac{\partial V(\mathbf{x})_k}{\partial x_j} \right)_{j,k=1}^d$, where

$$\frac{\partial V(\mathbf{x})_j}{\partial x_k} = 2 \sum_{i=1}^d x_i p_{ik,j}. \tag{2.6}$$

Thus, $d_{\mathbf{x}}V$ is the restriction to $\Delta_0^d$ of $2M_{\mathbf{x}}$, where $M_{\mathbf{x}}$ is the multiplication operator given by the matrix

$$(M_{\mathbf{x}})_{jk} = \sum_{i=1}^{d} x_i p_{ij,k}.$$

**Lemma 2.1.1 ([139])** *For a matrix* $A = (a_{jk})_{j,k=1}^{d}$ *the subspace* $\Delta_0^d$ *is invariant if and only if the linear functional* $s$ *(see (2.5)) is an eigenvector of* $A^*$, *or, equivalently*

$$\sum_{j=1}^{d} a_{1j} = \sum_{j=1}^{d} a_{2j} = \cdots = \sum_{j=1}^{d} a_{dj}. \tag{2.7}$$

*Assuming such invariance, then the restriction of* $A$ *to* $\Delta_0^d$, *which is denoted by* $A_0$, *is nonsingular if and only if*

$$\begin{vmatrix} a_{11} & \cdots & a_{1,d} \\ \vdots & \ddots & \vdots \\ a_{d-1,1} & \cdots & a_{d-1,d} \\ 1 & \cdots & 1 \end{vmatrix} \neq 0. \tag{2.8}$$

*Remark 2.1.3* In (2.8) the first $d - 1$ rows of the matrix $A$ can be replaced by any $d - 1$ rows.

*Remark 2.1.4* One can show that the left-hand side of (2.8) is precisely the determinant of the restriction of $A$ to $\Delta_0^n$. We denote this by $det_0 A$, which is well defined only if the subspace $\Delta_0^d$ is $A$ invariant. If $det_0 A \neq 0$, then rank $A \geq d - 1$, and rank $A = d - 1$ if and only if there exists a row whose elements are equal.

Observe that for each $\mathbf{x} \in S^{d-1}$ the multiplication operator $M_{\mathbf{x}}$ is stochastic, i.e. $s$ is a fixed point for $M_{\mathbf{x}}^*$.

**Theorem 2.1.2** *For each* $\mathbf{x} \in S^{d-1}$, $s$ *is an eigenvector of* $M_{\mathbf{x}}$ *with eigenvalue 1.* $M_{\mathbf{x}}$ *has an eigenvector* $\tilde{\mathbf{x}} \in S^{d-1}$ *with eigenvalue 1 and the entire spectrum of* $M_{\mathbf{x}}$ *is contained in the unit circle. Moreover, a point* $\mathbf{x}$ *is fixed for* $V$ *if and only if* $\mathbf{x}$ *is an eigenvector of* $M_{\mathbf{x}}$ *with eigenvalue 1. Such a fixed point is non-degenerate if and only if* $\frac{1}{2}$ *is not an eigenvalue of* $M_{\mathbf{x}}^*$, *i.e.*

$$det_0(2M_{\mathbf{x}} - I) \neq 0. \tag{2.9}$$

*Proof* It is clear that for each $\mathbf{x} \in S^{d-1}$ the stochasticity of $M_{\mathbf{x}}$ implies that it is nonnegative and the common value in (2.7) is 1. Here we have used $a_{jk} = \sum_{i=1}^{d} x_i p_{ij,k}$. Thus, $M_{\mathbf{x}}$ is a nonnegative matrix with dominant eigenvalue 1. The existence of $\tilde{\mathbf{x}}$ and the spectral property follow from the Perron–Frobenius theory of such operators.

From $V(\mathbf{x}) = M_{\mathbf{x}}(\mathbf{x})$, one can see that $\mathbf{x}$ is a fixed point of $V$ precisely when $M_{\mathbf{x}}(\mathbf{x}) = \mathbf{x}$. Since the linearization of $d_{\mathbf{x}}V$ is $2M_{\mathbf{x}}$, such a fixed point is non-degenerate when (2.9) holds, i.e. when $\frac{1}{2}$ is not an eigenvalue of $M_{\mathbf{x}}$.

*Remark 2.1.5* This theorem allows us to investigate the structure of the fixed points of a q.s.o. For more details we refer the reader to [139, Chap. 8].

Recall that the *Lipschitz constant* of an operator $V$ is defined by

$$l(V) = \sup_{\mathbf{x} \neq \mathbf{y}} \frac{\|V\mathbf{x} - V\mathbf{y}\|_1}{\|\mathbf{x} - \mathbf{y}\|_1}.$$

If one has $l(V) < 1$, then $V$ is a strict contraction. Moreover, it has a unique fixed point, and all trajectories of $V$ converge to this point with exponential rate.

**Lemma 2.1.3** *Let $\mathscr{C}$ be convex d-dimensional compact subset of $\mathbb{R}^d$, and $F : \mathscr{C} \to \mathscr{C}$ be a smooth map. Then $l(F) = l'(F) \equiv \max_{\mathbf{x} \in \mathscr{C}} \|d_{\mathbf{x}}F\|_1$.*

*Proof* From the Mean Value Theorem we infer that $\|F_{\mathbf{x}} - F_{\mathbf{y}}\|_1 \leq l'(F)\|\mathbf{x} - \mathbf{y}\|_1$, i.e. $l(F) \leq l'(F)$. On the other hand, $\mathscr{C}$ is the closure of its interior, so for any $\epsilon > 0$ there exists an $\mathbf{x}_0 \in Int\mathscr{C}$ such that $\|d_{\mathbf{x}_0}F\|_1 > l'(F) - \epsilon$. Hence, there exist a vector $\mathbf{z}$ such that $\|\mathbf{z}\|_1 = 1$ and $\|d_{\mathbf{x}_0}F(\mathbf{z})\|_1 > l'(F) - \epsilon$. From the definition

$$F(\mathbf{x}_0 + h\mathbf{z}) - F(\mathbf{x}_0) = hd_{\mathbf{x}_0}F\mathbf{z} + o(h) \text{ as } h \to 0,$$

we find $l(F)h > (l'(F) - \epsilon)h - o(h)$, which in the limit gives $l(F) \geq l'(F) - \epsilon$ and hence $l(F) \geq l'(F)$.

**Lemma 2.1.4 ([139])** *Let a matrix $A = \left(a_{ij}\right)_{i,j=1}^{d}$ satisfy (2.7). Then*

$$\|A|\Delta_0^d\| = \frac{1}{2} \max_{i_1 \neq i_2} \sum_{j=1}^{d} |a_{i_1j} - a_{i_2j}|. \tag{2.10}$$

*Proof* Let us rewrite (2.10) as

$$\|A|\Delta_0^d\| = \max_{i_1 \neq i_2} \|A((\mathbf{e}_{i_1} - \mathbf{e}_{i_2})/2)\|_1,$$

where, as before, $\{\mathbf{e}_i\}$ is the standard basis in $\mathbb{R}^d$.

To prove the last equality, it is enough to show that $\epsilon_{i_1 i_2} = (\mathbf{e}_{i_1} - \mathbf{e}_{i_2})/2$, $(1 \leq i_1, i_2 \leq d; i_1 \neq i_2)$ is the set of all extremal points of the intersection of the unit ball $\|\mathbf{z}\|_1 \leq 1$ with the hyperplane $\Delta_0^d$.

Assume that

$$\epsilon_{i_1 i_2} = \frac{1}{2} \left( \sum_{i=1}^{d} \alpha_i \mathbf{e}_i + \sum_{i=1}^{d} \beta_i \mathbf{e}_i \right), \tag{2.11}$$

where $\sum_{i=1}^{d} \alpha_i = \sum_{i=1}^{d} \beta_i = 0$ and $\sum_{i=1}^{d} |\alpha_i|, \sum_{i=1}^{d} |\beta_i| \le 1$. Then $\alpha_{i_1} + \beta_{i_1} = 1$, $\alpha_{i_2} + \beta_{i_2} = -1$ and $\alpha_i + \beta_i = 0$ for $i \ne i_1, i_2$. Hence,

$$\alpha_{i_1} + \beta_{i_1} - \alpha_{i_2} - \beta_{i_2} = 2$$

while

$$|\alpha_{i_1}| + |\beta_{i_1}| + |\alpha_{i_2}| + |\beta_{i_2}| \le 2.$$

So, $\alpha_{i_1}, \beta_{i_1} \ge 0$, $\alpha_{i_2}, \beta_{i_2} \le 0$ and

$$|\alpha_{i_1}| + |\beta_{i_1}| + |\alpha_{i_2}| + |\beta_{i_2}| = 2.$$

These imply $\alpha_i = \beta_i = 0$ for $i \ne i_1, i_2$ and

$$\alpha_{i_1} = \beta_{i_1} = -\alpha_{i_2} = -\beta_{i_2} = \frac{1}{2}$$

and the representation (2.11) is trivial. Consequently, all the $\epsilon_{i_1 i_2}$ are extremal.

If a point $\mathbf{z} \in \Delta_0^d$, ($\|\mathbf{z}\|_1 \le 1$) is not on the convex hull of the set $\{\epsilon_{i_1 i_2}\}$, then it is separated from it by some linear form $f$, i.e. $f(\mathbf{z}) > 0$ and $f(\epsilon_{i_1 i_2}) \le 0$. Since $\epsilon_{i_1 i_2} = -\epsilon_{i_1 i_2}$, we get $f(\epsilon_{i_1 i_2}) = 0$. Hence, $f(\mathbf{e}_i) = \cdots = f(\mathbf{e}_d)$, i.e. $f$ is proportional to $s$. But then $f(\mathbf{z}) = 0$, a contradiction. This completes the proof.

**Theorem 2.1.5 ([117])** *For any q.s.o. V given by (2.1) one has*

$$l(V) = \max_{i_1, i_2, k} \sum_{j=1}^{d} |p_{i_1 k, j} - p_{i_2 k, j}|.$$

*Proof* From (2.6) the derivative of $V$ on the simplex $S^{d-1}$ is

$$d_{\mathbf{x}} V = 2M_{\mathbf{x}} = 2 \sum_{k=1}^{n} x_k M_k,$$

where $M_k = M_{e_k}$ is the multiplication map with matrix $(P_{ik,j})_{i,j=1}^{d}$. By Lemma 2.1.3 we have

$$l(V) = 2 \max_{\mathbf{x} \in S^{d-1}} \|M_{\mathbf{x}}\|_1 = 2 \max_k \|M_k\|$$

and by Lemma 2.1.4, one finds

$$\|M_k\| = \frac{1}{2} \max_{i_1 i_2} \sum_{k=1}^{d} |p_{i_1 k, j} - p_{i_2 k, j}|,$$

which is the required equality. This completes the proof.

**Corollary 2.1.6 ([117])** *A q.s.o. V is a strict contraction if and only if*

$$\max_{i_1,i_2,k} \sum_{j=1}^{d} |p_{i_1 k,j} - p_{i_2 k,j}| < 1.$$

It is known that for a stochastic matrix $\mathbf{P} = (p_{ij})$ the condition $p_{ij} > 0$ (for all $i, j$) implies the regularity of $\mathbf{P}$, i.e. there is a $\mathbf{p} \in S^{d-1}$ such that

$$\mathbf{P}^n(\mathbf{x}) \to \mathbf{p} \quad \text{for all} \quad \mathbf{x} \in S^{d-1}.$$

Unfortunately, for the q.s.o. $V$ given by (2.1) the condition $p_{ij,k} > 0$ (for all $i, j, k$) does not imply its regularity. Let us provide an example.

*Example 2.1.1 ([139, p. 249],[126])* Let us consider the following operator $V_\varepsilon$ on $S^2$ depending on a parameter $\varepsilon > 0$:

$$x' = (1 - 4\varepsilon)x^2 + 2\varepsilon y^2 + 10\varepsilon z^2 + 4\varepsilon xy + (1 + 4\varepsilon)xz + 8\varepsilon yz,$$

$$y' = 2\varepsilon x^2 + (1 - 3\varepsilon)y^2 + \varepsilon z^2 + \left(\frac{1}{2} + 2\varepsilon\right)xy + 2\varepsilon xz + (1 - 12\varepsilon)yz,$$

$$z' = 2\varepsilon x^2 + \varepsilon y^2 + (1 - 11\varepsilon)z^2 + \left(\frac{3}{2} - 6\varepsilon\right)xy + (1 - 6\varepsilon)xz + (1 + 4\varepsilon)yz.$$

Here all the coefficients are positive for $\varepsilon < \frac{1}{12}$. One can check that if $\varepsilon < \frac{9-5\sqrt{2}}{124}$ then $V_\varepsilon$ has exactly three fixed points, which means $V_\varepsilon$ is not regular.

Therefore, one can ask: find the smallest $\alpha_d$ such that $p_{ij,k} > \alpha_d$ implies the regularity of $V$.

It is clear that if such a number $\alpha_d$ exists then one has $0 < \alpha_d < \frac{1}{d}$.

**Lemma 2.1.7** *Let* $\mathbf{x} = (x_i)$, $\mathbf{y} = (y_i) \in S^{d-1}$ *and* $x_i \geq \frac{1}{2d}$, $y_i \geq \frac{1}{2d}$, $i = \overline{1, d}$. *Then*

$$\|\mathbf{x} - \mathbf{y}\|_1 \leq 1.$$

*Moreover, if* $x_i > \frac{1}{2d}$, $y_i > \frac{1}{2d}$, *then* $\|\mathbf{x} - \mathbf{y}\|_1 < 1$.

*Proof* It is evident that the set

$$K = \left\{\mathbf{x} \in S^{d-1} : x_i \geq \frac{1}{2d}, i = \overline{1, d}\right\}$$

is convex and compact. One can prove that $\max\limits_{\mathbf{x},\mathbf{y}\in K}\|\mathbf{x}-\mathbf{y}\|_1$ is attained at the extremal points of $K$. It is clear that the extremal points of $K$ are the following $n$ vectors:

$$\left(\frac{d-1}{2d},\frac{1}{2d},\ldots,\frac{1}{2d}\right),\left(\frac{1}{2d},\frac{d-1}{2d},\ldots,\frac{1}{2d}\right),\ldots,\left(\frac{1}{2d},\frac{1}{2d},\ldots,\frac{d-1}{2d}\right).$$

The distance between two of them is 1. Hence, if $\mathbf{x},\mathbf{y}\in K$, then $\|\mathbf{x}-\mathbf{y}\|_1\leq 1$.

**Theorem 2.1.8** *Let $V$ be a q.s.o. If $p_{ij,k}>\frac{1}{2d}$ for all $i,j,k$, then $V$ is regular.*

*Proof* From Lemma 2.1.7 we immediately find that

$$\max\limits_{i_1 i_2 j}\sum_{k=1}^{d}|p_{i_1j,k}-p_{i_2j,k}|<1.$$

Therefore, due to Corollary 2.1.6, the operator $V$ is regular.

The above theorem yields that $\alpha_d\leq\frac{1}{2d}$.

*Remark 2.1.6* Note that there are also irregular q.s.o.s, the first of which was constructed in [251, 254]. Some generalizations have been investigated in [52, 69, 130, 131, 222]. The investigated (non-regular) q.s.o. is called a *Volterra operator*, i.e. $p_{ij,k}=0$ if $k\notin\{i,j\}$. These operators, in general, are studied in [74]. Non-Volterra q.s.o.s have also been studied intensively (see for example, [214–216]).

## 2.2 One-dimensional q.s.o.s

In this section we follow [138] to study the dynamics of an arbitrary q.s.o. on a one-dimensional simplex.

It is clear that in this setting any q.s.o. has the following form:

$$\begin{cases} x_1' = ax_1^2 + 2bx_1x_2 + cx_2^2 \\ x_2' = (1-a)x_1^2 + 2(1-b)x_1x_2 + (1-c)x_2^2 \end{cases},$$

where $x_1\geq 0$, $x_2\geq 0$, $x_1+x_2=1$ and $a=p_{11,1}$, $b=p_{12,1}=p_{21,1}$, $c=p_{22,1}$ are arbitrary coefficients with $0\leq a,b,c\leq 1$.

Assuming $x_1=x$, $x_1'=y$ and substituting for the second coordinate $x_2=1-x$, we obtain

$$y = ax^2 + 2bx(1-x) + c(1-x)^2 = (a-2b+c)x^2 + 2(b-c)x + c. \tag{2.12}$$

It is evident that a function (2.12) maps the segment $[0, 1]$ (one-dimensional simplex) into itself with

$$y\mid_{x=0} = c, \quad y\mid_{x=1} = a.$$

In order to avoid the analysis of particularities, in what follows, we suppose that $a < 1$ and $c > 0$.

**Lemma 2.2.1** *A fixed point of the transformation* (2.12) *is unique and belongs to* $(0, 1)$.

*Proof* In fact, the equation

$$x = (a - 2b + c)x^2 + 2(b - c)x + c \tag{2.13}$$

has a root in the interval $(1, \infty)$ when $a - 2b + c > 0$ and has a root in $(-\infty, 0)$ when $a - 2b + c < 0$. If $a - 2b + c = 0$, then the equation becomes a linear one with $c > 0$. Thus, for all cases a root in $[0, 1]$ is unique. It is clear that this root differs from 0 and 1.

Let us consider the discriminant of the quadratic equation (2.13) to investigate the local character of the fixed point:

$$\Delta = 4(1 - a)c + (1 - 2b)^2. \tag{2.14}$$

One can see that $0 < \Delta < 5$ and $\Delta$ takes all values in this interval.

**Lemma 2.2.2** *If* $0 < \Delta < 4$, *then the fixed point is attractive, and if* $4 < \Delta < 5$ *then it is repelling.*

*Proof* Let $\xi$ be a fixed point. Its character is defined by $f'(\xi)$, where $f(x)$ is the right-hand side of the equation (2.13) and $f'(x)$ is its derivative. Let $\lambda = f'(\xi)$, where

$$\lambda = 2(a - 2b + c)\xi + 2(b - c).$$

If $\mid \lambda \mid < 1$, then $\xi$ is an *attractive point*, and if $\mid \lambda \mid > 1$, then $\xi$ is a *repelling point* (see [35] for more information about the theory of dynamical systems). It is easy to check that

$$\lambda = 1 - \sqrt{\Delta}, \tag{2.15}$$

and the statement of the lemma follows from this equality.

**Theorem 2.2.3** *If* $0 < \Delta < 4$, *then all trajectories converge to a fixed point.*

*Proof* Let us decompose the function $f(x)$ by powers of $x - \xi$, i.e.

$$f(x) = \xi + \lambda(x - \xi) + A(x - \xi)^2,$$

where $A = a - 2b + c$.

If $\{x_m\}_0^\infty$ is a trajectory, then the differences $u_m = x_m - \xi$ are defined by the iterations

$$u_{m+1} = \lambda u_m + A u_m^2. \tag{2.16}$$

It is enough to prove that if $|\lambda| < 1$, then $\lim_{m \to \infty} u_m = 0$. We know that the sequence $\{u_m\}$ is located in the segment $[-\xi, 1 - \xi]$ which is invariant with respect to the transformation

$$\varphi(u) = \lambda u + A u^2.$$

Moreover, the last transformation has a unique fixed point $u = 0$ on this segment. According to the uniqueness of the fixed point we have

$$\lambda + Au \neq 1 \quad \text{for all} \ \ u \in [-\xi, 1 - \xi]$$

and due to $(\lambda + Au)\,|_{u=0} = \lambda < 1$, one gets $\lambda + Au < 1$. Here $u \in [-\xi, 1 - \xi]$.

If, in addition $\lambda + Au > -1$ for all $u \in [-\xi, 1 - \xi]$, then $q \equiv \max |\lambda + Au| < 1$. According to $|u_{n+1}| \leq q|u_n|$, in this case, we find $\lim_{m \to \infty} u_m = 0$.

Now assume that $\lambda + Au \leq -1$ for some $u \in [-\xi, 1 - \xi]$. Then the root $\bar{u} = -\frac{1+\lambda}{A}$ of the equation $\lambda + Au = -1$ belongs to $[-\xi, 1 - \xi]$. For the sake of definiteness we suppose $A > 0$, then $\bar{u} < 0$.

Let us consider the segment $I = [\bar{u}, 1 - \xi]$. Note that a point $u = 0$ is an inner point of $I$, and $\varphi(I) \subset I$, since $|\lambda + Au| \leq 1$ on $I$. Hence,

$$|\varphi(u)| \leq |u| \quad \text{for all} \ u \in I.$$

So, the segment $I$ is invariant with respect to the transformation $\varphi$.

On the other hand, the image of the whole segment $[-\xi, 1 - \xi]$ lies in $I$, since if $u < \bar{u}$, then $u < 0$ and $\lambda + Au < -1$, therefore $\varphi(u) > 0$.

From this we infer that any trajectory $\{x_m\}_0^\infty$ will enter into the segment $I$ no later than $m = 1$ and remain in there. Moreover, we have

$$|u_{m+1}| \leq |u_m| \quad (m = 1, 2, 3, \cdots),$$

which yields the existence of the limit $\gamma = \lim_{m \to \infty} |u_m|$.

If $\gamma > 0$, then $\lim_{m \to \infty} |\lambda + Au_m| = 0$, i.e. all limit points of the trajectory satisfy the equation $|\lambda + Au| = 1$. But $\bar{u}$ is a unique root of this equation on $[-\xi, 1 - \xi]$. Therefore, the trajectory should converge to $\bar{u}$, but this is impossible since $\bar{u}$ is not a fixed point. Hence, $\gamma = 0$ i.e. $\lim_{m \to \infty} u_m = 0$. This completes the proof.

*Remark 2.2.1* From the above calculations one can derive the asymptotic

$$u_m \sim C(u_0)\lambda^m.$$

**Theorem 2.2.4** *If* $4 < \Delta < 5$, *then there exists a cycle of second order and all trajectories tend to this cycle except the stationary trajectory starting with the fixed point.*

*Proof* We consider the previous equation (2.16), keeping in mind that $|\lambda| > 1$. Due to (2.15) one can see that $\lambda < -1$. As before, the segment $[-\xi, 1 - \xi]$ is invariant with respect to the transformation $\varphi$, and the fixed point $u = 0$ is unique. Therefore, $A \neq 0$ and $\lambda + Au < -1$, for all $u \in [-\xi, 1 - \xi]$.

Let us consider the fixed points of the transformation $\psi(v) = \varphi(\varphi(v))$, which are the roots of the equation

$$v = v(\lambda + Av)(\lambda + Av(\lambda + Av)).$$

One of these roots, $v = 0$, is a fixed point for $\varphi$. The second one is a root of $\lambda + Av = 1$, which lies outside of the segment $[-\xi, 1 - \xi]$. So, there remain two more roots:

$$v_\pm = \frac{|\lambda + 1| \pm \sqrt{|\lambda + 1|(3 - \lambda)}}{2A}.$$

They belong to $[-\xi, 1 - \xi]$ (of opposite signs) and form the second order cycle for the transformation $\varphi$.

Now let us separate the segment $[-\xi, 1 - \xi]$ into the following five parts:

$$I_{-2} = [-\xi, v_-); \ I_{-1} = [v_-, 0); \ I_0 = \{0\}; \ I_1 = (0, v_+]; \ I_2 = (v_+, 1 - \xi].$$

One can easily check the following statements:

(a) $\psi(I_1) \subset I_1$, with $\psi(v) \geq v \ (v \in I_1)$;
(b) $\psi(I_2) \subset I_1 \cup I_2$, moreover, if $v \in I_2$, then $\psi(v) < v$;
(c) $\psi(I_{-1}) \subset I_{-1}$, with $\psi(v) \leq v \ (v \in I_{-1})$;
(d) $\psi(I_2) \subset I_{-1} \cup I_{-2}$, moreover if $v \in I_{-2}$, then $\psi(v) > v$.

All these statements can easily be seen in the graphic (see Fig. 2.1).

Let $v_{m+1} = \psi(v_m)(m = 0, 1, 2, \cdots)$ and $v_0 > 0$. As soon as the trajectory falls into $I_1$, it does not leave it and $v_{m+1} \geq v_m(v_m \in I_1)$. In this case, we have $\lim\limits_{m \to \infty} v_m = v_+$. If the trajectory does not leave $I_2$, then $v_{m+1} \leq v_m(m = 0, 1, 2, \cdots)$, and again one finds $\lim\limits_{m \to \infty} v_m = v_+$.

In the case $v_0 < 0$, using the same argument, one gets $\lim\limits_{m \to \infty} v_m = v_-$.

Now it is clear that the trajectory

$$u_{m+1} = \varphi(u_m) \ (m = 0, 1, 2, \cdots)$$

behaves itself as announced above. In fact,

$$u_{m+2} = \varphi(u_m) \ (m = 0, 1, 2, \cdots)$$

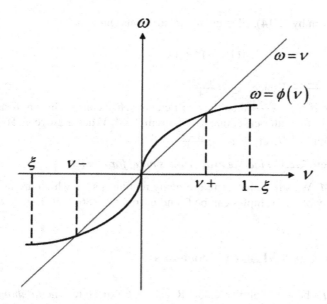

**Fig. 2.1** Graphic of the function $\psi$

and $\mathrm{sgn}(u_1) = -\mathrm{sgn}(u_0)$ (since $\lambda + Au_0 < 0$). If, for example, $u_0 > 0$, then $\lim\limits_{k\to\infty} u_{2k} = v_+$, and $\lim\limits_{k\to\infty} u_{2k+1} = v_-$. This completes the proof.

*Remark 2.2.2* The rate of convergence is again exponential, namely,

$$v_m - v_+ \sim C(v_0)(5 - \Delta)^m, \ \text{if} \ v_0 > 0,$$

$$v_m - v_- \sim C(v_0)(5 - \Delta)^m \ \text{if} \ v_0 < 0.$$

*Remark 2.2.3* Note that the existence of the second order cycle in the case $4 < \Delta < 5$ follows from a general theorem proved by Kesten [117]. However, he conjectured the convergence of the trajectories to the cycle which were confirmed by numerical experiments in [154].

Based on the above results, it turns out that in the one-dimensional setting one can calculate $\alpha_d$, which was introduced in the previous section.

**Theorem 2.2.5** *One has* $\alpha_2 = \frac{3-\sqrt{7}}{4}$.

*Proof* Due to Theorem 2.2.3 we know that if $0 < \Delta < 4$ then $V$ is regular.
Let $\alpha = \min\{a, b, c, 1 - a, 1 - b, 1 - c\} \le \frac{1}{4}$. Then

$$\Delta = 4(1 - a)c + (1 - 2b)^2 < 4(1 - \alpha)^2 + (1 - 2\alpha)^2.$$

Here $\Delta$ is given by (2.14). Elementary calculations show that

$$4(1 - \alpha)^2 + (1 - 2\alpha)^2 \leq 4$$

holds at $\alpha \geq \frac{3-\sqrt{7}}{4}$. Hence, $\alpha_2 \leq \frac{3-\sqrt{7}}{4}$.

Moreover, if $\alpha < \frac{3-\sqrt{7}}{4}$ then the coefficients $a, b, c$ can be chosen in such a way that $\Delta > 4$. In the latter case, due to Theorem 2.2.4, $V$ has a 2-cycle. Therefore, $V$ cannot be regular. Therefore, $\alpha_2 = \frac{3-\sqrt{7}}{4}$.

**Open problem 2.2.1** *Find the exact values of $\alpha_d$ for any $d \geq 3$.*

*Remark 2.2.4* We stress that there are many regular q.s.o.s which are not contractions. Some explicit examples can be found in [66, 67, 186, 189, 194, 224].

## 2.3  Q.s.o.s and Markov Processes

Let $(\Omega, \mathscr{F}, P)$ be a probability space. Recall that a discrete time *nonhomogeneous Markov chain* is defined as a stochastic process $\{\xi_n\}_{n \geq 0}$ on a discrete phase space $E$ (finite, when $E = \{1, 2, \ldots, d\}$) such that for any time $n \geq 0$ and any sequence of states $s, r, r_{n-1}, \ldots, r_0$ one has

$$P\left(\xi_{n+1} = s | \xi_n = r, \xi_{n-1} = r_{n-1}, \ldots, \xi_0 = r_0\right) = P\left(\xi_{n+1} = s | \xi_n = r\right) =: p_{r,s}^{n,n+1}.$$

The transition probability matrix $(p_{r,s}^{n,n+1})_{r,s \geq 1}$ at the instant $n$ is denoted by $P^{[n,n+1]}$ and the sequence of such matrices $\mathbf{P} := (P^{[n,n+1]})_{n \geq 0}$ is called a *nonhomogeneous chain* of stochastic operators (i.e. positive linear operators on $\mathfrak{X}$ which preserve $S^{d-1}$). Put simply, the matrix $P^{[n,n+1]}$ acts on $\mathfrak{X}$ by

$$(P^{[n,n+1]}\mathbf{x})_s = \sum_{r=1} x_r p_{r,s}^{n,n+1}.$$

For any natural numbers $n > m \geq 0$ we set

$$P^{[m,n]} = P^{[m,m+1]} \circ P^{[m+1,m+2]} \circ \cdots \circ P^{[n-1,n]},$$

which corresponds to transition probabilities in $n - m$ steps (here $\circ$ stands for the composition of linear operators, i.e. multiplication of matrices). If for each $n \geq 0$ one has $P^{[n,n+1]} = P$, then $\mathbf{P} = (P)_{n \geq 0}$ is called a *homogeneous Markov chain*. The set of all nonhomogeneous (including homogeneous) chains of stochastic operators $\mathbf{P} = (P^{[n,n+1]})_{n \geq 0}$ is denoted by $\mathfrak{S}$.

The space $\mathfrak{S}$ may be endowed with several natural metric topologies. Their geometric structures have recently been comprehensively studied in [206].

**Definition 2.3.1** A nonhomogeneous chain of stochastic operators **P** is called *asymptotically stable* if there exists a probability vector $\mathbf{p}^* \in S^{d-1}$ such that for all $m \geq 0$ and $\mathbf{x} \in S^{d-1}$ one has

$$\lim_{n \to \infty} \left\| P^{[m,n]}\mathbf{x} - \mathbf{p}^* \right\|_1 = 0.$$

The reader should be warned that authors do not always use the same names for the same notions, e.g. in [103] strong ergodicity is called norm mixing. Another example is [98], where a detailed classification of different asymptotic behaviors of nonhomogeneous Markov chains has been carried out (see also [32, 46, 97, 104, 106, 174, 177, 206]).

The structure of quadratic stochastic operators is much more complex than the structure of nonhomogeneous Markov chains. Namely, given any sequence of probability vectors $(\mathbf{x}_n)_{n \geq 0}$ one may take $p_{r,s}^{n,n+1} = \mathbf{Q}(\mathbf{x}_n, \mathbf{e}_r)_s$, where $r, s \geq 1$. It follows from (2.2) that $(p_{r,s}^{n,n+1})_{r,s \geq 1} = P^{[n,n+1]}$ is a transition probability matrix. The idea of studying the dynamics of a quadratic stochastic operator through the so-called associated nonhomogeneous Markov chain has a longer history (cf. [46, 84]).

**Definition 2.3.2** Given a quadratic stochastic operator **Q** and any initial distribution $\mathbf{y} \in S^{d-1}$, a *nonhomogeneous Markov chain* associated with **Q** and a seed $\mathbf{y} \in S^{d-1}$ is defined by

$$P_{\mathbf{y}}^{[n,n+1]} = \left( p_{\mathbf{y};j,k}^{n,n+1} \right)_{j,k \geq 1} = \left( \mathbf{Q}(V^n(\mathbf{y}), \mathbf{e}_j)_k \right)_{j,k \geq 1}. \tag{2.17}$$

The following lemma is obvious and its proof follows directly from the previous definition. In fact it states that the iterates of the quadratic stochastic operator $V$ can be defined as a nonhomogeneous Markov chain with transition probability matrix $(p_{\mathbf{y};j,k}^{n,n+1})_{j,k \geq 1}$ at the instant $n$.

**Lemma 2.3.1** *Let $\mathbf{P}_{\mathbf{y}}$ be a nonhomogeneous Markov chain associated with a quadratic stochastic operator $V$ and a seed $\mathbf{y} \in S^{d-1}$. Then for every $n \geq 0$ we have*

$$V^n(\mathbf{y}) \circ P_{\mathbf{y}}^{[n,n+1]} = V^{n+1}(\mathbf{y}) = P_{\mathbf{y}}^{[0,n+1]}(\mathbf{y}) \left( = \mathbf{y} \circ P_{\mathbf{y}}^{[0,n+1]} \right).$$

*Proof* The proof immediately follows from $\sum_{j=1}^{d} V^n(\mathbf{y})_j p_{\mathbf{y};j,k}^{n,n+1} = V^{n+1}(\mathbf{y})_k$.

Let us note that if the seed $\mathbf{y} = \mathbf{p}^* \in S^{d-1}$ is $V$-invariant (i.e. if $V(\mathbf{p}^*) = \mathbf{Q}(\mathbf{p}^*, \mathbf{p}^*) = \mathbf{p}^*$), then the associated Markov chain $\mathbf{P}_{\mathbf{p}^*}$ is homogeneous. Indeed,

$$P_{\mathbf{p}^*}^{[n,n+1]} = (p_{\mathbf{p}^*;j,k}^{n,n+1})_{j,k \geq 1} = \left( \mathbf{Q}(V^n(\mathbf{p}^*), \mathbf{e}_j)_k \right)_{j,k \geq 1} = \left( \mathbf{Q}(\mathbf{p}^*, \mathbf{e}_j)_k \right)_{j,k \geq 1}$$

does not depend on $n$. Then we write $P_{\mathbf{p}^*}^{[n,n+1]} =: P_{\mathbf{p}^*}$ and $P_{\mathbf{p}^*}^{[0,n]} =: P_{\mathbf{p}^*}^n$.

Now we are going to show that, translating the relevant results for Markov (stochastic) operators to the theory of quadratic stochastic operators, the structure of invariant subsets and invariant vectors differs from what we would expect. We begin with

**Definition 2.3.3** Given a quadratic stochastic operator $V$ we say that a subset $D \subseteq E$ is *V-invariant (or absorbing)* if $p_{ij,k} = 0$ for all $i, j \in D$ and $k \notin D$.

The most natural $V$-invariant sets are supports of $V$-invariant probability vectors $\mathbf{p}^* \in S^{d-1}$. Recall that by a support of a vector $\mathbf{x}$ we mean supp$(\mathbf{x}) = \{k : x_k \neq 0\} (= S_{\mathbf{x}}$ for short). It follows directly from

$$\sum_{k=1}^{d} p_k^* = \sum_{k \in S_{\mathbf{p}^*}} \sum_{i,j \in S_{\mathbf{p}^*}} p_i^* p_j^* p_{ij,k} = 1$$

that $S_{\mathbf{p}^*}$ is $\mathbf{Q}$-invariant. It is well known that invariant subsets of a stochastic operator $P$ (homogeneous Markov chain) may have a nontrivial structure (in the literature on Markov chains invariant sets are sometimes called absorbing [14, 255]). In particular, supports of invariant probability measures $\mathbf{p}^*$ are unions of disjoint minimal invariant sets, each being the support of an ergodic invariant measure. On each such minimal set the Markov chain is irreducible and all states belonging to the same minimal set have a common period $d_D$. In particular, if the period $d_D$ is 1 (so the Markov chain restricted to $D$ is aperiodic and positive recurrent), then for any probability vector $\mathbf{x} \in S^{d-1}$ supported on $D$ we have $\|P^n(\mathbf{x}) - \mathbf{p}_D^*\|_1 \to 0$, where $\mathbf{p}_D^*$ is a unique invariant probability on $D$. If the period $d_D > 1$, then the chain has a periodic structure and $P^{nd_D}(\mathbf{x})$ converges. Of course, in the infinite-dimensional case it may happen that there is no common period for all minimal sets $D$ (cf. [104, 106]). It is well known that if a stochastic operator $P$ possesses an invariant strictly positive density $\mathbf{p}^*$ and it overlaps supports (i.e. for any pair of densities $\mathbf{u}, \mathbf{v} \in S^{d-1}$ there exists an $n$ such that $P^n(\mathbf{u}) \wedge P^n(\mathbf{v}) \neq 0$), then its ergodic structure is trivial (it is irreducible and aperiodic) and $P^n(\mathbf{x}) \to \mathbf{p}^*$ in the norm for all $\mathbf{x} \in S^{d-1}$ (cf. [14]). We will show that for a quadratic stochastic operator, the associated stochastic operator $P_{\mathbf{p}^*}$ restricted to the support of a $V$-invariant probability measure $\mathbf{p}^*$ always overlaps supports.

In what follows, for any subset $A \subseteq E$ we define $\mathfrak{X}(A) := \{\mathbf{x} \in \mathfrak{X} : \text{supp}(\mathbf{x}) \subseteq A\}$.

**Lemma 2.3.2** *Let $\mathbf{p}^*$ be an invariant probability vector of a quadratic stochastic operator $V$. Then $\mathfrak{X}(S_{\mathbf{p}^*})$ is $P_{\mathbf{p}^*}$-invariant (i.e. $P_{\mathbf{p}^*}(\mathfrak{X}(S_{\mathbf{p}^*})) \subseteq \mathfrak{X}(S_{\mathbf{p}^*})$) for the associated (homogeneous) Markov operator $P_{\mathbf{p}^*}$ and*

$$\lim_{n \to \infty} \left\| P_{\mathbf{p}^*}^n(\mathbf{x}) - \left( \sum_{j=1}^{d} x_j \right) \mathbf{p}^* \right\|_1 = 0$$

*for all $\mathbf{x} \in \mathfrak{X}(S_{\mathbf{p}^*})$ (i.e. $P_{\mathbf{p}^*}$ is asymptotically stable on $\mathfrak{X}(S_{\mathbf{p}^*})$).*

*Proof* Since $S_{\mathbf{p}^*}$ is $V$-invariant, then $P_{\mathbf{p}^*}$ is well defined on $\mathfrak{X}(S_{\mathbf{p}^*})$. Moreover, $S_{\mathbf{p}^*}$ is also a $P_{\mathbf{p}^*}$-invariant subset, as obviously $\mathbf{p}^*$ is also a $P_{\mathbf{p}^*}$-invariant probability measure. Using condition (2.2) in the definition of a quadratic stochastic operator we infer that for any $r, s \in S_{\mathbf{p}^*}$ there exists a $k \in S_{\mathbf{p}^*}$ such that $p_{rs,k} > 0$. Hence

$$\left(P_{\mathbf{p}^*}(\mathbf{e}_r) \wedge P_{\mathbf{p}^*}(\mathbf{e}_s)\right)_k = \left(\sum_{i=1}^{d} p_i^* p_{ir,k}\right) \wedge \left(\sum_{i=1}^{d} p_i^* p_{is,k}\right)$$

$$\geq \min\left\{p_s^* p_{sr,k}, p_r^* p_{rs,k}\right\}$$

$$= \min\left\{p_r^*, p_s^*\right\} p_{rs,k} > 0$$

(here $\wedge$ stands for the pointwise minimum in $\mathfrak{X}$). In particular, the stochastic operator $P_{\mathbf{p}^*}$ overlaps supports. The rest follows from [14]. ∎

In the next theorem we use this lemma to obtain the convergence of associated Markov chains with arbitrarily fixed seeds. A short Example 2.3.1, following Corollary 2.3.4, will explain that an additional assumption concerning relations between supports may be necessary to assure the convergence of the associated Markov chains if $\mathfrak{X}$ is infinite-dimensional. However, the following theorem is valid both in the finite and infinite-dimensional cases.

**Theorem 2.3.3** *Let $V$ be a q.s.o. and let $\mathbf{x} \in S^{d-1}$ such that*

$$\lim_{n \to \infty} \|V^n(\mathbf{x}) - \mathbf{p}^*\|_1 = 0.$$

*Then*

$$\lim_{n \to \infty} \left\| P_{\mathbf{x}}^{[0,n]}(\mathbf{z}) - \left(\sum_{j=1}^{d} z_j\right) \mathbf{p}^* \right\|_1 = 0$$

*for any $\mathbf{z}$ satisfying $S_{\mathbf{z}} \subseteq S_{\mathbf{x}}$.*

*Proof* Let us fix $\mathbf{x} \in S^{d-1}$ such that the associated (nonhomogeneous) Markov chain satisfies $P_{\mathbf{x}}^{[0,n]}(\mathbf{x}) = V^n(\mathbf{x}) \to \mathbf{p}^*$.

If $0 \leq \mathbf{u} \leq \mathbf{x}$ (obviously $\mathbf{u} \notin S^{d-1}$ in general), then $0 \leq P_{\mathbf{x}}^{[0,n]}(\mathbf{u}) \leq P_{\mathbf{x}}^{[0,n]}(\mathbf{x})$. Since the iterates $P_{\mathbf{x}}^{[0,n]}(\mathbf{x})$ converge in the norm, the set

$$\left\{P_{\mathbf{x}}^{[0,n]}(\mathbf{u}) \; : \; 0 \leq \mathbf{u} \leq \mathbf{x} \text{ and } n \geq 1\right\}$$

is norm relatively compact. It follows that there exists a convergent subsequence $P_\mathbf{x}^{[0,n_\iota]}(\mathbf{u})$. Given another $\mathbf{y} \in S^{d-1}$, it follows from the inequality (2.4) that for any $\mathbf{z} \in \mathfrak{X}$

$$\left\| P_\mathbf{x}^{[n,n+1]}(\mathbf{z}) - P_\mathbf{y}^{[n,n+1]}(\mathbf{z}) \right\|_1 = \left\| \mathbf{Q}(V^n(\mathbf{x}), \mathbf{z}) - \mathbf{Q}(V^n(\mathbf{y}), \mathbf{z}) \right\|_1$$

$$\leq \left\| V^n(\mathbf{x}) - V^n(\mathbf{y}) \right\|_1 \left\| \mathbf{z} \right\|_1$$

holds true. Thus, denoting the operator norm by $\|| \cdot \||$, we get

$$\left\|\left| P_\mathbf{x}^{[n,n+1]} - P_\mathbf{y}^{[n,n+1]} \right|\right\| \leq \left\| V^n(\mathbf{x}) - V^n(\mathbf{y}) \right\|_1 .$$

Substituting $\mathbf{y} = \mathbf{p}^*$, iterating the above estimate and applying the triangle inequality several times, for a fixed $m$ and arbitrary $\mathbf{z} \in \mathfrak{X}$ we obtain

$$\lim_{n \to \infty} \left\| P_\mathbf{x}^{[0,n+m]}(\mathbf{z}) - P_{\mathbf{p}^*}^m(P_\mathbf{x}^{[0,n]}(\mathbf{z})) \right\|_1 = 0.$$

In particular, denoting the limit point $\lim_{\iota \to \infty} P_\mathbf{x}^{[0,n_\iota]}(\mathbf{u}) = \mathbf{w}$ (as before $0 \leq \mathbf{u} \leq \mathbf{x}$), we have $\lim_{\iota \to \infty} P_\mathbf{x}^{[0,n_\iota+1]}(\mathbf{u}) = P_{\mathbf{p}^*}(\mathbf{w})$. Now, let $\omega(\mathbf{u})$ be the $\omega$-limit (closed) set

$$\omega(\mathbf{u}) = \left\{ \mathbf{w} = \lim_{\iota \to \infty} P_\mathbf{x}^{[0,n_\iota]}(\mathbf{u}) \; : \; \text{for all subsequences } n_\iota \text{ such that the limit exists} \right\}.$$

Clearly, $\omega(\mathbf{u}) \subset \{\theta \; : \; 0 \leq \theta \leq \mathbf{p}^*\}$, and therefore $S_\mathbf{w} \subseteq S_{\mathbf{p}^*}$. Hence $\omega(\mathbf{u})$ is compact and obviously it is a $P_{\mathbf{p}^*}$-invariant subset of $\mathfrak{X}$. Manipulating subsequences $n_\iota$ (if necessary) it follows from the norm compactness of $\omega(\mathbf{u})$ that for each natural $m$ and any $\mathbf{w} \in \omega(\mathbf{u})$ there exists a $\mathbf{w}_m \in \omega(\mathbf{u})$ such that $P_{\mathbf{p}^*}^m(\mathbf{w}_m) = \mathbf{w}$. Choosing a subsequence $\mathbf{w}_{m_\iota}$ we have $\lim_{\iota \to \infty} \mathbf{w}_{m_\iota} = \mathbf{w}_\star \in \omega(\mathbf{u})$. We have obtained

$$\left\| P_{\mathbf{p}^*}^{m_\iota}(\mathbf{w}_\star) - \mathbf{w} \right\|_1 = \left\| P_{\mathbf{p}^*}^{m_\iota}(\mathbf{w}_\star) - P_{\mathbf{p}^*}^{m_\iota}(\mathbf{w}_{m_\iota}) \right\|_1 \leq \left\| \mathbf{w}_\star - \mathbf{w}_{m_\iota} \right\|_1 \to 0.$$

It follows from $S_{\mathbf{w}_\star} \subseteq S_{\mathbf{p}^*}$ and Lemma 2.3.2 that $\lim_{m \to \infty} P_{\mathbf{p}^*}^m(\mathbf{w}_\star) = \|\mathbf{w}_\star\|_1 \mathbf{p}^*$. We get $\mathbf{w} = \|\mathbf{w}_\star\|_1 \mathbf{p}^* = \|\mathbf{u}\|_1 \mathbf{p}^*$. Thus $\omega(\mathbf{u})$ is a singleton and the limit $\lim_{n \to \infty} P_\mathbf{x}^{[0,n]}(\mathbf{u}) = \|\mathbf{u}\|_1 \mathbf{p}^*$ exists.

Now let us apply the fact that all operators $P_\mathbf{x}^{[0,n]}$ are linear and the vectors $\mathbf{u}$ from the order interval $[0, \mathbf{x}]$ span the whole $\mathfrak{X}(S_\mathbf{x})$. Thus $\lim_{n \to \infty} P_\mathbf{x}^{[0,n]}(\mathbf{z}) = (\sum_{k=1}^d z_k)\mathbf{p}^*$ holds for all $\mathbf{z} \in \mathfrak{X}(S_\mathbf{x})$. Finally, if $\mathbf{z} \in S^{d-1}$ with $S_\mathbf{z} \subseteq S_\mathbf{x}$, then $\lim_{n \to \infty} P_\mathbf{x}^{[0,n]}(\mathbf{z}) = \mathbf{p}^*$.

Directly from the above theorem we obtain

**Corollary 2.3.4** *Let V be a q.s.o. Given* $\mathbf{x} \in S^{d-1}$, *the iterates* $V^n(\mathbf{x})$ *converge to a (V-invariant) probability vector* $\mathbf{p}^*$ *if and only if*

$$\lim_{n \to \infty} \left\| P_{\mathbf{x}}^{[0,n]}(\mathbf{z}) - \mathbf{p}^* \right\|_1 = 0$$

*for all* $\mathbf{z} \in S^{d-1}$ *satisfying* $S_{\mathbf{z}} \subseteq S_{\mathbf{x}}$.

If $S_{\mathbf{z}} \subseteq S_{\mathbf{x}}$ fails, then the trajectories of the associated Markov chain $P_{\mathbf{x}}^{[0,n]}(\mathbf{z})$ may behave differently than in the above statement. For instance, if $\mathcal{X}$ is infinite-dimensional we have

*Example 2.3.1* Let $V$ be a q.s.o. defined on $\ell^1$ by $p_{11,1} = p_{22,2} = p_{12,2} = 1$ and $p_{ij,i+j} = 1$ if $\max\{i,j\} \geq 3$. Clearly, $\mathbf{p}^* = \mathbf{e}_1$ is $V$-invariant and $P_{\mathbf{e}_1}^n(\mathbf{e}_2) = \mathbf{e}_2 \neq \mathbf{e}_1$ for all $n$. Moreover, the sequence $P_{\mathbf{e}_1}^n(\mathbf{z})$ diverges if $\mathbf{z}$ is supported on $\{3, 4, \dots\}$.

## 2.4 Asymptotic Stability of q.s.o.s and Markov Processes

In this section we focus on the asymptotic stability (regularity) of quadratic stochastic operators and associated Markov process. Applying the results of the previous section we show that the asymptotic stability of a nonlinear transformation **Q** may be expressed in terms of the convergence of the associated nonhomogeneous Markov chains.

**Theorem 2.4.1** *Let V be a q.s.o. The following statements are equivalent:*

*(i) V is asymptotically stable;*
*(ii) there exists a* $\mathbf{p}^* \in S^{d-1}$ *such that for all* $\mathbf{x} \in S^{d-1}$ *and* $\mathbf{z} \in S^{d-1}$ *with* $S_{\mathbf{z}} \subseteq S_{\mathbf{x}}$ *we have*

$$\lim_{n \to \infty} \left\| P_{\mathbf{x}}^{[0,n]}(\mathbf{z}) - \mathbf{p}^* \right\|_1 = 0;$$

*(iii) there exists a* $\mathbf{p}^* \in S^{d-1}$ *such that for all* $m \geq 0$ *and all* $\mathbf{x} \in S^{d-1}$, $\mathbf{z} \in S^{d-1}$ *with* $S_{\mathbf{z}} \subseteq S_{V^m(\mathbf{x})}$ *we have*

$$\lim_{n \to \infty} \left\| P_{\mathbf{x}}^{[m,n]}(\mathbf{z}) - \mathbf{p}^* \right\|_1 = 0.$$

*Proof* The implication (i) $\Rightarrow$ (ii) follows directly from Theorem 2.3.3. In order to prove (ii) $\Rightarrow$ (iii) notice that $P_{\mathbf{x}}^{[m,n]}(\mathbf{z}) = P_{V^m(\mathbf{x})}^{[0,n-m]}(\mathbf{z})$. Applying (ii) we get

$$\lim_{n \to \infty} P_{\mathbf{x}}^{[m,n]}(\mathbf{z}) = \mathbf{p}^*.$$

The implication (iii) $\Rightarrow$ (i) is trivial. It is sufficient to put $m = 0$ and $\mathbf{z} = \mathbf{x}$.

The following example shows that in the above theorem the assumptions $S_\mathbf{z} \subseteq S_\mathbf{x}$ and $S_\mathbf{z} \subseteq S_{V^m(\mathbf{x})}$, respectively, may perhaps be removed.

*Example 2.4.1* Let

$$p_{11,1} = p_{22,1} = 1,$$

$$p_{12,1} = p_{21,1} = p_{12,2} = p_{21,2} = \frac{1}{2}.$$

We notice that $\mathbf{Q} = [q_{ij,k}]$ is a properly defined q.s.o. on $\mathbb{R}^2$. Clearly, $\mathbf{e}_1$ is $V$-invariant and

$$V^{n+1}(\mathbf{x})_2 = V^n(\mathbf{x})_1 V^n(\mathbf{x})_2 = V^n(\mathbf{x})_2(1 - V^n(\mathbf{x})_2).$$

The iterates of the function $[0, 1] \ni t \mapsto \varphi(t) = t(1-t) \in [0, \frac{1}{4}]$ tend to 0. It follows that $\lim_{n\to\infty} V^n(\mathbf{x})_2 = 0$. We infer that $\lim_{n\to\infty} V^n(\mathbf{x})_1 = 1$. Hence $V$ is asymptotically stable with the unique invariant distribution $\mathbf{e}_1 = (1, 0)$. As in the thesis of Theorem 2.4.1

$$P_{\mathbf{e}_1}^n(x_1, x_2) = (x_1 + (1 - 0.5^n)x_2, 0.5^n x_2) \to \mathbf{e}_1$$

for all $(x_1, x_2) \in S^{d-1}$.

**Question** Are the assumptions $S_\mathbf{z} \subseteq S_\mathbf{x}$ in (2) and $S_\mathbf{z} \subseteq S_{V^m(\mathbf{x})}$ in (3) in Theorem 2.4.1 redundant in general?

A partial answer is included in the next theorem.

**Theorem 2.4.2** *Given a q.s.o. $V$ such that for all $\mathbf{x}, \mathbf{z} \in S^{d-1}$ the trajectories $\{P_\mathbf{x}^{[0,n]}(\mathbf{z}) : n = 1, 2, \ldots\}$ are norm relatively compact, the following statements are equivalent:*

(i) *$V$ is asymptotically stable;*
(ii) *there exists a $\mathbf{p}^* \in S^{d-1}$ such that for all $\mathbf{x} \in S^{d-1}$ and $\mathbf{z} \in S^{d-1}$ we have*

$$\lim_{n\to\infty} \left\| P_\mathbf{x}^{[0,n]}(\mathbf{z}) - \mathbf{p}^* \right\|_1 = 0;$$

(iii) *there exists a $\mathbf{p}^* \in S^{d-1}$ such that for all $m \geq 0$ and all $\mathbf{x} \in S^{d-1}$, $\mathbf{z} \in S^{d-1}$ we have*

$$\lim_{n\to\infty} \left\| P_\mathbf{x}^{[m,n]}(\mathbf{z}) - \mathbf{p}^* \right\|_1 = 0$$

*(in particular, independently of a seed $\mathbf{x} \in S^{d-1}$, all nonhomogeneous Markov chains $\mathbf{P}_\mathbf{x} = (P_\mathbf{x}^{[n,n+1]})_{n\geq 0}$ are asymptotically stable with a common limit distribution $\mathbf{p}^*$).*

*Proof* Since $P_{\mathbf{x}}^{[m,n]}(\mathbf{z}) = P_{V^m(\mathbf{x})}^{[0,n-m]}(\mathbf{z})$, it follows that it is sufficient to prove the implication $(i) \Rightarrow (ii)$. For arbitrarily fixed $\mathbf{x}, \mathbf{z} \in S^{d-1}$ let

$$\omega(\mathbf{x}, \mathbf{z}) = \overline{\left\{ \mathbf{w} : \exists_{n_j \nearrow \infty} \ P_{\mathbf{x}}^{[0,n_j]}(\mathbf{z}) \to \mathbf{w} \right\}}^{\|\cdot\|_1}$$

be a closed $\omega$-limit set. By the assumption of the relative compactness of trajectories, the set $\omega(\mathbf{x}, \mathbf{z})$ is nonempty and norm compact. Moreover, it is contained in $S^{d-1}$. It follows from the asymptotic stability $(V^{n_j-1}(\mathbf{x}) \to \mathbf{p}^*)$ and $P_{\mathbf{x}}^{[0,n_j]}(\mathbf{z}) = P_{V^{n_j-1}(\mathbf{x})}^{[0,1]}(P_{\mathbf{x}}^{[0,n_j-1]}(\mathbf{z}))$ that, choosing a subsequence $n_{j_k}$ if necessary, any limit point $\mathbf{w} \in \omega(\mathbf{x}, \mathbf{z})$ has a representation $\mathbf{w} = P_{\mathbf{p}^*}(\mathbf{v})$ for some $\mathbf{v} \in \omega(\mathbf{x}, \mathbf{z})$. Considering $P_{\mathbf{x}}^{[0,n_j+1]}(\mathbf{z}) = P_{V^{n_j}(\mathbf{x})}^{[0,1]}(P_{\mathbf{x}}^{[0,n_j]}(\mathbf{z}))$ and passing to the limit as $j \to \infty$ we get $P_{\mathbf{p}^*}(\omega(\mathbf{x}, \mathbf{z})) \subseteq \omega(\mathbf{x}, \mathbf{z})$. It follows that $P_{\mathbf{p}^*}(\omega(\mathbf{x}, \mathbf{z})) = \omega(\mathbf{x}, \mathbf{z})$. Now for $\mathbf{z} \in S^{d-1}$ let us introduce

$$L(\mathbf{z}) = \limsup_{n \to \infty} (P_{\mathbf{x}}^{[0,n]}(\mathbf{z}))(S_{\mathbf{p}^*}) = \limsup_{n \to \infty} \sum_{j \in S_{\mathbf{p}^*}} (P_{\mathbf{x}}^{[0,n]}(\mathbf{z}))_j = \sup_{\mathbf{w} \in \omega(\mathbf{x},\mathbf{z})} \mathbf{w}(S_{\mathbf{p}^*}).$$

It follows from the compactness of $\omega(\mathbf{x}, \mathbf{z})$ that there exists a $\mathbf{w}_* \in \omega(\mathbf{x}, \mathbf{z})$ such that $L(\mathbf{z}) = \mathbf{w}_*(S_{\mathbf{p}^*})$. Since $S_{\mathbf{p}^*}$ is $V$-invariant, then $\sum_k p_{ij,k} 1_{S_{\mathbf{p}^*}}(k) = 1$ whenever $i, j \in S_{\mathbf{p}^*}$. Now for any $i \in S_{\mathbf{p}^*}$ we get

$$P_{\mathbf{p}^*}^*(1_{S_{\mathbf{p}^*}})(i) = \sum_k [P_{\mathbf{p}^*}]_{i,k} 1_{S_{\mathbf{p}^*}}(k) = \sum_k \sum_j p_j^* p_{ij,k} 1_{S_{\mathbf{p}^*}}(k) = \sum_j p_j^* = 1,$$

where $P_{\mathbf{p}^*}^* : \mathcal{X}^* \to \mathcal{X}^*$ stands for the adjoint operator. We get $P_{\mathbf{p}^*}^*(1_{S_{\mathbf{p}^*}}) \geq 1_{S_{\mathbf{p}^*}}$. From

$$P_{\mathbf{p}^*}(\mathbf{w}_*)(S_{\mathbf{p}^*}) = \langle P_{\mathbf{p}^*} \mathbf{w}_*, 1_{S_{\mathbf{p}^*}} \rangle = \langle \mathbf{w}_*, P_{\mathbf{p}^*}^* 1_{S_{\mathbf{p}^*}} \rangle \geq \langle \mathbf{w}_*, 1_{S_{\mathbf{p}^*}} \rangle = \mathbf{w}_*(S_{\mathbf{p}^*}) = L(\mathbf{z})$$

we infer that $P_{\mathbf{p}^*}(\mathbf{w}_*)(S_{\mathbf{p}^*}) = L(\mathbf{z})$ as $P_{\mathbf{p}^*}(\mathbf{w}_*) \in \omega(\mathbf{x}, \mathbf{z})$. Applying the induction method one gets $P_{\mathbf{p}^*}^n(\mathbf{w}_*)(S_{\mathbf{p}^*}) = L(\mathbf{z})$ for all $n = 0, 1, \ldots$.

If $L(\mathbf{z}) = 1$, then the limit point $\mathbf{w}_*$ is supported on the set $S_{\mathbf{p}^*}$. Hence by Theorem 2.3.3 we have $\lim_{n \to \infty} \|P_{\mathbf{p}^*}^n(\mathbf{w}_*) - \mathbf{p}^*\|_1 = 0$. In particular, $\mathbf{p}^* \in \omega(\mathbf{x}, \mathbf{z})$.

Let $n_j \nearrow \infty$ be any sequence of natural numbers such that $\|P_{\mathbf{x}}^{[0,n_j]}(\mathbf{z}) - \mathbf{p}^*\|_1 \to 0$. For any $\mathbf{w} \in \omega(\mathbf{x}, \mathbf{z})$ there exists a $k_j \nearrow \infty$ such that $\lim_{j \to \infty} \|P_{\mathbf{x}}^{[0,n_j+k_j]}(\mathbf{z}) - \mathbf{w}\|_1 = 0$.

Hence,

$$
\begin{aligned}
\|\mathbf{w} - \mathbf{p}^*\|_1 &= \lim_{j \to \infty} \left\| P_{\mathbf{x}}^{[0,n_j+k_j]}(\mathbf{z}) - V^{n_j+k_j}(\mathbf{x}) \right\|_1 \\
&= \lim_{j \to \infty} \left\| P_{V^{n_j}(\mathbf{x})}^{[0,k_j]}(P_{\mathbf{x}}^{[0,n_j]}(\mathbf{z})) - P_{V^{n_j}(\mathbf{x})}^{[0,k_j]}(V^{n_j}(\mathbf{x})) \right\|_1 \\
&= \lim_{j \to \infty} \left\| P_{V^{n_j}(\mathbf{x})}^{[0,k_j]}(P_{\mathbf{x}}^{[0,n_j]}(\mathbf{z}) - V^{n_j}(\mathbf{x})) \right\|_1 \\
&\leq \lim_{j \to \infty} \left\| P_{\mathbf{x}}^{[0,n_j]}(\mathbf{z}) - V^{n_j}(\mathbf{x}) \right\|_1 \\
&= \lim_{j \to \infty} \left\| P_{\mathbf{x}}^{[0,n_j]}(\mathbf{z}) - \mathbf{p}^* \right\|_1 = 0.
\end{aligned}
$$

In particular, $L(\mathbf{z}) = 1$ implies that $\omega(\mathbf{x}, \mathbf{z}) = \{\mathbf{p}^*\}$ and the convergence

$$
\lim_{n \to \infty} P_{\mathbf{x}}^{[0,n]}(\mathbf{z}) = \mathbf{p}^*
$$

follows.

Now let us suppose that $0 < L(\mathbf{z}) < 1$. We have a representation

$$
\mathbf{w}_* = L(\mathbf{z}) \mathbf{1}_{S_{\mathbf{p}^*}} \frac{\mathbf{w}_*}{L(\mathbf{z})} + (1 - L(\mathbf{z})) \mathbf{u}_*,
$$

where $\mathbf{u}_* = \mathbf{1}_{S_{\mathbf{p}^*}^{\mathbb{C}}} \frac{\mathbf{w}_*}{1-L(\mathbf{z})}$. We easily find that for any natural $j$ the iterates $P_{\mathbf{p}^*}^{j}(\mathbf{u}_*)$ are supported on $S_{\mathbf{p}^*}^{\mathbb{C}}$ (otherwise $\limsup_{n \to \infty} P_{\mathbf{x}}^{[0,n+j]}(\mathbf{z})(S_{\mathbf{p}^*}) \geq L(\mathbf{z}) + P_{\mathbf{p}^*}^{j}(\mathbf{u}_*)(S_{\mathbf{p}^*}) > L(\mathbf{z})$). Applying the Eberlein mean ergodic theorem the Cesàro means $\frac{1}{N} \sum_{n=1}^{N} P_{\mathbf{p}^*}^{n}(\mathbf{u}_*)$ converge in the norm to a $P_{\mathbf{p}^*}$-invariant vector $\mathbf{a}$ which is lattice orthogonal to $\mathbf{p}^*$. We have $\sum_{i,j=1} p_i^* a_j p_{ij,k} = a_k$. Therefore, $p_{ij,k} = 0$ if $i \in S_{\mathbf{p}^*}$, $j \in S_{\mathbf{a}}$ and $k \in S_{\mathbf{a}}^{\mathbb{C}}$. Given $\mathbf{r} \in S^{d-1}$, we put $A(\mathbf{r}) = \mathbf{r}(S_{\mathbf{a}})$ and $B(\mathbf{r}) = \mathbf{r}(S_{\mathbf{p}^*})$, i.e.

$$
A(\mathbf{r}) = \sum_{j \in S_{\mathbf{a}}} r_j, \quad B(\mathbf{r}) = \sum_{i \in S_{\mathbf{p}^*}} r_i.
$$

Clearly, $A(\mathbf{r}) + B(\mathbf{r}) \leq 1$. It can easily be seen that $B(V(\mathbf{r})) \geq B^2(\mathbf{r})$ and $A(V(\mathbf{r})) \geq 2A(\mathbf{r})B(\mathbf{r})$ for any $\mathbf{r} \in S^{d-1}$. Now let us consider a $P_{\mathbf{p}^*}$-invariant vector $\mathbf{s} = L(\mathbf{z})\mathbf{p}^* + (1 - L(\mathbf{z}))\mathbf{a}$. Substituting $\mathbf{r} = \mathbf{s}$ it follows that $A(V^n(\mathbf{s})) > 0$ and $B(V^n(\mathbf{s})) > 0$ for all natural $n$. By our assumption (i) the convergence $\lim_{n \to \infty} V^n(\mathbf{s}) = \mathbf{p}^*$ holds true. Hence $\lim_{n \to \infty} B(V^n(\mathbf{s})) = 1$. For $n$ large enough we have $B(V^n(\mathbf{s})) \geq \frac{3}{4}$.

Then

$$\limsup_{k \to \infty} A(V^{n+k}(\mathbf{s})) \geq \limsup_{k \to \infty} 2A(V^{n+k-1}(\mathbf{s}))B(V^{n+k-1}(\mathbf{s}))$$

$$\geq \limsup_{k \to \infty} \left(\frac{3}{2}\right)^k A(V^n(\mathbf{s})) = \infty,$$

a contradiction.

It remains to discuss the case $L(\mathbf{z}) = 0$. We have $P_{\mathbf{p}^*}^n(\mathbf{w})(S_{\mathbf{p}^*}) = 0$ for each $\mathbf{w} \in \omega(\mathbf{x}, \mathbf{z})$, where $n = 0, 1, \ldots$. The Cesàro limit $\lim_{N \to \infty} \frac{1}{N} \sum_{n=1}^{N} P_{\mathbf{p}^*}^n(\mathbf{w}) = \mathbf{a}$ is invariant and disjoint from $\mathbf{p}^*$. Defining the $P_{\mathbf{p}^*}$-invariant vector $\mathbf{s} = \frac{1}{2}\mathbf{p}^* + \frac{1}{2}\mathbf{a}$ we obtain, as before, $A(V^n(\mathbf{s})) \to \infty$, which contradicts our assumption.

## 2.5 Comments and References

We note that some parts of Sect. 2.1 are taken from [139] and [93]. Section 2.2 was published in [138]. Sections 2.3 and 2.4 are taken from [15]. Some generalizations of these sections have already been published in [16].

The considered q.s.o. has a direct connection to nonlinear Markov evolution (see [257] and [50]). Nonlinear Markov evolution (see [123, 124]) has become a subject of interest due to its immense range of applications, which include population and disease dynamics, statistical mechanics, evolutionary biology and economic and social systems. The fundamental issue is the study of the limit behavior of such processes, but because of nonlinearity the problem is not easily tractable. However, it appears that the theory of nonlinear Markov processes is rooted in the study of linear Markov semigroups and processes, whose theory is a well-developed field of mathematics.

It is known that Lotka–Volterra (LV) systems typically model the time evolution of conflicting species in biology [112, 113, 134, 252]. On the other hand, the use of LV discrete-time systems is a well-known subject of applied mathematics [99, 139]. They were first introduced in a bio-mathematical context by Moran [157], and later popularized in [150–153, 243, 245, 249]. Since then, LV systems have proved to be a rich source of analysis for the investigation of dynamical properties and modeling in different domains (see for example, [100, 135, 207]). Typically in all these applications, the LV systems are taken to be quadratic. The dynamical properties of some LV systems are studied in [13, 192, 204].

The main problem in nonlinear operator theory is to study the behavior of nonlinear operators. Even in the simplest case (q.s.o.s), this study is not complete [37].

In [72, 86, 88, 117, 138, 140, 243] the stability of trajectories of q.s.o.s are investigated, i.e. the uniqueness of fixed points are studied. Note that Theorems 2.1.8 and 2.2.5 were proved in [87]. The results of Sects. 2.3 and 2.4 have been taken from [138] and [15], respectively. A quadratic stochastic process such that the dynamics of a q.s.o. is described by such a process was first introduced in [73]. For this kind of process a regularity condition was studied. In the next chapter we will deal with the mentioned processes.

In [38, 70, 71, 83, 126, 127, 130, 225, 228] the structures of fixed and periodic points of q.s.o.s are studied. In [119, 120, 257] the limiting behaviors of q.s.o.s are investigated. Certain ergodic type theorems for q.s.o.s have been proved in [86, 146, 147, 160, 229, 230, 233, 257].

We stress that the asymptotic behavior of quadratic stochastic operators is complicated even on a low dimensional simplex.

Volterra quadratic operators, i.e. q.s.o.s with the constraint $p_{ij,k} = 0$ if $k \notin \{i,j\}$, were introduced in [74]. Particular cases of such operators (in low dimensions) are discussed in [251, 254]. There it is shown that the dynamics of Volterra operators might not be regular, i.e. even the Cesàro averages $\frac{1}{n}\sum_{k=1}^{n} V^n$ may not converge. Such operators are called *non-ergodic*. More systematic studies of Volterra operators are carried out in [74, 76, 80, 130–132, 191, 227, 231]. In [52, 54, 55, 63, 69, 109, 131] sufficient conditions are found for Volterra operators to be non-ergodic. It turns out that such operators may have different sort of behaviors such as Li–Yorke chaos (see [221, 222]). Other properties of Volterra operators have been discussed in [65, 66, 90, 108, 110, 198, 217].

Furthermore, in order to understand the dynamics of q.s.o.s many researchers have focussed on a certain class of q.s.o.s and studied their behavior. So, in [77–79, 81, 83] permutations of Volterra operators are discussed and it is shown that only these kind of operators (among q.s.o.s) form automorphisms of the simplex. In [90, 213, 214] $\ell$-Volterra operators were introduced and their dynamics were studied. Recently, in [186, 189, 193–195] the limiting behavior of $\ell$-Volterra and permuted Volterra operators were investigated. These works show that the dynamics of the mentioned operators can be very complicated. There are many classes of q.s.o.s, such as quasi-Volterra q.s.o.s [64], non Volterra operators generated by a product measure [212], $F$-q.s.o.s [23, 215], strictly non-Volterra [216] and bistochastic q.s.o.s [75, 82, 85, 89, 224]. However, these classes of q.s.o.s together do not cover all q.s.o.s. There are many classes of q.s.o.s which have not yet been studied.

In [155, 156] the behavior of q.s.o.s defined by means of Ising and Potts models was investigated. The study of genetic processes given by q.s.o.s via the theory of Gibbs distributions is proposed in [47, 49, 68].

A self-contained exposition of recent achievements and open problems in the theory of q.s.o.s is given in [84].

# Chapter 3
# Quadratic Stochastic Processes

In this chapter we introduce quadratic stochastic processes (q.s.p.s) and give some examples of such processes. Furthermore, constructions of q.s.p.s are provided. Associated with a given q.s.p. are two kind of processes, called *marginal processes*, one of which is a Markov process. We prove that such processes uniquely determine a q.s.p. This allows us to construct a discrete q.s.p. from a given q.s.o. Moreover, we provide other constructions of nontrivial examples of q.s.p.s. The weak ergodicity of q.s.p.s is also studied in terms of the marginal processes.

## 3.1 Definition of Quadratic Processes

A Markov chain is completely defined by its one-step transition probability matrix $(P_{ij})_{i,j=1}$ and the specification of a probability distribution on the state of the process at time 0. The analysis of a Markov chain mainly concerns the calculation of the probabilities of the possible realizations of the process. Central to these calculations are the $n$-step transition probability matrices.

**Theorem 3.1.1** *The n-step transition probabilities of a homogeneous Markov chain satisfy*

$$P_{ij}^{(n)} = \sum P_{ik} P_{kj}^{(n-1)}, \tag{3.1}$$

*where*

$$P_{ij}^{(0)} = \begin{cases} 1, & \text{if } i = j, \\ 0, & \text{otherwise.} \end{cases}$$

© Springer International Publishing Switzerland 2015
F. Mukhamedov, N. Ganikhodjaev, *Quantum Quadratic Operators and Processes*,
Lecture Notes in Mathematics 2133, DOI 10.1007/978-3-319-22837-2_3

Quadratic processes arise naturally in the study of certain models with interactions, where interactions are described by quadratic stochastic operators. Namely, following the standard procedure, we first produce formulas for computing the $n$-step transition functions of a quadratic stochastic process when the set $E$ is finite and the time $t$ is discrete.

Let $E = \{1, \cdots, m\}$ and $S^{m-1}$ be the corresponding simplex, and let $\mathbf{x}^{(0)} = (x_1^{(0)}, \cdots, x_m^{(0)})$ be an initial distribution on $E$. For arbitrary moments of time $s$ and $t$ with $s \leq t$ the transition function $P(s, i, j, t, k)$ is defined as the probability of the following event: if states $i$ and $j$ interact at time $s$, then with probability $P(s, i, j, t, k)$ the state $k$ will be realized at time $t$. Assume that $P_{ij,k} = P(0, i, j, 1, k)$ and $P_{ij,k}^{[s,t]} = P(s, i, j, t, k)$. By $V^{[s,t]}$ we denote the corresponding q.s.o. defined by the family of transition functions $P_{ij,k}^{[s,t]}$, i.e.

$$(V^{[s,t]}\mathbf{x})_k = \sum_{i,j=1}^{m} P_{ij,k}^{[s,t]} x_i x_j, \quad (k = 1, \cdots, m), \quad \mathbf{x} \in S^{m-1}. \tag{3.2}$$

For the sake of simplicity, as in the Markov case, we assume that $V^{[t,t+1]} = V$, that is $P_{ij,k}^{[t,t+1]} = P_{ij,k}$ for any $i, j, k$ and for all $t \geq 1$. This property is called *homogeneity per unit of time*. Note that for Markov chains the homogeneity follows from the homogeneity per unit time.

Now due to (3.2) the distribution $\mathbf{x}^{(1)} = (x_1^{(1)}, \cdots, x_m^{(1)})$ at the moment of time $t = 1$ is defined as follows

$$x_k^{(1)} = \sum_{i,j=1}^{m} P_{ij,k} x_i^{(0)} x_j^{(0)}, \quad (k = 1, \cdots, m). \tag{3.3}$$

An operator $V^{[0,2]}$ is defined as

$$V^{[0,2]}\mathbf{x}^{(0)} = V\mathbf{x}^{(1)} = \mathbf{x}^{(2)}$$

or in a coordinate form as follows

$$x_k^{(2)} = \sum_{i,j=1}^{m} P_{ij,k} x_i^{(1)} x_j^{(1)} = \sum_{\alpha,\beta=1}^{m} P_{\alpha\beta,k}^{[0,2]} x_\alpha^{(0)} x_\beta^{(0)}, \quad (k = 1, \cdots, m). \tag{3.4}$$

According to (3.3) the equality (3.4) can be reduced to

$$x_k^{(2)} = \sum_{i,j,\gamma,\delta,n,l=1}^{m} P_{ij,k} P_{i\delta,k} P_{nl,j} x_\gamma^{(0)} x_\delta^{(0)} x_n^{(0)} x_l^{(0)}$$

$$= \sum_{\alpha,\beta=1}^{m} P_{\alpha\beta,k}^{[0,2]} x_\alpha^{(0)} x_\beta^{(0)}, \quad (k = 1, \cdots, m).$$

From this equality one concludes that the transition function $P_{\alpha\beta,k}^{[0,2]}$ can be found as follows:

I) If $\gamma = \alpha, \delta = \beta$, then

$$P_{\alpha\beta,k}^{[0,2]} = \sum_{i,j,n,l=1}^{m} P_{ij,k} P_{\alpha\beta,i} P_{nl,j} x_n^{(0)} x_l^{(0)} = \sum_{i,j=1}^{m} P_{\alpha\beta,i} P_{ij,k} x_j^{(1)};$$

II) If $\gamma = \alpha, n = \beta$, then

$$P_{\alpha\beta,k}^{[0,2]} = \sum_{i,j,\delta,l=1}^{m} P_{ij,k} P_{\alpha\delta,i} P_{\beta l,j} x_\delta^{(0)} x_l^{(0)};$$

III) If $\gamma = \alpha, l = \beta$, then

$$P_{\alpha\beta,k}^{[0,2]} = \sum_{i,j,\delta,n=1}^{m} P_{\alpha\delta,i} P_{\beta n,j} P_{ij,k} x_\delta^{(0)} x_n^{(0)};$$

IV) If $\delta = \alpha, n = \beta$, then

$$P_{\alpha\beta,k}^{[0,2]} = \sum_{i,j,\gamma,l=1}^{m} P_{\alpha\gamma,i} P_{\beta l,j} P_{ij,k} x_\gamma^{(0)} x_l^{(0)};$$

V) If $\delta = \alpha, l = \beta$, then

$$P_{\alpha\beta,k}^{[0,2]} = \sum_{i,j,\gamma,n=1}^{m} P_{\alpha\gamma,i} P_{\beta n,j} P_{ij,k} x_\gamma^{(0)} x_n^{(0)};$$

VI) If $n = \alpha, l = \beta$, then

$$P_{\alpha\beta,k}^{[0,2]} = \sum_{i,j,\gamma,\delta=1}^{m} P_{\alpha\beta,j} P_{ji,k} P_{\gamma\delta,i} x_\gamma^{(0)} x_\delta^{(0)} = \sum_{i,j=1}^{m} P_{\alpha\beta,j} P_{ji,k} x_i^{(1)}.$$

For all the other combinations of indexes we will have one of the equalities I)–VI) up to renumeration. Note that equality I) coincides with equality VI) and equalities II)–V) are the same. Hence, we obtain only two possible definitions of the transition function $P_{ij,k}^{[0,2]}$:

A) $P_{\alpha\beta,k}^{[0,2]} = \sum_{i,j=1}^{m} P_{\alpha\beta,j} P_{ji,k} x_i^{(1)};$

B) $P_{\alpha\beta,k}^{[0,2]} = \sum_{i,j,n,l=1}^{m} P_{\alpha i,n} P_{\beta j,l} P_{nl,k} x_i^{(0)} x_j^{(0)}.$

The equalities A) and B) can be interpreted as two different rules of reproduction of "grandchildren".

Now from the equalities

$$V^{[0,3]}\mathbf{x}^{(0)} = V^{[1,3]}\mathbf{x}^{(1)} = V\mathbf{x}^{(2)} = \mathbf{x}^{(3)}$$

we can find $P^{[0,3]}$ and $P^{[1,3]}$. As above, in a coordinate form, one obtains

$$x_k^{(3)} = \sum_{\alpha,\beta}^m P_{\alpha\beta,k}x_\alpha^{(2)}x_\beta^{(2)} = \sum_{i,j=1}^m P_{ij,k}^{[1,3]}x_i^{(1)}x_j^{(1)}$$

$$= \sum_{n,l=1}^m P_{nl,k}^{[0,3]}x_n^{(0)}x_l^{(0)}, \quad (k = 1,\cdots,m).$$

Then in case A) one finds

$$P_{ij,k}^{[1,3]} = \sum_{\alpha,\beta=1}^m P_{ij,\alpha}P_{\alpha\beta,k}x_\beta^{(2)}$$

and

$$P_{ij,k}^{[0,3]} = \sum_{\alpha,\beta=1}^m P_{ij,\alpha}P_{\alpha\beta,k}^{[1,3]}x_\beta^{(1)} = \sum_{\alpha,\beta=1}^m P_{ij,\alpha}^{[0,2]}P_{\alpha\beta,k}x_\beta^{(2)}$$

and in case B) we have

$$P_{ij,k}^{[1,3]} = \sum_{\alpha,\beta,\gamma,\delta=1}^m P_{i\alpha,\gamma}P_{j\beta,\delta}P_{\gamma\delta,k}x_\alpha^{(1)}x_\beta^{(1)}$$

and

$$P_{ij,k}^{[0,3]} = \sum_{\alpha,\beta,\gamma,\delta=1}^m P_{i\alpha,\gamma}P_{j\beta,\delta}P_{\gamma\delta,k}^{[1,3]}x_\alpha^{(0)}x_\beta^{(0)} = \sum_{\alpha,\beta,\gamma,\delta=1}^m P_{i\alpha,\gamma}^{[0,2]}P_{j\beta,\delta}^{[0,2]}P_{\gamma\delta,k}x_\alpha^{(0)}x_\beta^{(0)}.$$

Now by induction, one can show that in case A)

$$P_{ij,k}^{[s,t]} = \sum_{\alpha,\beta=1}^m P_{ij,\alpha}^{[s,\tau]}P_{\alpha\beta,k}^{[\tau,t]}x_\beta^{(\tau)}, s < \tau < t, \tag{3.5}$$

where

$$x_\beta^{(\tau)} = \sum_{i,j=1}^{m} P_{ij,\beta}^{[0,\tau]} x_i^{(0)} x_j^{(0)}. \tag{3.6}$$

Hence, in case B) we have

$$P_{ij,k}^{[s,t]} = \sum_{\alpha,\beta,\gamma,\delta=1}^{m} P_{i\alpha,\gamma}^{[s,\tau]} P_{j\beta,\delta}^{[\tau,t]} P_{\gamma\delta,k}^{[\tau,t]} x_\alpha^{(s)} x_\beta^{(s)}. \tag{3.7}$$

So, we are ready to define a quadratic process.

**Definition 3.1.1**  A family of functions $\{P_{ij,k}^{[s,t]} : i,j,k \in E, s,t \in \mathbb{R}_+, t-s \geq 1\}$ with an initial state $\mathbf{x}^{(0)} = (x_k^{(0)})_{k \in E} \in S^{m-1}$ is said to be a *quadratic stochastic process* (*q.s.p.*) if, for fixed $s,t \in \mathbb{R}_+$, it satisfies the following conditions:

(i)  $P_{ij,k}^{[s,t]} = P_{ji,k}^{[s,t]}$ for any $i,j,k \in E$;

(ii)  $P_{ij,k}^{[s,t]} \geq 0$ and $\sum_{k \in E} P_{ij,k}^{[s,t]} = 1$ for any $i,j,k \in E$;

(iii)  An analogue of the Kolmogorov–Chapman equation: there are two variants: for the initial point $\mathbf{x}^{(0)}$ and $s < r < t$ such that $t - r \geq 1, r - s \geq 1$ one has

(iii$_A$)

$$P_{ij,k}^{[s,t]} = \sum_{m,l,k} P_{ij,m}^{[s,r]} P_{ml,k}^{[r,t]} x_l^{(r)},$$

where $x_k^{(r)}$ is defined as follows:

$$x_k^{(r)} = \sum_{i,j} P_{ij,k}^{[0,r]} x_i^{[0]} x_j^{(0)};$$

(iii$_B$)

$$P_{ij,k}^{[s,t]} = \sum_{m,l,g,h} P_{im,l}^{[s,r]} P_{jg,h}^{[s,r]} P_{lh,k}^{[r,t]} x_m^{(s)} x_g^{(s)}.$$

In what follows, by $(E, P_{ij,k}^{[s,t]}, \mathbf{x}^{(0)})$ we denote the defined process. We say that the q.s.p. is of *type (A)* (resp. *type (B)*) if it satisfies the fundamental equations (*iii$_A$*) (resp. (*iii$_B$*)). The equations (*iii$_A$*) and (*iii$_B$*) can be interpreted as different laws of behavior of the "offspring".

Now we define quadratic stochastic processes in a general setting. Let $(E, \mathfrak{I})$ be a measurable space and $\mathfrak{M}$ be a collection of all probability measures on $(E, \mathfrak{I})$.

Assume that a family of transition functions $\{P(s, x, y, t, A) : x, y \in E, A \in \Im, s, t \in \mathbb{R}_+, t - s \geq 1\}$ is given and satisfies the following conditions:

   (I)   $P(s, x, y, t, A) = P(s, y, x, t, A)$, for all $x, y \in E$, and $A \in \Im$;
  (II)   $P(s, x, y, t, A) \in \mathfrak{M}$ for all $x, y \in E$, and $s, t \in \mathbb{R}_+$ with $t - s \geq 1$;
 (III)   For any fixed $A \in \Im$ and $s, t$ with $t - s \geq 1$ the function $P(s, x, y, t, A)$ as a function of two variables $x, y$ is measurable with respect to the $\sigma$-algebra $\Im \otimes \Im$ on $E \times E$;
 (IV)   For an initial measure $m_0 \in \mathfrak{M}$ and arbitrary $s, \tau, t \in \mathbb{R}_+$ such that $t - \tau \geq 1$ and $\tau - s \geq 1$, we have either

(IV)$_A$

$$P(s, x, y, t, A) = \int_E \int_E P(s, x, y, \tau, du) P(\tau, u, v, t, A) m_\tau(dv), \qquad (3.8)$$

where the measure $m_\tau$ on $(E, \Im)$ with $\tau \geq 1$ is given by

$$m_\tau(B) = \int_E \int_E P(0, x, y, \tau, B) dm_0(x) dm_0(y), \qquad (3.9)$$

or

(IV)$_B$

$$P(s, x, y, t, A) = \int_E \int_E \int_E \int_E P(s, x, z, \tau, du)$$
$$\times P(s, y, v, \tau, dw) P(\tau, u, w, t, A) m_s(dv). \qquad (3.10)$$

The first condition is determined by the symmetric property of quadratic stochastic operators, that is $p_{ij,k} = p_{ji,k}$.

The second and third conditions are reformulations of similar conditions for transition probabilities of Markov processes. The meaning of the condition $t - s \geq 1$ will be clarified below.

Condition IV) is the analogue of the Kolmogorov–Chapman equation. As we noted above there are two different forms of this condition.

Equations $(IV)_A$ and $(IV)_B$ can be interpreted as different rules for the appearance of the "grandchildren". These equations also have implications for chemical treatments. So, the appearance of particles in reactions occurring in ordinary chemical kinetics are described either by the equation of type $(IV)_A$ or by the equation of type $(IV)_B$ which reflect the appearance of particles in processes of catalysis. Then the triple $\{(E, \Im), P(s, x, y, t, A), m_0\}$ satisfying the conditions I)–III) is called a *quadratic stochastic process (q.s.p.) of type (A)* (resp. (B)) if $(IV)_A$ (resp. $(IV)_B$) is satisfied. In this definition, $P(s, x, y, t, A)$ is the probability of the following event: if $x$ and $y$ states in $E$ interact at time $s$, then one of the elements of the set $A \in \Im$ will be realized at time $t$. The realization of an interaction in a physical, chemical, or biological phenomena requires some time. We assume that the minimum of these values of time is equal to 1 (see Boltzmann's model [114] or the biological models in [137, 139]). Hence, $P(s, x, y, t, A)$ is defined for $t - s \geq 1$.

The set of quadratic stochastic processes can be decomposed into the following three classes:

(i) *homogeneous*, that is, the transition functions $P(s, x, y, t, A)$ depend only on $t - s$ for arbitrary $x, y \in E, A \in \Im, s, t \in R^+$ such that $t - s \geq 1$;

(ii) *homogeneous per unit time*, that is, the transition functions satisfy $P(t, x, y, t + 1, A) = P(0, x, y, 1, A)$ for any $t \geq 1$, but don't belong to the first class;

(iii) the processes which don't belong to the second class are called *nonhomogeneous*.

In general, homogeneity does not follow from homogeneity per unit time.

Thus, quadratic stochastic processes are related to quadratic transformations in the same way as Markov processes are related to linear transformations, and a number of concepts and problems in the theory of Markov processes can also be considered for quadratic processes.

## 3.2  Examples of Quadratic Stochastic Processes

In this subsection we are going to provide some examples of quadratic stochastic processes.

*Example 3.2.1* Let us consider a simple Mendelian inheritance for a single gene with two alleles $A$ and $a$. The rules of simple Mendelian inheritance indicate that the next generation will inherit either $A$ or $a$ with an equal frequency. For brevity let us rename the genes $A$ and $a$ by 1 and 2, respectively. Then this model of heredity is defined by a quadratic stochastic operator with $E = \{1, 2\}$ and the following transition functions (coefficients of heredity):

$$P_{11,1} = 1, P_{12,1} = P_{21,1} = \frac{1}{2}, \ P_{22,1} = 0,$$

$$P_{11,2} = 0, \ P_{12,2} = P_{21,2} = \frac{1}{2}, \ P_{22,2} = 1.$$

Let $x_1^{(0)} = x, x_2^{(0)} = 1 - x$ be an initial distribution on $E$ with $0 \leq x \leq 1$. Then the following system of transition functions

$$P_{11,1}^{[s,t]} = \frac{1}{2^{t-s-1}} \cdot [(2^{t-s-1} - 1)x + 1],$$

$$P_{12,1}^{[s,t]} = P_{21,1}^{[s,t]} = \frac{1}{2^{t-s-1}} \cdot [(2^{t-s-1} - 1)x + \frac{1}{2}],$$

$$P_{22,1}^{[s,t]} = \frac{1}{2^{t-s-1}} \cdot (2^{t-s-1} - 1)x,$$

with $P_{ij,2}^{[s,t]} = 1 - P_{ij,1}^{[s,t]}$ for all $i,j = 1,2$, satisfies all conditions I)–IV). The process determined by this collection of transition functions we call a *simple Mendel quadratic stochastic process*. Clearly, this process is homogeneous, and one has $x_1^{(t)} = x, x_2^{(t)} = 1 - x$ for any $t \geq 1$.

*Example 3.2.2* Let $E = \{1,2\}$, and a model of heredity be defined by

$$P_{11,1} = 1 - 2\varepsilon, \ P_{12,1} = P_{21,1} = \frac{1}{2} - \varepsilon, \ P_{22,1} = 0,$$

$$P_{11,2} = 2\varepsilon, \ P_{12,2} = P_{21,2} = \frac{1}{2} + \varepsilon, P_{22,2} = 1,$$

and $x_1^{(0)} = x, x_2^{(0)} = 1 - x$ ($x \in [0,1]$) be an initial distribution on $E$. Then the following system of transition functions

$$P_{11,1}^{[s,t]} = \frac{(1 - 2\varepsilon)^{t-s}}{2^{t-s-1}} \cdot [(2^{t-s-1} - 1)(1 - 2\varepsilon)^s x + 1], \tag{3.11}$$

$$P_{12,1}^{[s,t]} = P_{21,1}^{[s,t]} = \frac{(1 - 2\varepsilon)^{t-s}}{2^{t-s-1}} \cdot [(2^{t-s-1} - 1)(1 - 2\varepsilon)^s x + \frac{1}{2}], \tag{3.12}$$

$$P_{22,1}^{[s,t]} = \frac{(1 - 2\varepsilon)^{t-s}}{2^{t-s-1}} \cdot (2^{t-s-1} - 1)x, \tag{3.13}$$

with $P_{ij,2}^{[s,t]} = 1 - P_{ij,1}^{[s,t]}$ for all $i,j = 1,2$, satisfies all conditions I)–V) for $0 \leq \varepsilon \leq \frac{1}{2}$. Moreover, one has $x_1^{(t)} = (1 - 2\varepsilon)^t x, x_2^{(t)} = 1 - (1 - 2\varepsilon)^t x$ for any $t \geq 1$.

The determined process is called a *Mendel quadratic stochastic process*. If $\varepsilon = 0$, then such a process is homogeneous, otherwise (i.e. if $\varepsilon \neq 0$) it belongs to the second class.

*Example 3.2.3* Let $E = \{1,2\}$ and $x_1^{(0)} = x, x_2^{(0)} = 1 - x$ ($x \in [0,1]$) be an initial distribution on $E$. Then the following system of transition functions

$$P_{11,1}^{[s,t]} = \frac{\varepsilon^{t-s}}{2^{t-s-1}} \cdot \frac{s+1}{t+1} \cdot [(2^{t-s-1} - 1)\frac{\varepsilon^s}{s+1}x + 1],$$

$$P_{12,1}^{[s,t]} = P_{21,1}^{[s,t]} = \frac{\varepsilon^{t-s}}{2^{t-s-1}} \cdot \frac{s+1}{t+1} \cdot [(2^{t-s-1} - 1)\frac{\varepsilon^s}{s+1}x + \frac{1}{2}],$$

$$P_{22,1}^{[s,t]} = \frac{2^{t-s-1} - 1}{2^{t-s-1}} \cdot \frac{\varepsilon^t}{t+1}x,$$

with $P_{ij,2}^{[s,t]} = 1 - P_{ij,1}^{[s,t]}$ for all $i,j = 1,2$, satisfies all conditions I)–V) for $0 \leq \varepsilon \leq 1$. In addition, one has $x_1^{(t)} = \frac{\varepsilon^t}{t+1}x, x_2^{(t)} = 1 - \frac{\varepsilon^t}{t+1}x$ for any $t \geq 1$.

These processes are of both types A and B.

*Example 3.2.4* Let $E = \{1, 2\}$ and $x_1^{(0)} = x, x_2^{(0)} = 1 - x$ ($x \in [0, 1]$) be an initial distribution on $E$. Then the following system of transition probabilities

$$P_{11,1}^{[s,t]} = x^{2^t - 2^{s+1}}, \quad P_{12,1}^{[s,t]} = P_{21,1}^{[s,t]} = P_{22,1}^{[s,t]} = 0,$$

with $P_{ij,2}^{[s,t]} = 1 - P_{ij,1}^{[s,t]}$ for all $i, j = 1, 2$, satisfies all conditions I)–IV). Moreover, one has $x_1^{(t)} = x^{2^t}, x_2^{(t)} = 1 - x^{2^t}$ for any $t \geq 1$. Such a process is of type (A).

*Example 3.2.5* Let $E = \{1, 2, 3\}$ and $x_1^{(0)} = x_1, x_2^{(0)} = x_2, x_3^{(0)} = 1 - x_1 - x_2$ be an initial distribution on $E$ with $x_1 \geq 0, x_2 \geq 0$ and $x_1 + x_2 \leq 1$. Then the following system of transition functions

$$P_{11,1}^{[s,t]} = 2\varepsilon^{t-s} + \frac{2^{t-s-1} - 1}{2^{t-s-1}} x_1^{(t+1)},$$

$$P_{12,1}^{[s,t]} = P_{13,1}^{[s,t]} = \varepsilon^{t-s} + \frac{2^{t-s-1} - 1}{2^{t-s-1}} x_1^{(t+1)},$$

$$P_{22,1}^{[s,t]} = P_{23,1}^{[s,t]} = P_{33,1}^{[s,t]} = \frac{2^{t-s-1} - 1}{2^{t-s-1}} x_1^{(t+1)},$$

$$P_{11,2}^{[s,t]} = P_{13,2}^{[s,t]} = P_{33,2}^{[s,t]} = \frac{2^{t-s-1} - 1}{2^{t-s-1}} x_2^{(t+1)},$$

$$P_{22,2}^{[s,t]} = 2\varepsilon^{t-s} + \frac{2^{t-s-1} - 1}{2^{t-s-1}} x_2^{(t+1)},$$

$$P_{12,2}^{[s,t]} = P_{23,2}^{[s,t]} = \varepsilon^{t-s} + \frac{2^{t-s-1} - 1}{2^{t-s-1}} x_2^{(t+1)},$$

with $P_{ij,3}^{[s,t]} = 1 - P_{ij,1}^{[s,t]} - P_{ij,2}^{[s,t]}$ for all $i, j = 1, 2, 3$ satisfies conditions I)–IV$_A$) for $0 \leq \varepsilon \leq \frac{1}{2}$. In addition, one has $x_1^{(t)} = (2\varepsilon)^t x_1, x_2^{(t)} = (2\varepsilon)^t x_2$ for any $t \geq 1$. This process is a homogeneous q.s.p. of type (A), but not of type (B).

*Example 3.2.6* Let $E = \{1, 2, 3\}$ and $x_1^{(0)} = x_1, x_2^{(0)} = x_2, x_3^{(0)} = 1 - x_1 - x_2$ be an initial distribution on $E$ with $x_1 \geq 0, x_2 \geq 0$ and $x_1 + x_2 \leq 1$. Then the following system of transition functions

$$\tilde{P}_{11,1}^{[s,t]} = 2\varepsilon^{t-s} + \frac{2^{t-s-1} - 1}{2^{t-s-1}} x_1^{(t)},$$

$$\tilde{P}_{12,1}^{[s,t]} = \tilde{P}_{13,1}^{[s,t]} = \varepsilon^{t-s} + \frac{2^{t-s-1} - 1}{2^{t-s-1}} x_1^{(t)},$$

$$\tilde{P}^{[s,t]}_{22,1} = \tilde{P}^{[s,t]}_{23,1} = \tilde{P}^{[s,t]}_{33,1} = \frac{2^{t-s-1} - 1}{2^{t-s-1}} x_1^{(t)},$$

$$\tilde{P}^{[s,t]}_{11,2} = \tilde{P}^{[s,t]}_{13,2} = \tilde{P}^{[s,t]}_{33,2} = \frac{2^{t-s-1} - 1}{2^{t-s-1}} x_2^{(t)},$$

$$\tilde{P}^{[s,t]}_{22,2} = 2\varepsilon^{t-s} + \frac{2^{t-s-1} - 1}{2^{t-s-1}} x_2^{(t)},$$

$$\tilde{P}^{[s,t]}_{12,2} = \tilde{P}^{[s,t]}_{23,2} = \varepsilon^{t-s} + \frac{2^{t-s-1} - 1}{2^{t-s-1}} x_2^{(t)},$$

with $\tilde{P}^{[s,t]}_{ij,3} = 1 - \tilde{P}^{[s,t]}_{ij,1} - \tilde{P}^{[s,t]}_{ij,2}$ for all $i,j = 1,2,3$ satisfies conditions I)–IV$_B$) for $0 \le \varepsilon \le \frac{1}{2}$. Moreover, one has $x_1^{(t)} = (2\varepsilon)^t x_1$, $x_2^{(t)} = (2\varepsilon)^t x_2$ for any $t \ge 1$. This process is a homogeneous q.s.p. of type (B), but not of type (A).

It is easy to see that for the Examples 3.2.5 and 3.2.6 one has

$$P^{[t,t+1]}_{ij,k} = \tilde{P}^{[t,t+1]}_{ij,k},$$

however $P^{[s,t]}_{ij,k} \ne \tilde{P}^{[s,t]}_{ij,k}$ at $t - s > 1$ for all $i,j,k$.

Thus, the sets of quadratic stochastic processes are nonempty. Moreover, quadratic processes of type (A) and type (B) are different.

## 3.3   Marginal Markov Processes Related to q.s.p.s

In this section we are going to prove that every q.s.p. can be uniquely determined by two kinds of processes.

Let $E$ be an at most countable set. In this case, the set of probability measures coincides with

$$S = \{x \in \ell^1(E) : x_n \ge 0, \; n \in E; \; \|x\|_1 = 1\},$$

where

$$\ell^1(E) = \{x = (x_n)_{n \in E} : \|x\|_1 = \sum_{n \in E} |x_n| < \infty; \; x_n \in \mathbb{R}\}.$$

Recall that a matrix $(U_{ij})_{i,j \in E}$ is called *stochastic* if for any $i,j \in E$ one has

$$U_{ij} \ge 0, \quad \sum_{j \in E} U_{ij} = 1.$$

A family of stochastic matrices $\{(U_{ij}^{[s,t]})_{i,j \in E} : s,t \in \mathbb{R}_+, t - s \geq 1\}$ is called a *Markov process* if the following condition holds: for every $s < r < t$ one has

$$U_{ij}^{[s,t]} = \sum_{k \in E} U_{ik}^{[s,r]} U_{kj}^{[r,t]}. \tag{3.14}$$

This equation is known as the Kolmogorov–Chapman equation.

Let $(E, P_{ij,k}^{[s,t]}, \mathbf{x}^{(0)})$ be a q.s.p. Let us define

$$H_{ij}^{[s,t]} = \sum_{\ell} P_{i\ell,j}^{[s,t]} x_l^{(s)}, \tag{3.15}$$

$$Q_{(ij)(uv)}^{[s,t]} = P_{ij,u}^{[s,t]} x_v^{(t)}, \tag{3.16}$$

where $i, j, u, v \in E$.

It is clear that for each pair $s, t \in \mathbb{R}_+$ the matrix $H_{ij}^{[s,t]}$ is stochastic, and $Q_{(ij)(uv)}^{[s,t]}$ is also stochastic, in the following sense

$$Q_{(ij)(uv)}^{[s,t]} \geq 0, \quad \forall (i,j), (u,v) \in E \times E$$

$$\sum_{u,v \in E} Q_{(ij)(uv)}^{[s,t]} = 1, \quad \forall (i,j) \in E \times E.$$

**Theorem 3.3.1** *Let $(E, P_{ij,k}^{[s,t]}, x^{(0)})$ be a q.s.p. of type (A). Then the defined processes $(H_{ij}^{[s,t]})$ and $(Q_{(ij)(uv)}^{[s,t]})$ are Markov processes. Moreover, for any $s, r, t \in \mathbb{R}_+$ with $t - r \geq 1, r - s \geq 1$ one has*

$$P_{ij,k}^{[s,t]} = \sum_m P_{ij,m}^{[s,r]} H_{m,k}^{[r,t]}. \tag{3.17}$$

*Proof* Using (3.15), (iii$_A$) for any $s < r < t$ we have

$$\sum_k H_{ik}^{[s,r]} H_{kj}^{[r,t]} = \sum_k \left( \sum_\ell P_{i\ell,k}^{[s,r]} x_\ell^{(s)} \right) \left( \sum_m P_{km,j}^{[r,t]} x_m^{(r)} \right)$$

$$= \sum_\ell \left( \sum_{k,m} P_{i\ell,k}^{[s,r]} P_{km,j}^{[r,t]} x_m^{(r)} \right) x_\ell^{(s)}$$

$$= \sum_\ell P_{i\ell,j}^{[s,t]} x_\ell^{(s)}$$

$$= H_{ij}^{[s,t]},$$

hence $(H_{ij}^{[s,t]})_{i,j \in E}$ is a Markov process.

Similarly, from (3.16) and (iii$_A$) one gets

$$Q^{[s,t]}_{(ij)(uv)} = \left( \sum_{m,\ell} P^{[s,r]}_{ij,m} P^{[r,t]}_{m\ell,u} x^{(r)}_\ell \right) x^{(t)}_v$$

$$= \sum_{m,\ell} P^{[s,r]}_{ij,m} x^{(r)}_\ell P^{[r,t]}_{m\ell,u} x^{(t)}_v$$

$$= \sum_{m,\ell} Q^{[s,r]}_{(ij)(m\ell)} Q^{[r,t]}_{(m\ell)(uv)}.$$

So, $(Q^{[s,t]}_{(ij)(uv)})$ is also a Markov process.

The equality (3.17) immediately follows from (iii$_A$) with (3.15). This completes the proof.

**Theorem 3.3.2** *Let $(E, P^{[s,t]}_{ij,k}, \mathbf{x}^{(0)})$ be a q.s.p. of type (B). Then the process $(H^{[s,t]}_{ij})$ defined by (3.15) is Markov. Moreover, the process $(Q^{[s,t]}_{(ij)(uv)})$ defined by (3.16) satisfies the following equation*

$$Q^{[s,t]}_{(ij)(uv)} = \sum_{\ell,k} H^{[s,\tau]}_{i\ell} H^{[s,\tau]}_{jk} Q^{[\tau,t]}_{(\ell k)(uv)}$$

*for any $t - \tau \geq 1$, $\tau - s \geq 1$. Moreover, one has*

$$P^{[s,t]}_{ij,k} = \sum_{m,\ell} H^{[s,\tau]}_{i,m} H^{[s,\tau]}_{j,\ell} P^{[\tau,t]}_{m\ell,k}. \tag{3.18}$$

*Proof* First we want to show that

$$x^{(t)}_k = \sum_{i,j} P^{[s,t]}_{ij,k} x^{(s)}_i x^{(s)}_j.$$

Indeed, from

$$x^{(s)}_k = \sum_{i,j} P^{[0,s]}_{ij,k} x^{(0)}_i x^{(0)}_j$$

and (see (iii$_B$))

$$P^{[0,t]}_{ij,k} = \sum_{m,\ell,g,u} P^{[0,s]}_{im,\ell} P^{[0,s]}_{jg,u} P^{[s,t]}_{\ell u,k} x^{(0)}_m x^{(0)}_g,$$

one finds

$$x_k^{(t)} = \sum_{i,j} P_{ij,k}^{[0,t]} x_i^{(0)} x_j^{(0)}$$

$$= \sum_{i,j} \left( \sum_{m,\ell,g,u} P_{im,\ell}^{[0,s]} P_{jg,u}^{[0,s]} P_{\ell u,k}^{[s,t]} x_m^{(0)} x_g^{(0)} \right) x_i^{(0)} x_j^{(0)}$$

$$= \sum_{\ell,u} P_{\ell u,k}^{[s,t]} \left( \sum_{i,m} P_{im,\ell}^{[0,s]} x_i^{(0)} x_m^{(0)} \right) \left( \sum_{j,g} P_{jg,u}^{[0,s]} x_j^{(0)} x_g^{(0)} \right)$$

$$= \sum_{\ell,u} P_{\ell u,k}^{[s,t]} x_\ell^{(s)} x_u^{(s)}.$$

Now, using the last equality with (iii$_B$), we obtain

$$H_{ij}^{[s,t]} = \sum_\ell \left( \sum_{m,u,g,v} P_{im,u}^{[s,\tau]} P_{\ell g,v}^{[s,\tau]} P_{uv,j}^{[\tau,t]} x_m^{(s)} x_g^{(s)} \right) x_\ell^{(s)}$$

$$= \sum_{u,v} \left( \sum_m P_{im,u}^{[s,\tau]} x_m^{(s)} \right) \left( \sum_{\ell,g} P_{\ell g,v}^{[s,\tau]} x_g^{(s)} x_\ell^{(s)} \right) P_{uv,j}^{[\tau,t]}$$

$$= \sum_{u,v} H_{i,u}^{[s,\tau]} P_{uv,j}^{[\tau,t]} x_v^{(\tau)}$$

$$= \sum_u H_{i,u}^{[s,\tau]} H_{u,j}^{[\tau,t]}.$$

This shows $(H_{ik}^{[s,t]})$ is a Markov process.

From (iii$_B$) we immediately find

$$Q_{(ij)(uv)}^{[s,t]} = P_{ij,u}^{[s,t]} x_v^{(t)} = \left( \sum_{m,\ell,g,k} P_{im,\ell}^{[s,\tau]} P_{jg,k}^{[s,\tau]} P_{\ell k,u}^{[\tau,t]} x_m^{(s)} x_g^{(s)} \right) x_v^{(t)}$$

$$= \sum_{\ell,k} \left( \sum_m P_{im,\ell}^{[s,\tau]} x_m^{(s)} \right) \left( \sum_g P_{jg,k}^{[s,\tau]} x_g^{(s)} \right) P_{\ell k,u}^{[\tau,t]} x_v^{(t)}$$

$$= \sum_{\ell,k} H_{i\ell}^{[s,\tau]} H_{jk}^{[s,\tau]} Q_{(lk)(uv)}^{[\tau,t]}.$$

The equality (3.18) immediately follows from the last equality. This completes the proof.

*Remark 3.3.1*  Theorems 3.3.1 and 3.3.2 have several applications. In [254, 257] the ergodic properties of quadratic mappings are studied. By the mentioned theorems, the sequence of trajectories of the quadratic mapping can be considered as a sequence of one-dimensional distributions of some nonhomogeneous Markovian chain, therefore one can apply the theory of nonhomogeneous Markov chains. In Sect. 3.6 we will apply this to study the weak ergodicity of q.s.p.s. Moreover, in Chap. 4 we will apply these theorems to investigate differential equations for q.s.p.s by means of Kolmogorov's differential equations for Markov processes [121].

Now we have two processes associated with a q.s.p. Note that the defined processes are related to each other by the following formula

$$\sum_j Q^{[s,t]}_{(ij)(uv)} x^{(s)}_j = H^{[s,t]}_{iu} x^{(t)}_v.$$

We are interested in the reverse problem. Namely, assume that $E$ is an at most countable set. Let us suppose that we are given two non-homogeneous processes $(\mathscr{Q}^{[s,t]}_{(ij)(uv)})_{i,j,u,v \in E}$ and $(\mathscr{H}^{[s,t]}_{ij})_{i,j \in E}$. Under what conditions do these two processes uniquely determine some q.s.p.? To answer this question, we first fix an initial state $\mathbf{x}^{(0)} = (x^0_i) \in S$, and define

$$z^{(t)}_k = \sum_i \mathscr{H}^{[0,t]}_{ik} x^{(0)}_i, \quad y^{(t)}_k = \sum_{i,j,\ell} \mathscr{Q}^{[0,t]}_{(ij)(k\ell)} x^{(0)}_i x^{(0)}_j, \quad k \in E.$$

**Theorem 3.3.3** *Let $E$ be a finite or countable set. Let $(\mathscr{Q}^{[s,t]}_{(ij)(uv)})$ and $(\mathscr{H}^{[s,t]}_{ij})$ be two stochastic processes on $E \times E$ and $E$, respectively. Assume that*

*(a)* $\mathscr{Q}^{[s,t]}_{(ij)(uv)} = \mathscr{Q}^{[s,t]}_{(ji)(uv)}$ *for any* $i, j, u, v \in E$.

*(b)* $\sum_j \mathscr{Q}^{[s,t]}_{(ij)(uv)} y^{(s)}_j = \mathscr{H}^{[s,t]}_{iu} z^{(t)}_v$ *for any* $i, u, v \in E$;

*(c)* $\mathscr{Q}^{[s,t]}_{(ij)(uv)} = \sum_\ell \mathscr{Q}^{[s,t]}_{(ij)(u\ell)} y^{(t)}_v$ *for any* $i, j, u, v \in E$.

*Let*

$$P^{[s,t]}_{ij,k} = \sum_\ell \mathscr{Q}^{[s,t]}_{(ij)(k\ell)}. \tag{3.19}$$

*Then the following assertions hold true:*

*(i)* $z^{(t)}_k = y^{(t)}_k$ *for any* $k \in E$;

*(ii) If $(\mathscr{Q}^{[s,t]}_{(ij)(uv)})$ and $(\mathscr{H}^{[s,t]}_{ij})$ are Markov processes, then $(E, P^{[s,t]}_{ij,k}, \mathbf{x}^{(0)})$ is a q.s.p. of type (A).*

*(iii) If $(\mathscr{H}^{[s,t]}_{ij})$ is a Markov process and $(\mathscr{Q}^{[s,t]}_{(ij)(uv)})$ satisfies*

$$\mathscr{Q}^{[s,t]}_{(ij)(uv)} = \sum_{m,\ell} \mathscr{H}^{[s,\tau]}_{im} \mathscr{H}^{[s,\tau]}_{jl} \mathscr{Q}^{[\tau,t]}_{(m\ell)(uv)} \tag{3.20}$$

*then $(E, P^{[s,t]}_{ij,k}, \mathbf{x}^{(0)})$ is a q.s.p. of type (B).*

*Moreover, one has* $x_k^{(t)} = z_k^{(t)}$ *($k \in E$) and*

$$\mathcal{H}_{ik}^{[s,t]} = \sum_{j \in E} P_{ij,k}^{[s,t]} z_j^{(s)}. \tag{3.21}$$

*Proof*

(i) Let us show that $z_k^{(t)} = y_k^{(t)}$, ($k \in E$). Indeed, from condition (c) and the stochasticity of $\mathscr{Q}_{(ij)(uv)}^{[s,t]}$, one finds

$$\sum_m \mathscr{Q}_{(ij)(mk)}^{[s,t]} = \sum_{m,\ell} \mathscr{Q}_{(ij)(m\ell)}^{[s,t]} y_k^{(t)} = y_k^{(t)},$$

for any $k \in E$. Hence, the last equality with (b) yields

$$y_k^{(t)} = \sum_{m,j} \mathscr{Q}_{(ij)(mk)}^{[s,t]} y_j^{(s)} = \sum_m \left( \sum_j \mathscr{Q}_{(ij)(mk)}^{[s,t]} y_j^{(s)} \right) = \sum_m \mathcal{H}_{im}^{[s,t]} z_k^{(t)} = z_k^{(t)},$$

which is the required assertion.

(ii) From (3.19) one can see that $(P_{ij,k}^{[s,t]})$ satisfies conditions (i) and (ii) of the definition of a q.s.p. We need to check the equality $(iii_A)$. Let

$$x_k^{(t)} = \sum_{i,j} P_{ij}^{[0.t]} x_i^{(0)} x_j^{(0)}.$$

Then we find

$$x_k^{(t)} = \sum_{i,j} \left( \sum_\ell \mathscr{Q}_{(ij)(k\ell)}^{[s,t]} \right) x_i^{(0)} x_j^{(0)} = y_k^{(t)}.$$

Hence, due to (i), one has

$$x_k^{(t)} = y_k^{(t)} = z_k^{(t)},$$

for any $k \in E$. Let us directly check the fundamental equation $(iii_A)$. Indeed, for $s, \tau, t \in \mathbb{R}_+$ with $\tau - s \geq 1$, $t - \tau \geq 1$ and due to the Markovianity of $(\mathscr{Q}_{(ij)(uv)}^{[s,t]})$, we have

$$P_{ij,k}^{[s,t]} = \sum_\ell \mathscr{Q}_{(ij)(k\ell)}^{[s,t]}$$

$$= \sum_\ell \left( \sum_{m,h=1} \mathscr{Q}_{(ij)(mh)}^{[s,\tau]} \mathscr{Q}_{(mh)(k\ell)}^{[\tau,t]} \right)$$

$$= \sum_{\ell} \left( \sum_{m,h} \left( \sum_{g} \mathscr{Q}^{[s,\tau]}_{(ij)(mg)} y^{(\tau)}_h \right) \mathscr{Q}^{[\tau,t]}_{(mh)(k\ell)} \right)$$

$$= \sum_{m,h} P^{[s,\tau]}_{ij,m} P^{[\tau,t]}_{mh,k} x^{(\tau)}_h .$$

(iii)  From (b) one gets

$$\sum_{j,\ell} \mathscr{Q}^{[s,t]}_{(ij)(k\ell)} y^{(s)}_j = \mathscr{H}^{[s,t]}_{ik}$$

for any $i, k \in E$. Then using the last equality and (3.20) we obtain

$$\sum_{m,\ell,h,g} P^{[s,\tau]}_{im,\ell} P^{[s,\tau]}_{jh,g} P^{[\tau,t]}_{\ell g,k} x^{(s)}_m x^{(s)}_h$$

$$= \sum_{m,\ell,h,g} \left( \sum_{u} \mathscr{Q}^{[s,\tau]}_{(im)(\ell u)} \right) \left( \sum_{v} \mathscr{Q}^{[s,\tau]}_{(jh)(gv)} \right) \left( \sum_{a} \mathscr{Q}^{[\tau,t]}_{(\ell g)(ka)} \right) x^{(s)}_m x^{(s)}_h$$

$$= \sum_{a,\ell,g} \mathscr{H}^{[s,\tau]}_{i\ell} \mathscr{H}^{[s,\tau]}_{jg} \mathscr{Q}^{[\tau,t]}_{(\ell g)(ka)}$$

$$= \sum_{a} \mathscr{Q}^{[s,t]}_{(ij)(ka)} = P^{[s,t]}_{ij,k} ,$$

which means that the q.s.p. satisfies the equation $(iii_B)$.

Note that from (b) we immediately find (3.21). This completes the proof.

From the proved Theorems 3.3.1, 3.3.2 and 3.3.3 we conclude that any q.s.p. can be uniquely defined by two kinds of processes. Such processes are called *marginal Markov processes* associated with the q.s.p. The marginal Markov processes allow us to investigate q.s.p.s in terms of Markov processes. An application of this result will be illustrated in Sect. 3.6.

## 3.4  Quadratic Stochastic Operators and Discrete Time q.s.p.s

In this section, as an application of Theorem 3.3.3, we are going to demonstrate how to produce a discrete time q.s.p. by means of a given q.s.o. In what follows, for the sake of simplicity, we will consider such operators and processes with a discrete state space.

As before, let $E = \{1, \ldots, d\}$ and $S^{d-1}$ denotes the simplex of probability measures on $E$. Assume that we are given a q.s.o. $V$ (see (2.1)) defined by heredity coefficients $\{p_{ij,k}\}$.

Fix any initial distribution $\mathbf{x}_0^{(0)} \in S^{d-1}$ and consider

$$\mathbf{x}^{(n)} = V^n \mathbf{x}^{(0)}, \quad n \in \mathbb{N}.$$

Clearly, $\mathbf{x}^{(n)} \in S^{d-1}$, and $\mathbf{x}^{(n)} = (x_1^{(n)}, \ldots, x_d^{(n)})$.

For each $n \in \mathbb{N}$ define stochastic matrices by

$$H_{ij}^{[n,n+1]} = \sum_{j=1}^{d} p_{ij,k} x_j^{(n)}, \quad Q_{(ij)(uv)}^{[n,n+1]} = p_{ij,u} x_v^{(n+1)}. \tag{3.22}$$

Note that the defined stochastic matrix $H_{ij}^{[n,n+1]}$ is the same as (2.17).

For any $k, n \in \mathbb{N}$ ($k < n$) define a nonhomogeneous Markov process generated by $H^{[n,n+1]}$ as follows

$$H_{ij}^{[k,n]} = \sum_{m_1, m_2, \ldots, m_{n-k-1}} H_{im_1}^{[k,k+1]} H_{m_1 m_2}^{[k+1,k+2]} \cdots H_{m_{n-k-1}j}^{[n-1,n]}. \tag{3.23}$$

From (3.22) we find

$$\sum_i H_{ij}^{[k,k+1]} x_i^{(k)} = x_j^{(k+1)}$$

which with (3.23) implies

$$\sum_i H_{ij}^{[k,n]} x_j^{(k)} = x_j^{(n)}. \tag{3.24}$$

Similarly, again we define a nonhomogeneous Markov process generated by $Q^{[n,n+1]}$ as follows

$$Q_{(ij)(uv)}^{[k,n]} = \sum_{m_1, \ell_1} \cdots \sum_{m_{n-k-1}, \ell_{n-k-1}} Q_{(ij)(m_1 \ell_1)}^{[k,k+1]} \cdots \cdots Q_{(m_{n-k-1} \ell_{n-k-1})(uv)}^{[n-1,n]}. \tag{3.25}$$

So, we have a pair $(H_{ij}^{[k,n]}, Q_{(ij)(uv)}^{[k,n]})$ of nonhomogeneous Markov processes. One can find (see (3.24)) that

$$x_k^{(n)} = \sum_i H_{ik}^{[0,n]} x_i^{(0)} = \sum_{i,j,\ell} Q_{(ij)(k\ell)}^{[0,n]} x_i^{(0)} x_j^{(0)}, \quad k \in E.$$

**Proposition 3.4.1** *The pair* $(H_{ij}^{[k,n]}, Q_{(ij)(uv)}^{[k,n]})$ *of nonhomogeneous Markov processes generates a q.s.p.* $(E, P_{ij,k}^{[k,n]}, \mathbf{x}^{(0)})$ *of type (A). Moreover, one has* $P_{ij,k}^{[n,n+1]} = p_{ij,k}$ *for all* $n \in \mathbb{N}$, *and*

$$P_{ij,k}^{[k,n]} = \sum_m p_{ij,m} H_{m,k}^{[k+1,n]}. \tag{3.26}$$

*Proof* Due to Theorem 3.3.3, to prove the required statement it is enough to check that the pair $(H_{ij}^{[k,n]}, Q_{(ij)(uv)}^{[k,n]})$ satisfies conditions (a)–(b) of the mentioned theorem. Indeed, (a) is evident. Therefore, let us check (b). Note that from (3.22) for every $k \in \mathbb{N}$ one finds

$$\sum_j Q_{(ij)(uv)}^{[k,k+1]} x_j^{(k)} = \sum_j p_{ij,u} x_v^{(k+1)} x_j^{(k)} = H_{iu}^{[k,k+1]} x_v^{(k+1)}. \tag{3.27}$$

Hence, using (3.27) with (3.25) and (3.23) for any $k, n$ we obtain

$$\sum_j Q_{(ij)(uv)}^{[k,n]} x_j^{(k)}$$

$$= \sum_{m_1, \ell_1} \cdots \sum_{m_{n-k-1}, \ell_{n-k-1}} \sum_j Q_{(ij)(m_1 \ell_1)}^{[k,k+1]} x_j^{(k)} \cdots Q_{(m_{n-k-1} \ell_{n-k-1})(uv)}^{[n-1,n]}$$

$$= \sum_{m_1, \ell_1} \cdots \sum_{m_{n-k-1}, \ell_{n-k-1}} H_{im_1}^{[k,k+1]} Q_{(m_1 \ell_1)(m_2 \ell_2)}^{[k+1,k+2]} x_{\ell_1}^{(k+1)} \cdots Q_{(m_{n-k-1} \ell_{n-k-1})(uv)}^{[n-1,n]}$$

$$\cdots$$

$$= \sum_{m_1, \ldots, m_{n-k-1}} H_{im_1}^{[k,k+1]} \cdots H_{m_{n-k-1} u}^{[n-1,n]} x_v^{(n)}$$

$$= H_{iu}^{[k,n]} x_v^{(n)}.$$

This implies (b).

Now using (3.25) the equality (c) is obtained as follows

$$\sum_\ell Q_{(ij)(u\ell)}^{[k,n]} x_v^{(n)} = \sum_{m_1, \ell_1} \cdots \sum_{m_{n-k-1}, \ell_{n-k-1}} Q_{(ij)(m_1 \ell_1)}^{[k,k+1]} \cdots \sum_\ell Q_{(m_{n-k-1} \ell_{n-k-1})(u\ell)}^{[n-1,n]} x_v^{(n)}$$

$$= \sum_{m_1, \ell_1} \cdots \sum_{m_{n-k-1}, \ell_{n-k-1}} Q_{(ij)(m_1 \ell_1)}^{[k,k+1]} \cdots \sum_\ell p_{m_{n-k-1} \ell_{n-k-1}, u} x_\ell^{(n)} x_v^{(n)}$$

$$= \sum_{m_1, \ell_1} \cdots \sum_{m_{n-k-1}, \ell_{n-k-1}} Q_{(ij)(m_1 \ell_1)}^{[k,k+1]} \cdots Q_{(m_{n-k-1} \ell_{n-k-1})(uv)}^{[n-1,n]}$$

$$= Q_{(ij)(uv)}^{[k,n]}.$$

Hence, the required q.s.p. is defined by (3.19). So, we get $P_{ij,m}^{[n,n+1]} = p_{ij,m}$ for all $n \in \mathbb{N}$. The equality (3.26) immediately follows from (3.17). This completes the proof.

Now let us define another process $\tilde{Q}_{(ij)(uv)}^{[k,n]}$ generated by $H^{[k,n]}$ and $Q^{[n,n+1]}$ as follows

$$\tilde{Q}_{(ij)(uv)}^{[k,n]} = \sum_{m,\ell} H_{im}^{[k,n-1]} H_{j\ell}^{[k,n-1]} Q_{(m\ell)(uv)}^{[n-1,n]}. \qquad (3.28)$$

**Proposition 3.4.2** *The pair* $(H_{ij}^{[k,n]}, \tilde{Q}_{(ij)(uv)}^{[k,n]})$ *defines a q.s.p.* $(E, P_{ij,k}^{[k,n]}, \mathbf{x}^{(0)})$ *of type (B). Moreover, one has* $\tilde{P}_{ij,k}^{[n,n+1]} = p_{ij,k}$ *for all* $n \in \mathbb{N}$, *and*

$$\tilde{P}_{ij,k}^{[k,n]} = \sum_{m,\ell} H_{i,m}^{[k,n-1]} H_{j,\ell}^{[k,n-1]} p_{m\ell,k}. \qquad (3.29)$$

*Proof* To prove the required statement, we will check that the pair $(H_{ij}^{[k,n]}, \tilde{Q}_{(ij)(uv)}^{[k,n]})$ satisfies conditions (a)–(b) of Theorem 3.3.3. Condition (a) is evident. Using (3.27) and (3.24) from (3.28) one obtains condition (b) as follows

$$\sum_j \tilde{Q}_{(ij)(uv)}^{[k,n]} x_j^{(k)} = \sum_{m,\ell} \left( \sum_j H_{j\ell}^{[k,n-1]} x_j^{(k)} \right) H_{im}^{[k,n-1]} Q_{(m\ell)(uv)}^{[n-1,n]}$$

$$= \sum_m H_{im}^{[k,n-1]} \left( \sum_\ell Q_{(m\ell)(uv)}^{[n-1,n]} x_\ell^{(n-1)} \right)$$

$$= \sum_m H_{im}^{[k,n-1]} H_{mu}^{[n-1,n]} x_v^{(n)}$$

$$= H_{iu}^{[k,n]} x_v^{(n)}.$$

The equality (c) is similarly obtained as in the previous Proposition 3.4.1. Hence, the required statement immediately follows from Theorem 3.3.3 and (3.18). The proof is complete.

*Remark 3.4.1* From the proved propositions we infer that the nonhomogeneous Markov process defined by (2.17) naturally appears in the construction of a q.s.p. This construction was first proposed in [232].

Note that for both types of constructed q.s.p. one has

$$x_k^{(n)} = \sum_{i,j} P_{ij,k}^{[0,n]} x_i^{(0)} x_j^{(0)}$$

which implies that the study of the asymptotic behavior of the dynamics of a q.s.o.
$V$ is equivalent to the investigation of the limiting behavior of the q.s.p.

## 3.5 Construction of Quadratic Stochastic Processes

In the theory of quadratic stochastic processes, it is important to construct nontrivial
examples of such processes. In this section we are going to propose a construction
of q.s.p.s by means of two other given q.s.p.s. Using the proposed construction we
provide certain concrete examples of q.s.p.s.

**Theorem 3.5.1** *Let* $(E, P_{ij,k}^{[s,t]}, \mathbf{x}^{(0)})$ *and* $(F, G_{ij,k}^{[s,t]}, \mathbf{y}^{(0)})$ *be two q.s.p.s of type (A),*
*where E and F are at most countable sets. Then the process* $(E \times F, (P \otimes G)_{\bar{i}\bar{j},\bar{k}}^{[s,t]}, \mathbf{x}^{(0)} \times \mathbf{y}^{(0)})$ *defined by*

$$(P \otimes G)_{\bar{i}\bar{j},\bar{k}}^{[s,t]} = P_{i_1 j_1 k_1}^{[s,t]} G_{i_2 j_2 k_2}^{[s,t]}, \quad \mathbf{x}^{(0)} \times \mathbf{y}^{(0)} = (x_i^{(0)} y_j^{(0)})_{i \in E, j \in F},$$

*is a q.s.o. of type (A). Here* $\bar{i} = (i_1, i_2), \bar{j} = (j_1, j_2), \bar{k} = (k_1, k_2)$.

*Proof* It is clear that $(P \otimes G)_{\bar{i}\bar{j},\bar{k}}^{[s,t]} \geq 0$ and

$$\sum_{\bar{k}} (P \otimes G)_{\bar{i}\bar{j},\bar{k}}^{[s,t]} = \sum_{k_1, k_2} P_{i_1 j_1, k_1}^{[s,t]} G_{i_2 j_2, k_2}^{[s,t]} = \sum_{k_1} P_{i_1 j_1, k_1}^{[s,t]} \sum_{k_2} G_{i_2 j_2, k_2}^{[s,t]} = 1.$$

From

$$P_{i_1 j_1, k_1}^{[s,t]} G_{i_2 j_2, k_2}^{[s,t]} = G_{j_1 i_1, k_1}^{[s,t]} G_{j_2 i_2, k_2}^{[s,t]}$$

we immediately get $(P \otimes G)_{\bar{i}\bar{j},\bar{k}}^{[s,t]} = (P \otimes G)_{\bar{j}\bar{i},\bar{k}}^{[s,t]}$. Let us define $X_{\bar{i}}^{(0)} = x_{i_1}^{(0)} y_{i_2}^{(0)}$. Then,
one can see that $\mathbf{x}^{(0)} \times \mathbf{y}^{(0)} = (X_{\bar{i}}^{(0)})$. Let us find $X_{\bar{i}}^{(s)}$, which is given by

$$\begin{aligned}
X_{\bar{k}}^{(s)} &= \sum_{\bar{i}\bar{j}} (P \otimes G)_{\bar{i}\bar{j},\bar{k}}^{[0,s]} \bar{x}_{\bar{i}}^{(0)} \bar{x}_{\bar{j}}^{(0)} \\
&= \sum_{(i_1 i_2)(j_1 j_2)} P_{i_1 j_1, k_1}^{[0,s]} G_{i_2 j_2, k_2}^{[0,s]} x_{i_1}^{(0)} y_{i_2}^{(0)} x_{j_1}^{(0)} y_{j_2}^{(0)} \\
&= \left( \sum_{(i_1 i_2)} P_{i_1 j_1, k_1}^{[0,s]} x_{i_1}^{(0)} x_{j_1}^{(0)} \right) \left( \sum_{(j_1 j_2)} G_{i_2 j_2, k_2}^{[0,s]} y_{i_2}^{(0)} y_{j_2}^{(0)} \right) \\
&= x_{k_1}^{(s)} y_{k_2}^{(s)}.
\end{aligned} \tag{3.30}$$

Now let us check the equation $(iii_A)$.

Taking into account that the given processes $P_{ij,k}^{[s,t]}$ and $G_{ij,k}^{[s,t]}$ have type (A), then we have

$$\sum_{\bar{u},\bar{l}} (P \otimes G)_{\overline{ij},\bar{u}}^{[s,\tau]} (P \otimes G)_{\overline{ul},k}^{[\tau,t]} X_{\bar{l}}^{(s)}$$

$$= \sum_{(u_1 u_2)(l_1 l_2)} P_{i_1 j_1 u_1}^{[s,\tau]} G_{i_2 j_2 u_2}^{[s,\tau]} P_{u_1,l_1 k_1}^{[\tau,t]} G_{u_2 l_2 k_2}^{[\tau,t]} x_{l_1}^{(s)} y_{l_2}^{(s)}$$

$$= \left( \sum_{(u_1 l_1)} P_{i_1 j_1 u_1}^{[s,\tau]} P_{u_1,l_1 k_1}^{[\tau,t]} x_{l_1}^{(s)} \right) \left( \sum_{(u_2 l_2)} G_{i_2 j_2 u_2}^{[s,\tau]} G_{u_2 l_2 k_2}^{[\tau,t]} y_{l_2}^{(s)} \right)$$

$$= P_{i_1 j_1 k_1}^{[s,t]} G_{i_2 j_2 k_2}^{[s,t]}$$

$$= (P \otimes G)_{\overline{ij},\bar{k}}^{[s,t]}.$$

This completes the proof.

**Theorem 3.5.2** Let $(E, P_{ij,k}^{[s,t]}, \mathbf{x}^{(0)})$ and $(F, G_{ij,k}^{[s,t]}, \mathbf{y}^{(0)})$ be two q.s.p.s of type (B). Then the process $(E \times F, (P \otimes G)_{\overline{ij},\bar{k}}^{[s,t]}, \mathbf{x}^{(0)} \times \mathbf{y}^{(0)})$ defined by

$$(P \otimes G)_{\overline{ij},\bar{k}}^{[s,t]} = P_{i_1 j_1 k_1}^{[s,t]} G_{i_2 j_2 k_2}^{[s,t]}, \quad \mathbf{x}^{(0)} \times \mathbf{y}^{(0)} = (x_i^{(0)} y_j^{(0)})_{i \in E, j \in F},$$

is a q.s.o. of type (B). Here $\bar{i} = (i_1, i_2), \bar{j} = (j_1, j_2), \ \bar{k} = (k_1, k_2)$.

*Proof* To prove the theorem it is enough to check the equation $(iii_B)$. Noting that $X_{\bar{m}}^{(s)} = x_{m_1}^{(s)} y_{m_2}^{(s)}$, where $\bar{m} = (m_1, m_2)$, we get

$$\sum_{\bar{m},\bar{l},\bar{j},\bar{u}} (P \otimes G)_{\overline{i},\bar{m},\bar{l}}^{[s,\tau]} (P \otimes G)_{\overline{j},\bar{g},\bar{u}}^{[s,\tau]} (P \otimes G)_{\overline{l},\bar{u},k}^{[\tau,t]} X_{\bar{m}}^{(s)} X_{\bar{g}}^{(s)}$$

$$= \sum_{(m_1 m_2)(l_1 l_2)(j_1 j_2)(u_1 u_2)} P_{i_1,m_1,l_1}^{[s,\tau]} G_{j_2,m_2,l_2}^{[s,\tau]} P_{j_1,g_1,u_1}^{[s,\tau]} G_{j_2,g_2,u_2}^{[s,\tau]}$$

$$P_{l_1,u_1,k_1}^{[\tau,t]} G_{l_2,u_2,k_2}^{[\tau,t]} x_{m_1}^{(s)} y_{m_2}^{(s)} x_{g_1}^{(s)} y_{g_2}^{(s)}$$

$$= \left( \sum_{(m_1 l_1 j_1 u_1)} P_{i_1,m_1,l_1}^{[s,\tau]} P_{j_1,g_1,u_1}^{[s,\tau]} P_{l_1,u_1,k_1}^{[\tau,t]} x_{m_1}^{(s)} x_{g_1}^{(s)} \right)$$

$$\left( \sum_{(m_2 l_2 j_2 u_2)} G_{j_2,m_2,l_2}^{[s,\tau]} G_{j_2,g_2,u_2}^{[s,\tau]} G_{l_2,u_2,k_2}^{[\tau,t]} y_{m_2}^{(s)} y_{g_2}^{(s)} \right)$$

$$= P_{i_1 j_1 k_1}^{[s,t]} G_{i_2 j_2 k_2}^{[s,t]}$$

$$= (P \otimes G)_{\overline{ij},\bar{k}}^{[s,t]}.$$

This is the desired assertion.

From Theorems 3.5.1 and 3.5.2 we immediately get the following

**Corollary 3.5.3** *Let* $(E, P^{[s,t]}_{ij,k}, \mathbf{x}^{(0)})$, $(F, G^{[s,t]}_{ij,k}, \mathbf{y}^{(0)})$ *be two q.s.p.s, and* $(H^{[s,t]}_{ij,P})$ *and* $(H^{[s,t]}_{ij,G})$ *be their marginal Markov processes, respectively. Then the marginal Markov process* $(H^{[s,t]}_{\overline{ij},P\otimes G})$ *of* $(E \times F, (P \otimes G)^{[s,t]}_{\overline{ij},\overline{k}}, \mathbf{x}^{(0)} \times \mathbf{y}^{(0)})$ *is given by*

$$H^{[s,t]}_{\overline{ij},P\otimes G} = H^{[s,t]}_{i_1 j_1,P} H^{[s,t]}_{i_2 j_2,G}.$$

*Example 3.5.1* Now we are going to construct new examples of q.s.p.s. Let us consider the q.s.p. given in Example 3.2.2 (see (3.11)–(3.13)). In this case, we have $E = \{1, 2\}$ and $(x, 1 - x)$ is an initial distribution on $E$, where $x \in [0, 1]$. It has been shown that for $\epsilon \in [0, \frac{1}{2}]$ such a q.s.p. is of type (A) and type (B), simultaneously.

Now, we multiply the same process by itself according to the rule given in Theorem 3.5.1.

Let

$$F = E \times E = (1, 1), (1, 2), (2, 1), (2, 2)$$

and

$$X^0 = \big(xy, x(1 - y), (1 - x)y, (1 - x)(1 - y)\big)$$

where $x, y \in [0, 1]$ and define

$$\mathscr{P}^{[s,t]}_{(i_1 i_2)(j_1 j_2)(k_1 k_2)} = P^{[s,t]}_{i_1 j_1, k_1} P^{[s,t]}_{i_2 j_2 k_2}.$$

From Theorems 3.5.1 and 3.5.2 one finds that the defined process is a q.s.p. of type (A) and (B), simultaneously. Let us describe the process more precisely,

$$\mathscr{P}^{[s,t]}_{11,11,11} = \frac{1}{2^{2(t-s-1)}}[x^2(2^{t-s-1} - 1)^2 + 2x(2^{t-s-1} - 1) + 1]$$

$$\mathscr{P}^{[s,t]}_{11,11,12} = \frac{1}{2^{2(t-s-1)}}[-x^2(2^{t-s-1} - 1)^2 + x(2^{t-s-1} - 1)(2^{t-s-1} - 2) + 2^{t-s-1} - 1]$$

$$\mathscr{P}^{[s,t]}_{11,11,22} = \frac{1}{2^{2(t-s-1)}}[x^2(2^{t-s-1} - 1)^2 - 2x(2^{t-s-1} - 1)^2 + 2^2 t - s - 1 - 2^{t-s-1} + 1]$$

$$\mathscr{P}^{[s,t]}_{11,12,11} = \frac{1}{2^{2(t-s-1)}}[x^2(2^{t-s-1} - 1)^2 + \frac{3}{2}x(2^{t-s-1} - 1) + \frac{1}{2}]$$

$$\mathscr{P}^{[s,t]}_{11,12,12} = \frac{1}{2^{2(t-s-1)}}[x^2(2^{t-s-1} - 1)^2 - x(2^{t-s-1} - \frac{3}{2}2^{t-s-1} - \frac{3}{2}) + 2^{t-s-1} + \frac{1}{2}]$$

$$\mathscr{P}^{[s,t]}_{11,12,21} = \frac{1}{2^{2(t-s-1)}}[-x^2(2^{t-s-1}-1)^2 + x(2^{t-s-1} - \frac{3}{2}2^{t-s-1} + \frac{3}{2}) + \frac{1}{4}2^{t-s} - \frac{1}{2}]$$

$$\mathscr{P}^{[s,t]}_{11,12,22} = \frac{1}{2^{2(t-s-1)}}[x^2(2^{t-s-1}+1)^2 - x(2(2^{2(t-s-1)}) - \frac{7}{2}2^{t-s-1} - \frac{3}{2})$$
$$+ (2^{2(t-s-1)} + 2^{t-s-1} + \frac{1}{2})]$$

$$\mathscr{P}^{[s,t]}_{11,22,11} = \frac{1}{2^{2(t-s-1)}}[x^2(2^{t-s-1}-1)^2 + x(2^{t-s-1}-1) + \frac{1}{4}]$$

$$\mathscr{P}^{[s,t]}_{11,22,12} = \frac{1}{2^{2(t-s-1)}}[x^2(1-2^{t-s-1})(1+2^{t-s-1}) + x(2^{2(t-s-1)} - 2^{t-s-1} - 1) + \frac{1}{4}2^{t-s} + \frac{1}{4}]$$

$$\mathscr{P}^{[s,t]}_{11,22,22} = \frac{1}{2^{2(t-s-1)}}[x^2(2^{t-s-1}+1)^2 - x(2(2^{t-s-1})+1)(2^{t-s-1}+1) + (2^{t-s-1}+\frac{1}{2})^2]$$

$$\mathscr{P}^{[s,t]}_{22,12,11} = \frac{1}{2^{2(t-s-1)}}[x^2(2^{t-s-1}-1)^2 + \frac{1}{4}2^{t-s}x - \frac{1}{2}x]$$

$$\mathscr{P}^{[s,t]}_{22,12,12} = \frac{1}{2^{2(t-s-1)}}[-x^2(2^{t-s-1}-2^{t-s-1}-1) + x(2^{2(t-s-1)} - \frac{1}{2}2^{t-s-1} + \frac{1}{2}) + \frac{1}{4}2^{t-s}]$$

$$\mathscr{P}^{[s,t]}_{22,12,21} = \frac{1}{2^{2(t-s-1)}}[-x^2(2^{t-s-1}-1)^2 + x((2^{t-s-1}-1) + 1)(2(2^{t-s-1})-1)]$$

$$\mathscr{P}^{[s,t]}_{22,12,22} = \frac{1}{2^{2(t-s-1)}}[x^2(2^{t-s-1}-1)^2 - x(4(2^{t-s-1}))(2^{t-s-1}-1) + 2^{t-s-1}(2^{t-s-1}-\frac{1}{2})]$$

$$\mathscr{P}^{[s,t]}_{12,12,11} = \frac{1}{2^{2(t-s-1)}}[x^2(2^{2(t-s-1)} - 2^{t-s-1} + 1) - x]$$

$$\mathscr{P}^{[s,t]}_{12,12,12} = \frac{1}{2^{2(t-s-1)}}[-x^2(2^{2(t-s-1)}-2^{t-s-1}+1) + x(2^{2(t-s-1)}-2^{t-s-1}+1) + 2^{t-s-1}]$$

$$\mathscr{P}^{[s,t]}_{12,12,21} = \frac{1}{2^{2(t-s-1)}}[-x^2(2^{t-s-1}-1)^2 - x(1-2^{t-s-1})(1+2^{t-s-1})]$$

$$\mathscr{P}^{[s,t]}_{12,12,22} = \frac{1}{2^{2(t-s-1)}}[x^2(2^{t-s-1}-1)^2 - x(2(2^{t-s-1})-1)(2^{t-s-1}-1) + 2^{t-s-1}(2^{t-s-1}-1)]$$

$$\mathscr{P}^{[s,t]}_{12,22,11} = \frac{1}{2^{2(t-s-1)}}[x^2(2^{t-s-1}-1)^2 + \frac{1}{2}x(2^{t-s-1}-1)]$$

$$\mathscr{P}^{[s,t]}_{12,22,12} = \frac{1}{2^{2(t-s-1)}}[-x^2(2^{t-s-1}-1)^2 + x(2^{2(t-s-1)} - \frac{1}{4}2^{t-s} + \frac{1}{2}) + \frac{1}{4}2^{t-s}]$$

$$\mathscr{P}^{[s,t]}_{12,22,21} = \frac{1}{2^{2(t-s-1)}}[-x^2(2^{t-s-1} - 1)^2 + x(4(2^{t-s-1} - 2)(2^{t-s-1} - 1)]$$

$$\mathscr{P}^{[s,t]}_{12,22,22} = \frac{1}{2^{2(t-s-1)}}[x^2(2^{t-s-1} - 1)^2 - x(4(2^{t-s-1} - 1)(2^{t-s-1} - 1) + 2^{t-s-1}(2^{t-s-1} - \frac{1}{2})]$$

$$\mathscr{P}^{[s,t]}_{22,22,11} = \frac{1}{2^{t-s-1}}[x^2 - \frac{x^2}{2^{t-s-1}}]$$

$$\mathscr{P}^{[s,t]}_{22,22,12} = \frac{x}{2^{t-s-1}}[2^{t-s-1} - 1 - x + \frac{x}{2^{t-s-1}}]$$

$$\mathscr{P}^{[s,t]}_{22,22,22} = \frac{1}{2^{2(t-s-1)}}[x^2(2^{t-s-1} + 1)^2 - 2(2^{t-s-1}x)(2^{t-s-1} + 1) + 2^{2(t-s-1)}]$$

## 3.6   Weak Ergodicity of Quadratic Stochastic Processes

In this section we are going to study the weak ergodicity of q.s.o.s in terms of marginal processes.

Let $E$ be as before, i.e. $E$ is an at most countable set. We first recall that a Markov process $\left(U^{[s,t]}_{ij}\right)_{ij\in E}$ satisfies *weak ergodicity* if for every $s \in \mathbb{R}_+$ one has

$$\sup_{\varphi,\psi\in S} \left\|\mathbb{U}^{[s,t]}\varphi - \mathbb{U}^{[s,t]}\psi\right\|_1 \to 0 \text{ as } t \to \infty,$$

where

$$(\mathbb{U}^{[s,t]}\varphi)_k = \sum_{j\in E} U^{[s,t]}_{ik}\varphi_i, \quad \varphi = (\varphi_i) \in S.$$

**Proposition 3.6.1** *Let $\left(U^{[s,t]}_{ij}\right)$ be a Markov process. Then the following assertions are equivalent:*

*(i)  $\left(U^{[s,t]}_{ij}\right)$ is weakly ergodic;*
*(ii)  For every $s \in \mathbb{R}_+$ one has*

$$\lim_{t\to\infty} \sup_{i,j\in E} \left\|\mathbb{U}^{[s,t]}(e^{(i)}) - \mathbb{U}^{[s,t]}(e^{(j)})\right\|_1 = 0,$$

*where $e^{(i)} = (0,0,\ldots,\underbrace{1}_{i},0,\ldots)$, $i \in E$.*

The proof is evident.

The weak ergodicity of Markov processes is well studied in the literature (see for example, [97, 103, 104, 106, 141, 247, 256]).

**Theorem 3.6.2** *Let* $\left(U_{ij}^{[s,t]}\right)_{ij\in E}$ *and* $\left(V_{ij}^{[s,t]}\right)_{ij\in F}$ *be two weakly ergodic Markov processes. Then the Markov process* $(U\otimes V)_{\tilde{i}\tilde{j}}^{s,t} = U_{i_1 j_1}^{[s,t]} V_{i_2 j_2}^{[s,t]}$ *is also weakly ergodic.*

*Proof* First let us define $f^{(k)} = (0, 0, \ldots, \underbrace{1}_{k}, 0, \ldots), \ k \in F$. Then one can see that

$$(U\otimes V)^{[s,t]}(e^{(i)} \otimes f^{(j)}) = U^{[s,t]}(e^{(i)}) \otimes V^{[s,t]}(f^{(j)}). \tag{3.31}$$

Now take any $i_1, j_1 \in E$, $i_2, j_2 \in F$. Then from (3.31) one gets

$$\|(U\otimes V)^{[s,t]}(e^{(i_1)} \otimes f^{(i_2)}) - (U\otimes V)^{[s,t]}(e^{(j_1)} \otimes f^{(j_2)})\|_1$$
$$\leq \|U^{[s,t]}(e^{(i_1)}) - U^{[s,t]}(e^{(j_1)})\|_1 + \|V^{[s,t]}(f^{(i_2)}) - V^{[s,t]}(f^{(j_2)})\|_1.$$

Hence, using Proposition 3.6.1 we get the desired assertion.

Let $(E, P_{ij,k}^{[s,t]}, \mathbf{x}^{(0)})$ be a q.s.p. Then we define a mapping $\mathbb{P}^{[s,t]} : \ell^1(E) \times \ell^1(E) \to \ell^1(E)$ as follows

$$\left(\mathbb{P}^{[s,t]}(\varphi, \psi)\right)_k = \sum_{i,j\in E} P_{ij,k}^{[s,t]} \varphi_i \psi_j, \quad \varphi, \psi \in \ell^1(E).$$

Using the definition of a q.s.p. one finds

(a) $\mathbb{P}^{[s,t]}(\varphi, \psi) = \mathbb{P}^{[s,t]}(\psi, \varphi), \quad \varphi, \psi \in \ell^1(E)$;
(b) $\left\|\mathbb{P}^{[s,t]}(\varphi, \psi) - \mathbb{P}^{[s,t]}(\varphi, \psi_1)\right\|_1 \leq \|\psi - \psi_1\|_1, \quad \varphi \in S, \ \psi, \psi_1 \in \ell_1(E)$.

We say a q.s.p. $(E, P^{[s,t]}, x^{(0)})$ satisfies *weak ergodicity* if for every $s \in \mathbb{R}_+$ one has

$$\lim_{t\to\infty} \sup_{\varphi,\psi,\varphi_1,\psi_1\in S} \left\|\mathbb{P}^{[s,t]}(\varphi, \psi) - \mathbb{P}^{[s,t]}(\varphi_1, \psi_1)\right\|_1 = 0.$$

**Theorem 3.6.3** *Let* $\left(E, P_{ij,k}^{[s,t]}, \mathbf{x}^{(0)}\right)$ *be a q.s.p. and* $(H_{ij}^{[s,t]})$ *be its marginal Markov process. Then the following statements are equivalent:*

*(i) For every $s \in \mathbb{R}_+$ one has*

$$\lim_{t\to\infty} \sup_{i,j,k\in E} \left\|\mathbb{P}^{[s,t]}(e^{(i)}, e^{(k)}) - \mathbb{P}^{[s,t]}(e^{(j)}, e^{(k)})\right\|_1 = 0;$$

*(ii) For every $s \in \mathbb{R}_+$ one has*

$$\lim_{t\to\infty} \sup_{i,j,k,u\in E} \left\|\mathbb{P}^{[s,t]}(e^{(i)}, e^{(j)}) - \mathbb{P}^{[s,t]}(e^{(k)}, e^{(u)})\right\|_1 = 0;$$

*(iii)* $\left(E, P_{ij,k}^{[s,t]}, \mathbf{x}^{(0)}\right)$ *is weakly ergodic;*

*(iv)* $\left(H_{ij}^{[s,t]}\right)$ *is weakly ergodic.*

*Proof (i)* $\Rightarrow$ *(ii)*. From

$$\left\| \mathbb{P}^{[s,t]}\left(e^{(i)}, e^{(j)}\right) - \mathbb{P}^{[s,t]}\left(e^{(k)}, e^{(u)}\right) \right\|_1 \le \left\| \mathbb{P}^{[s,t]}\left(e^{(i)}, e^{(j)}\right) - \mathbb{P}^{[s,t]}\left(e^{(k)}, e^{(j)}\right) \right\|_1$$

$$+ \left\| \mathbb{P}^{[s,t]}\left(e^{(k)}, e^{(j)}\right) - \mathbb{P}^{[s,t]}\left(e^{(k)}, e^{(j)}\right) \right\|_1$$

with (a), (i) we get the desired assertion.

*(ii)* $\Rightarrow$ *(iii)*. Consider the following elements:

$$\xi = \sum_{i=1}^{M} \alpha_i e^{(i)}, \quad \eta = \sum_{j=1}^{N} \beta_j e^{(j)},$$

where $\alpha_i, \beta_j \ge 0$, $\sum_{i=1}^{M} \alpha_i = \sum_{j=1}^{N} \beta_j = 1$.

Then from (ii) we infer that for any $\varepsilon > 0$ there exists a $t_0 > 0$ such that for all $t \ge t_0$ and every $u, v \in E$ one has

$$\left\| \mathbb{P}^{[s,t]}\left(e^{(u)}, e^{(i)}\right) - \mathbb{P}^{[s,t]}\left(e^{(v)}, e^{(j)}\right) \right\|_1 < \frac{\varepsilon}{3}, \quad \forall i, j \in E.$$

Hence, we get

$$\left\| \mathbb{P}^{[s,t]}\left(e^{(u)}, \xi\right) - \mathbb{P}^{[s,t]}\left(e^{(v)}, \eta\right) \right\|_1 \le \sum_{i,j=1}^{M,N} \alpha_i \beta_j \left\| \mathbb{P}^{[s,t]}\left(e^{(u)}, e^{(i)}\right) - \mathbb{P}^{[s,t]}\left(e^{(v)}, e^{(j)}\right) \right\|_1$$

$$< \frac{\varepsilon}{3}. \tag{3.32}$$

Define

$$G = \left\{ \sum_{i=1}^{M} \alpha_i e^{(i)} \,\middle|\, \alpha_i \ge 0, \sum_{i=1}^{M} \alpha_i = 1, M \in \mathbb{N} \right\}.$$

Let $\varphi, \psi \in S$, then one can find $\xi, \eta \in G$ such that $\|\varphi - \xi\|_1 < \frac{\varepsilon}{3}$, $\|\psi - \eta\|_1 < \frac{\varepsilon}{3}$. Now using (b) and (3.32) we obtain

$$\left\| \mathbb{P}^{[s,t]}\left(e^{(u)}, \varphi\right) - \mathbb{P}^{[s,t]}\left(e^{(v)}, \psi\right) \right\|_1 \le \left\| \mathbb{P}^{[s,t]}\left(e^{(u)}, \varphi\right) - \mathbb{P}^{[s,t]}\left(e^{(u)}, \xi\right) \right\|_1$$

$$+ \left\| \mathbb{P}^{[s,t]}\left(e^{(v)}, \psi\right) - \mathbb{P}^{[s,t]}\left(e^{(v)}, \eta\right) \right\|_1$$

$$+ \left\| \mathbb{P}^{[s,t]}\left(e^{(u)}, \xi\right) - \mathbb{P}^{[s,t]}\left(e^{(v)}, \eta\right) \right\|_1$$

$$< \frac{\varepsilon}{3} + \frac{\varepsilon}{3} + \frac{\varepsilon}{3} = \varepsilon, \quad \forall t \ge t_0.$$

Now using the same argument, one finds

$$\sup_{\varphi,\psi,\varphi_1,\psi_1 \in S} \left\| \mathbb{P}^{[s,t]}(\varphi,\psi) - \mathbb{P}^{[s,t]}(\varphi_1,\psi_1) \right\|_1 < \varepsilon \quad \text{for all } t \geq t_0.$$

This is the desired assertion.

$(iii) \Rightarrow (iv)$. By $\mathbb{H}^{[s,t]}$ we denote the linear operator associated with the matrix $(H_{ij}^{[s,t]})$. From

$$\mathbb{H}^{[s,t]}(e^{(i)}) = \mathbb{P}^{[s,t]}(e^{(i)}, x^{(s)})$$

and (iii) we immediately find that $(H_{ij}^{[s,t]})$ is weakly ergodic.

$(iv) \Rightarrow (i)$. We consider two separate cases w.r.t. the type of q.s.p.

CASE A. Assume that $(E, P_{ij,k}^{[s,t]}, \mathbf{x}^{(0)})$ has type (A). Then for any $i \in E$ and $\varphi \in S$ one finds

$$\left( \mathbb{P}^{[s,t]}(e^{(i)}, \varphi) \right)_k = \sum_e P_{ie,k}^{[s,t]} \varphi_e = \sum_e \left( \sum_{m,h} P_{ie,m}^{[s,s+1]} P_{mh,k}^{[s+1,t]} x_h^{(s+1)} \right) \varphi_e$$

$$= \sum_e \sum_m P_{ie,m}^{[s,s+1]} H_{m,k}^{s+1,t} \varphi_e = \sum_m H_{m,k}^{[s+1,t]} y_{\varphi,m}^{[i]} = \left( \mathbb{H}^{[s+1,t]} \mathbf{y}_\varphi^{[i]} \right)_k,$$

where $y_{\varphi,m}^{[i]} = \sum_e P_{ie,m}^{[s,s+1]} \varphi_e$.

Hence, we get

$$\left\| \mathbb{P}^{[s,t]}(e^{(i)}, \varphi) - \mathbb{P}^{[s,t]}(e^{(j)}, \psi) \right\|_1 = \left\| \mathbb{H}^{[s,t]} \mathbf{y}_\varphi^{[i]} - \mathbb{H}^{[s,t]} \mathbf{y}_\psi^{[j]} \right\|_1$$

which yields

$$\sup_{i,j,\varphi,\psi} \left\| \mathbb{P}^{[s,t]}(e^{(i)}, \varphi) - \mathbb{P}^{[s,t]}(e^{(j)}, \psi) \right\|_1 \leq \sup_{\varphi,\psi} \left\| \mathbb{H}^{[s,t]} \varphi - \mathbb{H}^{[s,t]} \psi \right\|_1 \to 0 \quad \text{as } t \to \infty.$$

This implies the desired assertion.

CASE B. Now suppose that $\left( E, P_{ij,k}^{[s,t]}, \mathbf{x}^{(0)} \right)$ has type (B). Then we find

$$\left( \mathbb{P}^{[s,t+1]}(e^{(i)}, \varphi) \right)_k = \sum_e \left( \sum_{a,b,c,d} P_{ia,b}^{[s,t]} P_{ec,d}^{[s,t]} P_{bd,k}^{[t,t+1]} x_a^{(s)} x_e^{(s)} \right) \varphi_e$$

$$= \sum_e \sum_{b,d} H_{ib}^{[s,t]} H_{ed}^{[s,t]} P_{bd,k}^{[t,t+1]} \varphi_e$$

$$= \mathbb{P}^{[t,t+1]} \left( \mathbb{H}^{[s,t]}(e^{(i)}), \mathbb{H}^{[s,t]}(\varphi) \right).$$

Therefore, using (b) one gets

$$\left\| \mathbb{P}^{[s,t]}\left(e^{(i)}, \varphi\right) - \mathbb{P}^{[s,t]}\left(e^{(j)}, \varphi\right) \right\|_1$$

$$= \left\| \mathbb{P}^{[t,t+1]}\left(\mathbb{H}^{[s,t]}\left(e^{(i)}\right), \mathbb{H}^{[s,t]}(\varphi)\right) - \mathbb{P}^{[t,t+1]}\left(\mathbb{H}^{[s,t]}\left(e^{(j)}\right), \mathbb{H}^{[s,t]}(\varphi)\right) \right\|_1$$

$$\leq \left\| \mathbb{H}^{[s,t]}\left(e^{(i)}\right) - \mathbb{H}^{[s,t]}\left(e^{(j)}\right) \right\|_1 .$$

Hence, we find the desired assertion. This completes the proof.

Hence, by means of this theorem with Corollary 3.5.3 and Theorem 3.6.2 we find the following

**Corollary 3.6.4** *Let* $(E, P_{ij,k}^{[s,t]}, \mathbf{x}^{(0)})$ *and* $(F, G_{ij,k}^{[s,t]}, \mathbf{y}^{(0)})$ *be two weakly ergodic q.s.p.s, and* $(H_{ij,P}^{[s,t]})$ *and* $(H_{ij,G}^{[s,t]})$ *be their marginal Markov processes, respectively. Then the q.s.p.* $(E \times F, (P \otimes G)_{\overline{ij},\overline{k}}^{[s,t]}, \mathbf{x}^{(0)} \times \mathbf{y}^{(0)})$ *is weakly ergodic.*

## 3.7   Comments and References

The notion of a q.s.p. was first introduced and studied in [73]. One of the motivations for considering such processes was that their dynamics well describes the trajectories of q.s.o.s. The theory of q.s.p.s was further developed in [46, 48, 234, 236]. Sections 3.1 and 3.2 are essentially taken from [46]. We note that Sects. 3.3 and 3.5 have been published in [197]. Some parts of Sect. 3.6 are taken from [50], but the main result of the last section is published in [196]. Note that Theorem 3.6.3 allows us to study the weak ergodicity of q.s.p.s by means of the weak ergodicity of Markov processes. In the literature there are a lot papers devoted to this problem for Markov processes. One of the powerful tools in this study is the so-called Dobrushin's ergodicity coefficient (see [32, 36, 91, 105, 106, 237, 238, 247, 256]). Using this coefficient and Theorem 3.6.3 one can prove the following

**Theorem 3.7.1** *Let* $\left(E, P_{ij,k}^{[s,t]}, \mathbf{x}^{(0)}\right)$ *be a q.s.p. If there is a number* $\lambda \in [0,1)$ *and for each* $s \in \mathbb{R}_+$ *one can find* $t_0 \geq s + 1$ *such that*

$$\frac{1}{2} \sup_{i,j,u,v \in E} \sum_k \left| P_{ij,k}^{[s,t_0]} - P_{uv,k}^{[s,t_0]} \right| \leq \lambda,$$

*then the q.s.p. is weakly ergodic.*

In [50] we studied a weaker condition than weak ergodicity, which was called the *ergodic principle*. Namely, a q.s.p. $(E, P^{[s,t]}, x^{(0)})$ is said to satisfy *weak ergodicity* (or $L^1$-*weak ergodicity*) if for every $s \in \mathbb{R}_+$ and $\varphi, \psi, \varphi_1, \psi_1 \in S$ one has

$$\lim_{t \to \infty} \left\| \mathbb{P}^{[s,t]}(\varphi, \psi) - \mathbb{P}^{[s,t]}(\varphi_1, \psi_1) \right\|_1 = 0.$$

In [50, 206] the following is proved

**Theorem 3.7.2** *Let* $\left(E, P_{ij,k}^{[s,t]}, \mathbf{x}^{(0)}\right)$ *be a q.s.p. If there exists a function* $\lambda(s)$, $s \in [1, \infty)$ $(l(s) \in [0, 1))$ *satisfying*

$$\liminf_{s \to \infty} \lambda(s) > 0 \tag{3.33}$$

*and such that for some states* $\{n_s\} \subset E$

$$P_{ij,n_s}^{[s-1,s]} \geq \lambda(s) \ \text{for all} \ i, j \in E, \ s \geq 1, \tag{3.34}$$

*then the q.s.p. satisfies the ergodic principle.*

Recently, a similar kind of theorem was proved in [177] for nonhomogeneous continuous-time q.s.o.s with a general state space. Furthermore, in [174] we have provided necessary and sufficient conditions for q.s.p.s to enjoy the ergodic principle. Earlier papers [73, 230, 236] have investigated certain sufficient conditions for the fulfilment of the ergodic principle. In [233] regularity was investigated for a certain class of q.s.p.s.

From Theorems 3.3.1 and 3.3.2 we infer that for a given q.s.p. one can define an associated Markov process. Central limit theorems for such Markov chains were investigated in [237, 238]. In [62] absolute continuity and singularity of Markov measures corresponding to different kinds of quadratic stochastic processes were investigated.

# Chapter 4
# Analytic Methods in the Theory of Quadratic Stochastic Processes

In this chapter we are going to develop analytical methods for q.s.p.s. We will follow the lines of Kolmogorov's [121] paper. Namely, we will derive partial differential equations with delaying argument for quadratic processes of type (A) and (B), respectively.

## 4.1 Quadratic Processes with a Finite Set of States

In this chapter, we will assume that quadratic stochastic processes are homogeneous per unit time, i.e. $P_{ij,k} = P_{ij,k}^{[t,t+1]}$ for all $t \geq 1$.

We are going to consider two cases with respect to the type of the q.s.p.

First we consider quadratic processes of type (A). Let $(E, P_{ij,k}^{[s,t]}, \mathbf{x}^{(0)})$ be a q.s.p. of type (A). Suppose that the functions $P_{ij,k}^{[s,t]}$ are continuous with respect to the variables $s$ and $t$ and the functions $P_{ij,k}^{[s,t]}$ are differentiable with respect to $s$ and $t$ with $t > s+1$. Then for $t > s+2$, using (3.5), we have

$$P_{ij,k}^{[s,t+h]} - P_{ij,k}^{[s,t]} = \sum_{m,l} P_{ij,m}^{[s,t-1]} P_{ml,k}^{[t-1,t+h]} x_l^{(t-1)} - \sum_{m,l} P_{ij,m}^{[s,t-1]} P_{ml,k}^{[t-1,t]} x_l^{(t-1)}$$

$$= \sum_{m,l} P_{ij,m}^{[s,t-1]} \left( P_{ml,k}^{[t-1,t+h]} - P_{ml,k}^{[t-1,t]} \right) x_l^{(t-1)}. \qquad (4.1)$$

Assume

$$a_{ml,k}(t) = \lim_{h \to 0+} \frac{P_{ml,k}^{[t-1,t+h]} - P_{ml,k}}{h}, \qquad (4.2)$$

© Springer International Publishing Switzerland 2015
F. Mukhamedov, N. Ganikhodjaev, *Quantum Quadratic Operators and Processes*,
Lecture Notes in Mathematics 2133, DOI 10.1007/978-3-319-22837-2_4

provided that the limit exists. Dividing both sides of equality (4.1) by $h$ and passing to the limit as $h \to 0$, we get the first system of differential equations

$$\frac{\partial P_{ij,k}^{[s,t]}}{\partial t} = \sum_{m,l} a_{ml,k}(t) x_l^{(t-1)} P_{ij,m}^{[s,t-1]}, \quad i,j,k = 1, \cdots, n. \tag{4.3}$$

By (3.5) we rewrite the equation (4.3) as follows

$$\frac{\partial P_{ij,k}^{[s,t]}}{\partial t} = \sum_{m,l,r,q} a_{ml,k}(t) x_r^{(0)} x_q^{(0)} P_{rq,l}^{[0,t-1]} P_{ij,m}^{[s,t-1]}. \tag{4.4}$$

Similarly for $t > s + 2$ one gets

$$P_{ij,k}^{[s,t]} - P_{ij,k}^{[s+h,t]}$$

$$= \sum_{m,l} P_{ij,m}^{[s,s+1+h]} P_{ml,k}^{[s+1+h,t]} x_l^{(s+1+h)} - \sum_{m,l} P_{ij,m}^{[s+h,s+1+h]} P_{ml,k}^{[s+1+h,t]} x_l^{(s+1+h)}$$

$$= \sum_{m,l} \left( P_{ij,m}^{[s,s+1+h]} - P_{ij,m}^{[s,s+1]} \right) x_l^{(s+1+h)} P_{ml,k}^{[s+1+h,t]} x_l^{(t-1)}.$$

Here we have used the equality $P_{ij,m}^{[s+h,s+1+h]} = P_{ij,m}^{[s,s+1]}$. Dividing both sides of this equality by $h$ and passing to the limit as $h \to 0$, one finds the second system of partial differential equations

$$\frac{\partial P_{ij,k}^{[s,t]}}{\partial s} = -\sum_{m,l} a_{ij,m}(s+1) x_l^{(s+1)} P_{ml,k}^{[s+1,t]}, \quad i,j,k = 1, \cdots, n. \tag{4.5}$$

Again due to (3.5) the Eq. (4.5) can be rewritten as follows

$$\frac{\partial P_{ij,k}^{[s,t]}}{\partial s} = -\sum_{m,l,r,q} a_{ij,m}(s+1) x_r^{(0)} x_q^{(0)} P_{rq,l}^{[0,s+1]} P_{ml,k}^{[s+1,t]}. \tag{4.6}$$

Now let us derive a differential equation for $x_k^{(t)}$. Since for $t > 2$

$$x_k^{(t+h)} = \sum_{i,j}^{m} P_{ij,k}^{[t-1,t+h]} x_i^{(t-1)} x_j^{(t-1)}$$

and

$$x_k^{(t)} = \sum_{i,j}^{m} P_{ij,k}^{[t-1,t]} x_i^{(t-1)} x_j^{(t-1)},$$

then

$$x_k^{(t+h)} - x_k^{(t)} = \sum_{i,j=1}^{m} (P_{ij,k}^{[t-1,t+h]} - P_{ij,k}^{[t-1,t]}) x_i^{(t-1)} x_j^{(t-1)},$$

and dividing both sides of this equality by $h$ and passing to the limit as $h \to 0$, we obtain the following system of differential equations

$$\dot{x}_k^{(t)} = \sum_{i,j}^{m} a_{ij,k}(t) x_i^{(t-1)} x_j^{(t-1)}, \quad k = 1, \cdots, n. \tag{4.7}$$

So, we have proved the following theorem.

**Theorem 4.1.1** *Let* $(E, P_{ij,k}^{[s,t]}, \mathbf{x}^{(0)})$ *be a q.s.p. of type (A). Then it satisfies the partial differential equations* (4.4) *and* (4.6).

Now assume that $(E, P_{ij,k}^{[s,t]}, \mathbf{x}^{(0)})$ is a q.s.p. of type (B), and

$$\tilde{a}_{ml,k}(t) = \lim_{h \to 0+} \frac{\tilde{P}_{ml,k}^{[t-1,t+h]} - \tilde{P}_{ml,k}}{h}, \tag{4.8}$$

provided that the limit exists. For $t > s + 2$, due to (3.7), one has

$$\tilde{P}_{ij,k}^{[s,t+h]} - \tilde{P}_{ij,k}^{[s,t]} = \sum_{m,l,r,q} \tilde{P}_{im,l}^{[s,t-1]} \tilde{P}_{ir,q}^{[s,t-1]} \tilde{P}_{lq,k}^{[t-1,t+h]} x_m^{(s)} x_r^{(s)}$$

$$- \sum_{m,l,r,q} \tilde{P}_{im,l}^{[s,t-1]} \tilde{P}_{ir,q}^{[s,t-1]} \tilde{P}_{lq,k}^{[t-1,t]} x_m^{(s)} x_r^{(s)}$$

$$= \sum_{m,l,r,q} \tilde{P}_{im,l}^{[s,t-1]} \tilde{P}_{ir,q}^{[s,t-1]} \left( \tilde{P}_{lq,k}^{[t-1,t+h]} - \tilde{P}_{lq,k}^{[t-1,t]} \right) x_m^{(s)} x_r^{(s)}.$$

Then dividing both sides of this equality by $h$ and passing to the limit as $h \to 0$, we find the first system of differential equations

$$\frac{\partial \tilde{P}_{ij,k}^{[s,t]}}{\partial t} = \sum_{m,l,r,q} \tilde{a}_{lq,k}(t) x_m^{(s)} x_r^{(s)} \tilde{P}_{im,l}^{[s,t-1]} \tilde{P}_{jr,q}^{[s,t-1]}, i, j, k = 1, \cdots, n. \tag{4.9}$$

For $t > s + 2$, from

$$\tilde{P}_{ij,k}^{[s,t]} = \sum_{m,l,r,q} \tilde{P}_{im,l}^{[s,s+1+h]} \tilde{P}_{ir,q}^{[s,s+1+h]} \tilde{P}_{lq,k}^{[s+1+h,t]} x_m^{(s)} x_r^{(s)}$$

$$\tilde{P}_{ij,k}^{[s+h,t]} = \sum_{m,l,r,q} \tilde{P}_{im,l}^{[s,s+1+h]} \tilde{P}_{ir,q}^{[s,s+1+h]} \tilde{P}_{lq,k}^{[s+1+h,t]} x_m^{(s+h)} x_r^{(s+h)}$$

one gets

$$\tilde{P}_{ij,k}^{[s,t]} - \tilde{P}_{ij,k}^{[s+h,t]} = \sum_{m,l,r,q} \tilde{P}_{im,l}^{[s,s+1+h]} x_r^{(s)} \left( \tilde{P}_{jr,q}^{[s,s+1+h]} x_m^{(s)} - \tilde{P}_{jr,q} x_m^{(s)} \right.$$

$$+ \tilde{P}_{jr,q} x_m^{(s)} - \tilde{P}_{jr,q} x_m^{(s+h)} \right) \tilde{P}_{lq,r}^{[s+1+h,t]}$$

$$+ \sum_{m,l,r,q} \tilde{P}_{jr,q} x_m^{(s+h)} \left( \tilde{P}_{im,l}^{[s,s+1+h]} x_r^{(s)} - \tilde{P}_{im,l} x_r^{(s)} \right.$$

$$+ \tilde{P}_{im,l} x_r^{(s)} - \tilde{P}_{im,l} x_r^{(s+h)} \right) \tilde{P}_{lq,r}^{[s+1+h,t]}.$$

Simplifying the last one we find

$$\tilde{P}_{ij,k}^{[s,t]} - \tilde{P}_{ij,k}^{[s+h,t]} = \sum_{m,l,r,q} \tilde{P}_{im,l}^{[s,s+1+h]} x_m^{(s)} x_r^{(s)} \left( \tilde{P}_{jr,q}^{[s,s+1+h]} - \tilde{P}_{jr,q} \right) \tilde{P}_{lq,r}^{[s+1+h,t]}$$

$$+ \sum_{m,l,r,q} \tilde{P}_{im,l}^{[s,s+1+h]} \tilde{P}_{jr,q} x_r^{(s)} \left( x_m^{(s)} - x_m^{(s+h)} \right) \tilde{P}_{lq,r}^{[s+1+h,t]}$$

$$+ \sum_{m,l,r,q} \tilde{P}_{jr,q} x_m^{(s+h)} x_r^{(s)} \left( \tilde{P}_{im,l}^{[s,s+1+h]} - \tilde{P}_{im,l} \right) \tilde{P}_{lq,r}^{[s+1+h,t]}$$

$$+ \sum_{m,l,r,q} \tilde{P}_{jr,q} \tilde{P}_{im,l} x_m^{(s+h)} \left( x_r^{(s)} - x_r^{(s+h)} \right) \tilde{P}_{lq,r}^{[s+1+h,t]},$$

where dividing both sides of this equality by $h$ and passing to the limit as $h \to 0$, we obtain the second system of partial differential equations

$$\frac{\partial \tilde{P}_{ij,k}^{[s,t]}}{\partial s} = \sum_{m,l,r,q} \left( \tilde{P}_{im,l} \tilde{P}_{jr,q} (x_m^{(s)} x_r^{(s)})' \right.$$

$$- \tilde{P}_{im,l} \tilde{a}_{jr,q}(s+1) + \tilde{P}_{jr,q} \tilde{a}_{im,l}(s+1)) x_m^{(s)} x_r^{(s)} \right) \tilde{P}_{lq,k}^{[s+1+h,t]}, \qquad (4.10)$$

where $(x_m^{(s)} x_r^{(s)})' = \frac{d(x_m^{(s)} x_r^{(s)})}{ds}$ and $i, j, k = 1, \cdots, n$.

**Theorem 4.1.2** *Let* $(E, P_{ij,k}^{[s,t]}, \mathbf{x}^{(0)})$ *be a q.s.p. of type (B). Then it satisfies the partial differential equations* (4.9) *and* (4.10).

Note that all the derived differential equations are equations with delaying argument [40].

*Example 4.1.1* Let $E = \{1, 2\}$, $x^{(0)} = (x, 1-x)$, be an initial distribution on $E$ and $a_{11,1} = (x-1)\ln 2$, $a_{12,1} = [(2x-1)/2]\ln 2$, $a_{22,1} = x\ln 2$. Since $a_{ij,2} = -a_{ij,1}$, and due to (4.7), we obtain the following system of differential equations:

$$\dot{x}_1^{(t)} = (x - x_1^{(t-1)}) \ln 2,$$
$$\dot{x}_2^{(t)} = -\dot{x}_1^t. \tag{4.11}$$

From $x_1^{(0)} = x$ we have that $\dot{x}_1^{(t)}|_{t=1} = 0$, i.e., $x_1^{(1)} = x$, therefore $x_1^{(t)} = x$, $x_2^{(t)} = 1 - x$. In this case systems (4.3) and (4.5) have the following form:

$$\frac{\partial P_{ij,1}^{[s,t]}}{\partial t} = \ln \sqrt{2}(x - P_{ij,1}^{[s,t-1]}),$$

$$\frac{\partial P_{ij,2}^{[s,t]}}{\partial t} = \ln \sqrt{2}(1 - x - P_{ij,2}^{[s,t-1]}), \tag{4.12}$$

and

$$\frac{\partial P_{11,k}^{[s,t]}}{\partial s} = -(x - 1) \ln 2 [x \cdot P_{11,k}^{[s+1,t]} + (1 - 2x)P_{12,k}^{[s+1,t]} + (1 - x)P_{22,k}^{[s+1,t]}],$$

$$\frac{\partial P_{12,k}^{[s,t]}}{\partial s} = -\frac{(2x - 1)}{2} \ln 2 [x \cdot P_{11,k}^{[s+1,t]} + (1 - 2x)P_{12,k}^{[s+1,t]} + (1 - x)P_{22,k}^{[s+1,t]}],$$

$$\frac{\partial P_{22,k}^{[s,t]}}{\partial s} = -x \ln 2 [x \cdot P_{11,k}^{[s+1,t]} + (1 - 2x)P_{12,k}^{[s+1,t]} + (1 - x)P_{22,k}^{[s+1,t]}]. \tag{4.13}$$

The process defined in Chap. 3, Sect. 3.2 (see Example 4.2), with $\varepsilon = 0$, is a solution of the systems (4.12) and (4.13).

*Example 4.1.2* Let $E = \{1, 2\}$, $x^{(0)} = (x, 1 - x)$, be an initial distribution and $a_{11,1}(t) = \ln 2 \ln x \cdot 2^{t+1}$, $a_{12,1} = a_{21,1} = a_{22,1} = 0$. Since $a_{ij,2} = -a_{ij,1}$, for the distribution $x^{(t)}$ we obtain the following system of equations:

$$\dot{x}_1^{(t)} = \ln 2 \ln x \cdot 2^{t+1}(x_1^{(t-1)})^2,$$
$$\dot{x}_2^{(t)} = -\ln 2 \ln x \cdot 2^{t+1}(1 - x_2^{(t-1)})^2.$$

It has the following solution

$$x_1^{(t)} = x^{2^t}, \quad x_2^{(t)} = 1 - x^{2^t}$$

for $x_1^{(0)} = x$ and $x_2^{(0)} = 1 - x$. In this case the systems (4.3) and (4.5) have the following form:

$$\frac{\partial P_{ij,1}^{[s,t]}}{\partial t} = \ln 2 \ln x x^{2^{t-1}} P_{ij,1}^{[s,t-1]},$$

$$\frac{\partial P_{ij,2}^{[s,t]}}{\partial t} = -\ln 2 \ln x x^{2^{t-1}} \left(1 - P_{ij,2}^{[s,t-1]}\right), \tag{4.14}$$

and

$$\frac{\partial P_{11,k}^{[s,t]}}{\partial s} = -\ln 2 \ln x 2^{s+1} \left( x^{2^{s+1}} \left( P_{11,k}^{[s+1,t]} - 2P_{12,k}^{[s+1,t]} + P_{22,k}^{[s+1,t]} \right), \right.$$

$$\left. -P_{12,k}^{[s+1,t]} - P_{22,k}^{[s+1,t]} \right),$$

$$\frac{\partial P_{ij,k}^{[s,t]}}{\partial s} = 0 \quad \text{in all other cases.} \tag{4.15}$$

The process defined in Chap. 3, Sect. 3.2 (see Example 3.2.4) is a solution of the systems (4.14) and (4.15).

## 4.2   Quadratic Processes with a Continuous Set of States

First let us give an example of a quadratic process with continuous set $E$.

*Example 4.2.1* Let $(E, \Im)$ be a measurable space and $m_0$ be an initial measure on $(E, \Im)$. A transition function

$$P(s, x, y, t, A) = \frac{1}{2^{t-s-1}} \left( \frac{\delta_x(A) + \delta_y(A)}{2} + (2^{t-s-1} - 1)m_0(A) \right),$$

defined at $t - s \geq 1$ for $x, y \in E$ and $A \in \Im$, satisfies all conditions $I, II, II, IV_A, IV_B$. Moreover, it determines a homogeneous quadratic stochastic process of both types A and B.

Now let us produce differential equations for q.s.p.s with continuous $E$. As before, we first consider processes of type (A). Let $\{(E, \Im), P(s, x, y, t, A), m_0\}$ be a q.s.p. of type (A). As shown above, according to condition $IV_A$ we have for $t > s+2$

$$P(s, x, y, t + h, A) - P(s, x, y, t, A)$$

$$= \int_E \int_E P(s, x, y, t - 1, du)[P(t - 1, u, v, t + h, A) - P(t - 1, u, v, t, A)]m_{t-1}(dv).$$

Let

$$c(t, u, v, A) = \lim_{h \to 0+} \frac{P(t-1, u, v, t+h, A) - P(0, u, v, 1, A)}{h}$$

provided the limit exists. Then from the previous equality we get the first integro-differential equation

$$\frac{\partial P(s, x, y, t, A)}{\partial t} = \int_E \int_E P(s, x, y, t-1, du)c(t-1, u, v, A)m_{t-1}(dv). \qquad (4.16)$$

Similarly as in (4.6) and (4.7) one can produce the following second integro-differential equation

$$\frac{\partial P(s, x, y, t, A)}{\partial s} = -\int_E \int_E c(s+1, x, y, du)P(s+1, u, v, t, A)m_{s+1}(dv). \qquad (4.17)$$

Using the same argument, for a q.s.p. $\{(E, \mathfrak{S}), \tilde{P}(s, x, y, t, A), m_0\}$ of type (B), we obtain similar kinds of equations

$$\frac{\partial \tilde{P}(s, x, y, t, A)}{\partial t} = \int_E \int_E \int_E \int_E \tilde{c}(t, u, w, A)\tilde{P}(s, x, z, t-1, du)$$

$$\times \tilde{P}(s, y, v, t-1, dw)m_s(dz)m_s(dv) \qquad (4.18)$$

and

$$\frac{\partial \tilde{P}(s, x, y, t, A)}{\partial s} = -\int_E \int_E \tilde{P}(x, u, dv)\tilde{P}(y, w, dz)(m_s(du)m_s(dw))'$$

$$- (\tilde{c}(s+1, y, w, dz)\tilde{P}(x, u, dv) + \tilde{c}(s+1, x, u, dv)\tilde{P}(y, w, dz))$$

$$\times \tilde{P}(s+1, v, z, t, A)m_s(dv)m_s(dz), \qquad (4.19)$$

where $\tilde{c}$ is defined similarly as $c$ for $\tilde{P}(s, x, y, t, A)$, and $\tilde{P}(x, y, A) \equiv \tilde{P}(0, x, y, 1, A)$.

Let us consider the following case: $E = \mathbb{R}$ and $A_z = (-\infty, z], z \in \mathbb{R}$. Assume that $F(s, x, y, t, z) = P(s, x, y, t, A_z)$ and $\tilde{F}(s, x, y, t, z) = \tilde{P}(s, x, y, t, z)$. It is evident that $F$ and $\tilde{F}$ are the distribution functions for q.s.p.s of type (A) and type (B), respectively. If the functions $F$ and $\tilde{F}$ are absolutely continuous with respect to the variable $z$, then there are nonnegative functions $f$ and $\tilde{f}$ such that

$$F(s, x, y, t, z) = \int_z^\infty f(s, x, y, t, u)du$$

and

$$\tilde{F}(s, x, y, t, z) = \int_z^\infty \tilde{f}(s, x, y, t, u) du,$$

where $du$ is the usual Lebesgue measure on $\mathbb{R}$.

Then one can rewrite the fundamental equations $IV_A$ and $IV_B$, and integro-differential equations (4.16)–(4.19) with respect to the density functions $f$ and $\tilde{f}$, respectively. Namely, the equations $IV_A$ and $IV_B$ are reduced to

$IV_A'$

$$f(s, x, y, t, z) = \int_{-\infty}^\infty \int_{-\infty}^\infty f(s, x, y, \tau, u) f(\tau, u, v, t, z) du m_\tau(dv),$$

$IV_B'$

$$\tilde{f}(s, x, y, t, z) = \int_{-\infty}^\infty \int_{-\infty}^\infty \int_{-\infty}^\infty \int_{-\infty}^\infty \tilde{f}(s, x, u, \tau, v) \tilde{f}(s, y, w, \tau, h)$$

$$\times \tilde{f}(\tau, v, h, t, z) dv dh m_s(du) m_s(dw),$$

respectively. Similarly the Eqs. (4.16)–(4.19) can be reduced to

$$\frac{\partial f(s, x, y, t, z)}{\partial t} = \int_{-\infty}^\infty \int_{-\infty}^\infty a(t, u, v, z) f(s, x, y, t-1, u) du m_{t-1}(dv), \qquad (4.20)$$

$$\frac{\partial f(s, x, y, t, z)}{\partial s} = \int_{-\infty}^\infty \int_{-\infty}^\infty a(s+1, x, y, u) f(s+1, u, v, t, z) du m_{s+1}(dv), \qquad (4.21)$$

$$\frac{\partial \tilde{f}(s, x, y, t, z)}{\partial t} = \int_{-\infty}^\infty \int_{-\infty}^\infty \int_{-\infty}^\infty \int_{-\infty}^\infty \tilde{a}(t, v, h, z) \tilde{f}(s, x, u, t-1, v)$$

$$\times \tilde{f}(s, y, w, t-1, h) dv dh m_s(du) m_s(dw), \qquad (4.22)$$

$$\frac{\partial \tilde{f}(s, x, y, t, A)}{\partial s} = \int_{-\infty}^\infty \int_{-\infty}^\infty \left( \int_{-\infty}^\infty \int_{-\infty}^\infty \tilde{f}(x, u, v) \tilde{f}(y, w, h) (m_s(du) m_s(dw))' \right.$$

$$- (\tilde{a}(s+1, y, w, h) \tilde{f}(x, u, v)$$

$$\left. + \tilde{a}(s+1, x, u, v) \tilde{f}(y, w, h)) m_s(du) m_s(dw) \right)$$

$$\times \tilde{f}(s+1, v, h, t, z) dv dh, \qquad (4.23)$$

where $f(x, y, z) \equiv f(0, x, y, 1, z), \tilde{f}(x, y, z) \equiv \tilde{f}(0, x, y, 1, z)$ and

$$a(t, x, y, z) = \lim_{h \to 0} \frac{f(t - 1, x, y, t + h, z) - f(x, y, z)}{h},$$

$$\tilde{a}(t, x, y, z) = \lim_{h \to 0} \frac{\tilde{f}(t - 1, x, y, t + h, z) - \tilde{f}(x, y, z)}{h},$$

respectively.

If $m_\tau(dv) = r_\tau(v)dv$, then the fundamental equations $IV_A'$ and $IV_B'$ can be rewritten as follows

$I\tilde{V}_A$

$$f(s, x, y, t, z) = \int_{-\infty}^{\infty} \int_{-\infty}^{\infty} f(s, x, y, \tau, u)f(\tau, u, v, t, z)r_\tau(v)dudv,$$

$I\tilde{V}_B$

$$\tilde{f}(s, x, y, t, z) = \int_{-\infty}^{\infty} \int_{-\infty}^{\infty} \int_{-\infty}^{\infty} \int_{-\infty}^{\infty} \tilde{f}(s, x, u, \tau, v)\tilde{f}(s, y, w, \tau, h)$$

$$\times \tilde{f}(\tau, v, h, t, z)r_s(u)r_s(w)dudwdhdv,$$

respectively.

*Example 4.2.2* Let us consider a family of functions

$$f(s, x, y, t, z) = \frac{\exp(-\frac{(z-x-y)^2}{2^{t+1}-2^{s+2}+1})}{\sqrt{(2^{t+1} - 2^{s+2} + 1)\pi}}$$

with

$$r_t(v) = \frac{\exp(-\frac{v^2}{2^{t+1}-1})}{\sqrt{(2^{t+1} - 1)\pi}}.$$

One can establish that such a family determines a quadratic stochastic process of type (A).

*Example 4.2.3* A family of functions

$$f(s, x, y, t, z) = \frac{\exp(-\frac{(z-x)^2}{t-s}) + \exp(-\frac{(z-y)^2}{t-s})}{2^{t-s}\sqrt{(t-s)\pi}} + \frac{2^{t-s-1} - 1}{2^{t-s-1}} \cdot \frac{\exp(-\frac{z^2}{t+1})}{\sqrt{(t+1)\pi}}$$

with

$$r_t(v) = \frac{1}{\sqrt{(t+1)\pi}} \exp(-\frac{v^2}{t+1})$$

determines a quadratic stochastic process of type (A).

*Example 4.2.4* A family of functions

$$f(s,x,y,t,z) = \frac{2^{t-s-1} \exp(-\frac{4^{t-s-1}}{2^{2(t-s)-1}-1}) \cdot (z - \frac{x+y}{2^{t-s}})^2}{\sqrt{(2^{2(t-s)-1}-1)\pi}}$$

with

$$r_t(v) = \frac{\exp(-\frac{v^2}{2})}{\sqrt{2\pi}}$$

determines a quadratic stochastic process of both types A and B.

Under some conditions on the density functions $f$ and $\tilde{f}$, one can reduce the derived integro-differential equations (4.20)–(4.23) to some partial differential equations. Indeed, let us consider (4.20). First define

$$\Delta(s,x,y,t,z,h) = \frac{f(s,x,y,t+h,z) - f(s,x,y,t,z)}{h}. \tag{4.24}$$

Then from $IV_{A'}$ we get the following equality:

$$\Delta(s,x,y,t,z) = \int_{-\infty}^{\infty} \int_{-\infty}^{\infty} f(s,x,y,t-1,u)\Delta(t-1,u,v,t,z)m_{t-1}(dv)du. \tag{4.25}$$

Assume that the function $f(s,x,y,t-1,u)$ has partial derivatives up to third order with respect to the argument $u$, and consider its Taylor expansion in a neighborhood of the point $z$:

$$f(s,x,y,t-1,u) = f(s,x,y,t-1,z) + \frac{\partial f(s,x,y,t-1,z)}{\partial z}(u-z)$$

$$+ \frac{\partial^2 f(s,x,y,t-1,z)}{\partial z^2} \cdot \frac{(u-z)^2}{2} + \theta \frac{(u-z)^3}{6}$$

and substitute this expansion into (4.25). Then one finds

$$\Delta(s, x, y, t, z)$$

$$= f(s, x, y, t-1, z) \int_{-\infty}^{\infty} \int_{-\infty}^{\infty} \Delta(t-1, u, v, t, z, h) m_{t-1}(dv) du$$

$$+ \frac{\partial f(s, x, y, t-1, z)}{\partial z} \int_{-\infty}^{\infty} \int_{-\infty}^{\infty} \Delta(t-1, u, v, t, z, h)(u-z) m_{t-1}(dv) du$$

$$+ \frac{\partial^2 f(s, x, y, t-1, z)}{\partial z^2} \int_{R} \int_{R} \frac{\Delta(t-1, u, v, t, z, h)}{2} (u-z)^2 m_{t-1}(dv) du$$

$$+ \theta \int_{-\infty}^{\infty} \int_{-\infty}^{\infty} \frac{\Delta(t-1, u, v, t, z, h)}{6} (u-z)^3 m_{t-1}(dv) du. \qquad (4.26)$$

Assume that the following limits exist:

$$\lim_{h \to 0+} \int_{-\infty}^{\infty} \int_{-\infty}^{\infty} \Delta(t-1, u, v, t, z, h) m_{t-1}(dv) du = \bar{N}(t, z);$$

$$\lim_{h \to 0+} \int_{-\infty}^{\infty} \int_{-\infty}^{\infty} \Delta(t-1, u, v, t, z, h)(u-z) m_{t-1}(dv) du = \bar{A}(t, z);$$

$$\lim_{h \to 0+} \int_{-\infty}^{\infty} \int_{-\infty}^{\infty} \frac{\Delta(t-1, u, v, t, z, h)}{2} (u-z)^2 m_{t-1}(dv) du = \bar{B}^2(t, z);$$

$$\lim_{h \to 0+} \int_{-\infty}^{\infty} \int_{-\infty}^{\infty} \Delta(t-1, u, v, t, z, h)|u-z|^3 m_{t-1}(dv) du = 0.$$

Then passing to the limit in (4.26) when $h \to 0$, we obtain the following partial differential equation with delaying argument

$$\frac{\partial f(s, x, y, t, z)}{\partial t} = \bar{N}(t, z) f(s, x, y, t-1, z) + \bar{A}(t, z) \frac{\partial f(s, x, y, t-1, z)}{\partial z}$$

$$+ \bar{B}^2(t, z) \frac{\partial^2 f(s, x, y, t, z)}{\partial z^2}. \qquad (4.27)$$

Now let us elaborate on the integro-differential equation (4.21). Define

$$\tilde{\Delta}(s, x, y, t, z, h) = f(s, x, y, t, z) - f(s+h, x, y, t, z). \qquad (4.28)$$

Then again using IV$_{A'}$ one finds

$$\tilde{\Delta}(s, x, y, t, z, h) = \int_{-\infty}^{\infty} \int_{-\infty}^{\infty} \tilde{\Delta}(s, x, y, s + 1 + h, u, h)$$

$$\times f(s + 1 + h, u, v, t, z) m_{s+1+h}(dv) du. \qquad (4.29)$$

Assuming that the function $f(s + 1 + h, u, v, t, z)$ has partial derivatives up to third order, we expand it into a Taylor series in a neighborhood of the point $(x, y)$:

$$f(s + 1 + h, u, v, t, z)$$

$$= f(s + 1 + h, x, y, t, z) + \frac{\partial f(s + 1 + h, x, y, t, z)}{\partial x}(u - x)$$

$$+ \frac{\partial f(s + 1 + h, x, y, t, z)}{\partial y}(v - y)$$

$$+ \frac{1}{2} \frac{\partial^2 f(s + 1 + h, x, y, t, z)}{\partial x^2}(u - x)^2$$

$$+ \frac{1}{2} \frac{\partial^2 f(s + 1 + h, x, y, t, z)}{\partial y^2}(v - y)^2$$

$$+ \frac{\partial^2 f(s + 1 + h, x, y, t, z)}{\partial x \partial y}(u - x)(v - y)$$

$$+ \frac{1}{6} \frac{\partial^3 f(s + 1 + h, x + \theta_3(u - x), y + \theta_3(v - y), t, z)}{\partial x^3}(u - x)^3$$

$$- \frac{1}{2} \frac{\partial^3 f}{\partial x^2 \partial y}(u - x)^2(v - y) + \frac{1}{2} \frac{\partial^3 f}{\partial x \partial y^2}(u - x)(v - y)^2$$

$$+ \frac{1}{6} \frac{\partial^3 f}{\partial y^3}(v - y)^3.$$

By substituting this expansion into (4.29) let us evaluate the integrals from each summand. Then one finds

$$\int_{-\infty}^{+\infty} \int_{-\infty}^{+\infty} \tilde{\Delta}(s, x, y, s + 1 + h, u, h) f(s + 1 + h, x, y, t, z) m_{s+1+h}(dv) du = 0;$$

$$\int_{-\infty}^{+\infty} \int_{-\infty}^{+\infty} \tilde{\Delta}(s, x, y, s + 1 + h, u, h) f(s + 1 + h, x, y, t, z) \frac{\partial f}{\partial x}(u - x) m_{s+1+h}(dv) du$$

$$= \frac{\partial f(s + 1 + h, x, y, t, z)}{\partial x} \cdot \int_{-\infty}^{\infty} \tilde{\Delta}(s, x, y, s + 1 + h, u, h)(u - x) du.$$

Let

$$a(s, x, y, h) = \int_{-\infty}^{\infty} \tilde{A}(s, x, y, s + 1 + h, u, h)(u - x)du$$

and assume that

$$\int_{-\infty}^{\infty} \tilde{A}(s, x, y, s + 1 + h, u, h)\frac{\partial f}{\partial y}(v - y)m_{s+1+h}(dv)du$$

$$= \frac{\partial f(s + 1 + h, x, y, t, z)}{\partial x} \int_{-\infty}^{\infty} \tilde{A}(s, x, y, s + 1 + h, u, h)du \int_{-\infty}^{\infty} (v - y)m_{s+1+h}(dv)$$

$$= 0.$$

Now consider the second moments

$$\int_{-\infty}^{\infty}\int_{-\infty}^{\infty} \tilde{A}(s, x, y, s + 1 + h, u, h)\frac{1}{2}\frac{\partial f(s + 1 + h, x, y, t, z)}{\partial x}(u - x)^2 m_{s+1+h}(dv)du$$

$$= \frac{1}{2} \cdot \frac{\partial^2 f(s + 1 + h, x, y, t, z)}{\partial x^2} \int_{-\infty}^{\infty} \tilde{A}(s, x, y, s + 1 + h, u, h)(u - x)^2 du.$$

Put

$$b^2(s, x, y, h) = \int_{-\infty}^{\infty} [f(s, x, y, s + 1 + h, u) - f(s + h, x, y, s + 1 + h, u)](u - x)^2 du.$$

Furthermore,

$$\int_{-\infty}^{\infty}\int_{-\infty}^{\infty} [f(s, x, y, s + 1 + h, u) - f(s + h, x, y, s + 1 + h, u)]$$

$$\times \frac{\partial^2 f(s + 1 + h, x, y, t, z)}{\partial x \partial y}(u - x)(v - y)m_{s+1+h}(dv)du$$

$$= \frac{\partial^2 f(s + 1 + h, x, y, t, z)}{\partial x \partial y} \times \int_{-\infty}^{\infty} [f(s, x, y, s + 1 + h, u)$$

$$-f(s + h, x, y, s + 1 + h, u)](u - x)du \cdot \int_{-\infty}^{\infty} (v - y)m_{s+1+h}(dv).$$

Let

$$d(s + 1, y, h) = \int_{-\infty}^{\infty} (v - y)m_{s+1+h}(dv).$$

It is evident that

$$\int_{-\infty}^{\infty} \int_{-\infty}^{\infty} [f(s, x, y, s + 1 + h, u) - f(s + h, x, y, s + 1 + h, u)]$$

$$\times \frac{1}{2} \frac{\partial^2 f(s + 1 + h, x, y, t, z)}{\partial y^2} (v - y)^2 m_{s+1+h}(dv) du = 0.$$

Assume that the other integrals tend to zero when $h \to 0$. Then we get

$$\frac{f(s, x, y, t, z) - f(s + h, x, y, t, z)}{h}$$

$$= \frac{a(s, x, y, h)}{h} \cdot \frac{\partial f(s + 1 + h, x, y, t, z)}{\partial x}$$

$$+ \frac{b^2(s, x, y, h)}{h} \cdot \frac{1}{2} \frac{\partial^2 f(s + 1 + h, x, y, t, z)}{\partial x^2}$$

$$+ \frac{a(s, x, y, h)}{h} \cdot d(s + 1, y, h) \cdot \frac{\partial^2 f(s + 1 + h, x, y, t, z)}{\partial x \partial y}. \qquad (4.30)$$

Letting

$$A(s, x, y) = \lim_{h \to 0} \frac{a(s, x, y, h)}{h},$$

$$B^2(s, x, y) = \lim_{h \to 0} \frac{b^2(s, x, y, h)}{h},$$

$$D(s + 1, y) = \lim_{h \to 0} d(s + 1, y, h),$$

provided that these limits exist, equality (4.30) is transformed into the following differential equation:

$$\frac{\partial f(s, x, y, t, z)}{\partial s} = -A(s, x, y) \frac{\partial f(s + 1, x, y, t, z)}{\partial x} - B^2(s, x, y) \frac{\partial^2 f(s + 1, x, y, t, z)}{\partial x^2}$$

$$-A(s, x, y) D(s + 1, y) \frac{\partial^2 f(s + 1, x, y, t, z)}{\partial x \partial y}. \qquad (4.31)$$

Thus the integro-differential equations (4.20) and (4.21) with delaying argument (with respect to $t$ and $s$ respectively) have reduced to differential equations (4.27) and (4.31), respectively. The latter are also equations with delaying argument. In the next chapter, these equations will be reduced to well-known differential equations that are not equations with delaying argument. The integro-differential equations (4.22) and (4.23) will also be reduced to differential equations using another approach.

## 4.3 Averaging of Quadratic Stochastic Processes

In this chapter we are going to consider continuous analogues of Theorems 3.3.1 and 3.3.2. Namely, we consider relations between quadratic and Markovian processes.

**Theorem 4.3.1** *Let* $\{(E, \Im), P(s, x, y, t, A), m_0\}$ *be a q.s.p. Then the function*

$$H(s, x, t, A) = \int_E P(s, x, y, t, A) m_s(dy) \tag{4.32}$$

*is the transition function for some Markovian process with initial distribution* $m_0$.

*Proof* We consider two cases with respect to the type of the q.s.p. Assume that the q.s.p. is of type (A). Then one gets

$$H(s, x, t, A) = \int_E P(s, x, y, t, A) m_s(dy)$$

$$= \int_E \left( \int_E \int_E P(s, x, y, \tau, du) P(\tau, u, v, t, A) m_\tau(dv) m_s(dy) \right)$$

$$= \int \left( \int_E P(s, x, y, \tau, du) m_s(dy) \right) \left( \int_E P(\tau, u, v, t, A) m_\tau(dv) \right)$$

$$= \int_E H(s, x, \tau, du) H(\tau, u, t, A).$$

Similarly, for a type (B) process, we obtain

$$\tilde{H}(s, x, t, A) = \int_E \tilde{P}(s, x, y, t, A) m_s(dy)$$

$$= \int_E \left( \int_E \int_E \int_E \int_E \tilde{P}(s, x, z, \tau, du) \tilde{P}(s, y, u, \tau, dw) \right.$$

$$\times \tilde{P}(\tau, u, w, t, A) m_s(dv) \Big) m_s(dy)$$

$$= \int \int_E \int_E \tilde{P}(s, x, z, \tau, du) m_s(dz) \int_E \int_E \tilde{P}(s, y, v, \tau, dw)$$

$$\times \tilde{P}(\tau, u, w, t, A) m_s(dv)$$

$$= \int_E \int_E \tilde{P}(s, x, z, du) m_s(dz) \int_E \tilde{P}(\tau, u, w, t, A) m_\tau(dw)$$

$$= \int_E \tilde{H}(s, x, \tau, du) \tilde{H}(\tau, u, t, A).$$

Hence, the theorem is proved.

*Remark 4.3.1* Note that a process generated by transition probabilities $\{H(s, x, t, A)\}$, in general, forms a non-homogenous Markov process. Thus, starting from a quadratic process one can construct a non-homogenous Markov process.

*Remark 4.3.2* This theorem allows us to simplify the obtained system of differential and integro-differential equations. We demonstrate this below.

### 4.3.1   The Set E Is Finite

In this subsection we consider the case when $E$ is a finite set. Let $(E, P_{ij,k}^{[s,t]}, \mathbf{x}^{(0)})$ and $(E, \tilde{P}_{ij,k}^{[s,t]}, \mathbf{x}^{(0)})$ be q.s.p.s of types A and B, respectively. As before, we assume that q.s.p.s are homogeneous per unit time. Then the corresponding Markov processes are

$$H_{ij}^{[s,t]} = \sum_{k=1}^{n} P_{ik,j}^{[s,t]} x_k^{(s)}, \tag{4.33}$$

(respectively

$$\tilde{H}_{ij}^{[s,t]} = \sum_{k=1}^{n} \tilde{P}_{ik,j}^{[s,t]} x_k^{(s)}).$$

*Remark 4.3.3* Note that with any q.s.p, one can connect a Markovian chain with the same initial distribution. It is necessary to note that although the function $P_{ij,k}^{[t,t+h]}$ is not defined for $0 \le h < 1$, the quantity $H_{ij}^{[t,t+h]}$ is defined by means of a Markovian property but it cannot be represented in the form $\sum_{k=1}^{n} \tilde{P}_{ik,j}^{[t,t+h]} x_k^{(t)}$.

From Theorems 3.3.1 and 3.3.2 one has

$$P_{ij,k}^{[s,t]} = \sum_{m=1}^{n} P_{ij,m}^{[s,\tau]} H_{mk}^{[\tau,t]},$$

(4.34)

(respectively

$$\tilde{P}_{ij,k}^{[s,t]} = \sum_{m,l=1}^{n} \tilde{H}_{im}^{[s,\tau]} \tilde{H}_{jl}^{[s,\tau]} \tilde{P}_{ml,k}^{[\tau,t]}),$$

(4.35)

where $\tau - s \geq 0$, and $t - \tau \geq 1$ and

$$x_k^{(\tau)} = \sum_{i=1}^{n} H_{ik}^{[0,\tau]} x_i^{(0)}.$$

(4.36)

Following [121] we assume

$$\lim_{h \to 0+} \frac{H_{ij}^{[t,t+h]} - H_{ij}^{[t,t]}}{h} = A_{ij}(t)$$

(4.37)

and

$$\lim_{h \to 0+} \frac{\tilde{H}_{ij}^{[t,t+h]} - \tilde{H}_{ij}^{[t,t]}}{h} = \tilde{A}_{ij}(t).$$

(4.38)

**Proposition 4.3.2** *Let $(E, P_{ij,k}^{[s,t]}, \mathbf{x}^{(0)})$ and $(E, \tilde{P}_{ij,k}^{[s,t]}, \mathbf{x}^{(0)})$ be q.s.p.s of types A and B, respectively. Then the following equalities hold:*

$$a_{ij,k}(t) = \sum_{m=1}^{n} P_{ij,m} A_{mk}(t)$$

(4.39)

*and*

$$\tilde{a}_{ij,k}(t) = \sum_{l=1}^{n} (\tilde{A}_{il}(t) \tilde{P}_{lj,k} + \tilde{A}_{jl}(t) \tilde{P}_{il,k}).$$

(4.40)

*Proof* Let us first consider the case when the q.s.p. has type (A). Then according to (4.34), one finds

$$P_{ij,k}^{[t,t+1+h]} = \sum_{m=1}^{n} P_{ij,m} H_{mk}^{[t+1,t+1+h]}$$

$$P_{ij,k}^{[t,t+1]} = \sum_{m=1}^{n} P_{ij,m} H_{mk}^{[t+1,t+1]}.$$

Therefore, we obtain

$$a_{ij,k}(t+1) = \lim_{h \to 0+} \frac{P_{ij,k}^{[t,t+1+h]} - P_{ij,k}^{[t,t+1]}}{h}$$

$$\times \lim_{h \to 0+} \frac{\sum_{m=1}^{n} P_{ij,m} \left( H_{mk}^{[t+1,t+1+h]} - H_{mk}^{[t+1,t+1]} \right)}{h}$$

$$= \sum_{m=1}^{n} P_{ij,m} A_{mk}(t+1).$$

Now let us turn to a q.s.p. of type (B). By putting

$$\tilde{H}_{il}^{[t,t]} = \begin{cases} 1 \text{ if } i = l \\ 0 \text{ if } i \neq l \end{cases} \tag{4.41}$$

from (4.35) one finds

$$\frac{\tilde{P}_{ij,k}^{[t,t+1+h]} - \tilde{P}_{ij,k}^{[t,t+1]}}{h} = \sum_{l,m=1}^{n} \tilde{H}_{im}^{[t,t+h]} \frac{\tilde{H}_{jl}^{[t,t+h]} - \tilde{H}_{jl}^{[t,t]}}{h} \cdot \tilde{P}_{lm,k}$$

$$+ \sum_{l,m=1}^{n} \tilde{H}_{jl}^{[t,t]} \frac{\tilde{H}_{im}^{[t,t+h]} - \tilde{H}_{im}^{[t,t]}}{h} \cdot \tilde{P}_{lm,k}.$$

Therefore, passing to the limit as $h \to 0$ and taking into account (4.41), we obtain the equality (4.40).

*Remark 4.3.4* In [121] it was proved that the continuity of $H_{ij}^{[s,t]}$ with respect to $s$ and $t$ is sufficient for the existence of limits (4.37) and (4.38). From (4.39) and (4.40) it follows that if $P_{ij,k}^{[s,t]}$ and $\tilde{P}_{ij,k}^{[s,t]}$ are continuous with respect to $s$ and $t$, then the limits (4.2) and (4.9) exist.

The equalities (4.39) and (4.40) allow us to simplify the system of equations produced in Sect. 4.1. Let us consider the first system of equations (4.3). According to (4.39) we have

$$\sum_{m,l=1}^{n} a_{ml,k}(t) x_l^{(t-1)} P_{ij,m}^{[s,t-1]} = \sum_{m,l,r=1}^{n} p_{ml,r} A_{r,k}(t) x_l^{(t-1)} P_{ij,m}^{[s,t-1]}$$

$$= \sum_{r=1}^{n} \left( \sum_{m,l=1}^{n} P_{ij,m}^{[s,t-1]} P_{ml,r} x_l^{(t-1)} \right) A_{rk}(t)$$

$$= \sum_{r=1}^{n} P_{ij,r}^{[s,t]} A_{r,k}(t).$$

Hence, (4.3) is reduced to

$$\frac{\partial P_{ij,k}^{[s,t]}}{\partial t} = \sum_{l=1}^{n} A_{lk}(t) P_{ij,l}^{[s,t]}. \tag{4.42}$$

This system (4.42) is similar to Kolmogorov's direct differential equations for Markov chains [121]:

$$\frac{\partial q_{ik}(s,t)}{\partial t} = \sum_{l=1}^{n} A_{lk}(t) q_{il}(s,t).$$

Hence, the system of differential equations with delaying argument (4.3) is reduced to the well-known system of equations (4.42).

Now let us consider the system of differential equations (4.7). From (4.39) we get

$$\sum_{i,j=1}^{n} a_{ij,k}(t) x_i^{(t-1)} x_j^{(t-1)} = \sum_{i,j=}^{n} \left( \sum_{l=1}^{n} P_{ij,l} A_{lk}(t) x_i^{(t-1)} x_j^{(t-1)} \right)$$

$$= \sum_{l=1}^{n} A_{lk}(t) \left( \sum_{i,j=1}^{n} P_{ij,l} x_i^{(t-1)} x_j^{(t-1)} \right)$$

$$= \sum_{l=1}^{n} A_{mk}(t) x_m^{(t)}.$$

Then (4.7) can be rewritten as

$$\dot{x}_k^{(t)} = \sum_{m=1}^{n} A_{mk}(t) x_m^{(t)}. \tag{4.43}$$

Similarly, the reverse equation (4.5), according to (4.39), can be rewritten as follows

$$\frac{\partial P_{ij,k}^{[s,t]}}{\partial s} = - \sum_{m,l,r=1}^{n} P_{ij,r} A_{rm}(s+1) x_l^{(s+1)} P_{mj,k}^{[s+1,t]} \tag{4.44}$$

i.e., in this case the Eq. (4.5) cannot be reduced to an ordinary differential equation.

Now let us consider the system of differential equations (4.9) and (4.10) produced for a q.s.p. of type (B). According to (4.40), the Eq. (4.9) is reduced to the following one:

$$\frac{\partial \tilde{P}_{ij,k}^{[s,t]}}{\partial t} = \sum_{m,l,r,q,u,v=1}^{n} (\tilde{A}_{qv}(t-1)\tilde{P}_{lv,k} + \tilde{A}_{lu}(t-1)\tilde{P}_{uq,k}) x_m^{(s)} x_r^{(s)} \tilde{P}_{im,l}^{[s,t-1]} \tilde{P}_{jr,q}^{[s,t-1]}. \tag{4.45}$$

As in the previous case, the delaying of the argument is preserved.

We are going to reproduce the reverse equations (4.10) by means of (4.35). Namely, from (4.35) we obtain

$$\frac{\tilde{P}_{ij,k}^{[s,t]} - \tilde{P}_{ij,k}^{[s+h,t]}}{h} = \sum_{l,m=1}^{n} \tilde{H}_{jm}^{[s,s+h]} \frac{\tilde{H}_{il}^{[s,s+h]} - \tilde{H}_{il}^{[s,s]}}{h} \cdot \tilde{P}_{lm,k}^{[s+h,t]}$$

$$+ \sum_{l,m=1}^{n} \tilde{H}_{il}^{[s,s]} \frac{\tilde{H}_{jm}^{[s,s+h]} - \tilde{H}_{jm}^{[s,s]}}{h} \cdot \tilde{P}_{lm,k}^{[s+h,t]}.$$

Therefore, passing to limit as $h \to 0$ and taking into account (4.41), one finds

$$\frac{\partial \tilde{P}_{ij,k}^{[s,t]}}{\partial s} = - \sum_{l=1}^{n} (\tilde{A}_{jl}(s)\tilde{P}_{il,k}^{[s,t]} + \tilde{A}_{il}(s)\tilde{P}_{lj,k}^{[s,t]}). \tag{4.46}$$

Hence, the reverse equations are reduced to ordinary differential equations, which differ from the reverse Kolmogorov's equation for Markov chains [121] only by a number of summands.

So, as in [121], one can establish the existence and the uniqueness of the solutions for given initial conditions.

Consequently, for quadratic stochastic processes of type (A), the direct system of differential equations are similar to Kolmogorov's direct system for Markov chains [124], and in the case of processes of type (B) the reverse system of differential equations are similar to Kolmogorov's reverse system for Markov chains [121].

**Definition 4.3.1** If a quadratic stochastic process satisfies both the fundamental equations $IV_A$ and $IV_B$, then it is called *simple*.

Examples 4.1 and 4.2 in Chap. 3 are simple quadratic processes. The above results can be interpreted in the following way.

**Theorem 4.3.3** *The analytic theory of simple quadratic processes coincides with Kolmogorov's analytic theory of Markov chains.*

Similarly, one can consider the case when $E$ is a countable set, but here we shall omit it.

### 4.3.2 The Set E Is a Continuum

Now we consider the case when the set $E$ is a continuum. Let $f$ and $\tilde{f}$ be the density functions. Then according to Theorem 4.3.1 the functions

$$g(s,x,t,z) = \int_{-\infty}^{\infty} f(s,x,y,t,z)m_s(dy) \tag{4.47}$$

and

$$\tilde{g}(s,x,t,z) = \int_{-\infty}^{\infty} \tilde{f}(s,x,y,t,z)m_s(dy) \tag{4.48}$$

are the density functions for some Markov process. Then (4.47) and (4.48) can be rewritten as follows:

$$f(s,x,y,t,z) = \int_{-\infty}^{\infty} f(s,x,y,\tau,u)g(\tau,u,t,z)du \tag{4.49}$$

and

$$\tilde{f}(s,x,y,t,z) = \int_{-\infty}^{\infty}\int_{-\infty}^{\infty} \tilde{g}(s,x,\tau,v)\tilde{g}(s,y,\tau,h)\tilde{f}(\tau,v,h,t,z)dvdh \tag{4.50}$$

where $\tau - s \geq 0, t - \tau \geq 1$.

Now using (4.49), (4.50), and the same argument as in the case when $E$ is finite, and following the lines of [121], we then reduce the Eqs. (4.20) and (4.23) to the following:

$$\frac{\partial f(s,x,y,t,z)}{\partial t} = N(t,z)f(s,x,y,t,z) + A(t,z)\frac{\partial f(s,x,y,t,z)}{\partial z}$$

$$+B^2(t,z)\frac{\partial^2 f(s,x,y,t,z)}{\partial z^2}, \tag{4.51}$$

where

$$N(t,z) = \lim_{h \to 0} \frac{\int_{-\infty}^{\infty} g(t,u,t+h,z)du - 1}{h};$$

$$A(t,z) = \lim_{h \to 0} \frac{\int_{-\infty}^{\infty} g(t,u,t+h,z)(u-z)du}{h};$$

$$B^2(t,z) = \lim_{h \to 0} \frac{\int_{-\infty}^{\infty} g(t,u,t+h,z)(u-z)^2 du}{2h},$$

and

$$\frac{\partial \tilde{f}(s,x,y,t,z)}{\partial s} = -\tilde{A}(s,x,z)\frac{\partial \tilde{f}(s,x,y,t,z)}{\partial x} - \tilde{A}(s,y,z)\frac{\partial \tilde{f}(s,x,y,t,z)}{\partial y}$$

$$-B^2(s,x,z)\frac{\partial^2 \tilde{f}(s,x,y,t,z)}{\partial x^2} - B^2(s,x,z)\frac{\partial^2 \tilde{f}(s,x,y,t,z)}{\partial y^2},$$

(4.52)

where

$$\tilde{A}(t,z) = \lim_{h \to 0} \frac{\int_{-\infty}^{\infty}\int_{-\infty}^{\infty} \tilde{g}(s-h,x,s,v)\tilde{g}(s-h,y,s,w)(v-z)dvdw}{h};$$

$$\tilde{B}^2(s,x,z) = \lim_{h \to 0} \frac{\int_{-\infty}^{\infty}\int_{-\infty}^{\infty} \tilde{g}(s-h,x,s,v)\tilde{g}(s-h,y,s,v)(v-z)^2 dvdw}{2h}.$$

The existence of all above mentioned limits can be proved as in [121]. The existence and uniqueness of solutions for Eqs. (4.51) and (4.52) can be established by the methods of [121] and [42].

## 4.4   Diffusion Quadratic Processes

**Definition 4.4.1** We call a quadratic stochastic process *Wiener* (respectively *diffusion*, *Poisson*, etc.) if its average is a Wiener (respectively diffusion, Poisson etc.) process.

Let us consider the following process (see Example 4.2.4):

$$f(s,x,y,t,z) = \frac{2^{t-s-1}\exp(-\frac{4^{t-s-1}}{2^{2(t-s)-1}-1}\cdot(z-\frac{x+y}{2^{t-s}})^2)}{\sqrt{(2^{2(t-s)-1}-1)\pi}}$$

(4.53)

with

$$r_t(v) = \frac{\exp(-\frac{v^2}{2})}{\sqrt{2\pi}}. \tag{4.54}$$

**Proposition 4.4.1** *The quadratic stochastic process generated by (4.53)–(4.54) is a diffusion process.*

*Proof* Let us compute the mean (average) of the process (4.53):

$$g(s, x, y, t, z) = \int_{-\infty}^{\infty} f(s, x, y, t, z) r_s(y) dy$$

$$= \int_{-\infty}^{\infty} \frac{2^{t-s-1} \exp(-\frac{4^{t-s-1}}{2^{2(t-s)-1}-1}(z - \frac{x+y}{2^{t-s}})^2)}{\sqrt{(2^{2(t-s)-1}-1)\pi}} \cdot \frac{\exp(-\frac{y^2}{2})}{\sqrt{2\pi}} dy.$$

We have

$$\int_{-\infty}^{\infty} e^{-r^2 x^2} dx = \frac{\sqrt{\pi}}{r} \quad (r > 0),$$

and simple but unwieldy calculations show that

$$g(s, x, t, z) = \frac{1}{\sqrt{2\pi}} \frac{2^{t-s}}{\sqrt{4^{t-s}-1}} \exp\left(-\frac{4^{t-s}}{2(4^{t-s}-1)}\right) \left(z - \frac{x}{2^{t-s}}\right)^2, \tag{4.55}$$

and (4.55) is a density of transition probabilities that defines the diffusion process.

As mentioned above, the process (4.53) is simple. In this case, the corresponding differential equations (4.51) and (4.52) have the following forms:

$$\frac{\partial f(s, x, y, t, z)}{\partial t} = \ln 2 \left( f(s, x, y, t, z) + z \frac{\partial f(s, x, y, t, z)}{\partial z} \right.$$

$$\left. + (1 + z)^2 \frac{\partial^2 f(s, x, y, t, z)}{\partial z^2} \right),$$

and

$$\frac{\partial f(s, x, y, t, z)}{\partial s} = \ln 2 \left( x \frac{\partial f(s, x, y, t, z)}{\partial x} + y \frac{\partial f(s, x, y, t, z)}{\partial y} \right.$$

$$\left. - \frac{\partial^2 f(s, x, y, t, z)}{\partial x^2} - \frac{\partial^2 f(s, x, y, t, z)}{\partial y^2} \right).$$

## 4.5   Comments and References

The motivation behind the study of q.s.p.s came from the dynamics of q.s.o.s, where the q.s.p. describes its trajectory (see Chap. 1). A theory of q.s.p.s has been developed in [46, 234–236]. With the exception of the last section, all material in this chapter has essentially been taken from [46, 234, 237]. Note that in [197], the direct and reverse equations have been derived for general q.s.p.s (i.e. the process is not necessarily homogeneous). The results of the last section are taken from [51]. If we consider the Eq. (4.4) and (4.6), then one can ask:

**Open problem 4.5.1** *Under what conditions on the coefficients do these equations produce a quadratic stochastic process?*

In [187] we have found some conditions on the coefficients $(a_{ij,k}(t))$ for homogeneous q.s.p.s of type (A) so that the equations produce q.s.p.s.

# Chapter 5
# Quantum Quadratic Stochastic Operators

It is known that there are many systems which are described by nonlinear operators. One of the simplest nonlinear cases is the quadratic one. In the previous chapters we have considered classical (i.e. commutative) quadratic operators. These operators were defined over commutative algebras. However, such operators do not cover the case of quantum systems. Therefore, in the present chapter, we are going to introduce a noncommutative analogue of a q.s.o., which is called a *quantum quadratic stochastic operator* (q.q.s.o.). We will show that the set of q.q.s.o.s is weakly compact. By means of q.q.s.o.s one can define a nonlinear operator which is called a *quadratic operator*. We also study the asymptotically stability of the dynamics of quadratic operators. Moreover, in this chapter we recall the definition of quantum Markov chains and establish that each q.q.s.o. defines a quantum Markov chain.

## 5.1 Markov Operators

Let $B(H)$ be the algebra of all bounded linear operators on a separable complex Hilbert space $H$. Let $M \subset B(H)$ be a von Neumann algebra with unit $\mathbb{1}$. By $M_+$ (resp. $M_{sa}$) we denote the set of all positive (resp. self-adjoint) elements of $M$. By $M_*$ and $M^*$, respectively, we denote predual and dual spaces of $M$. The $\sigma(M, M_*)$-topology on $M$ is called the *ultraweak* topology. Note that the set $M_*$ coincides with the set of all ultraweakly continuous functionals on $M$ (see [226]). Recall that a linear functional $f$ on $M$ is called *Hermitian* if $f(x^*) = \overline{f(x)}$ for all $x \in M$. A linear functional $f$ on $M$ is called *positive* if $f(x) \geq 0$ whenever $x \geq 0$. A positive functional $f$ is called a *state* if $f(\epsilon) = 1$. By $M_*^h$ (resp. $M_{*,+}$) we denote the set of all Hermitian (resp. positive) functionals taken from $M_*$. The set of all ultraweakly continuous states on $M$ is denoted by $S$. It is clear that $S \subset M_{*,+}$. A state $f$ is called *normal* if it satisfies $f(\sup_\alpha x_\alpha) = \sup_\alpha f(x_\alpha)$ for every uniformly bounded increasing net $\{x_\alpha\}$ of positive elements of $M$. It is well known (see [25, 226]) that a state is normal if and only if it is ultraweakly continuous. In what follows, by $\|\cdot\|_1$ we denote the norm on $M_*$. It is known that for any $f \in M_{*,+}$ one has $\|f\|_1 = f(\epsilon)$. Moreover, for any Hermitian $f \in M_*^h$, there are uniquely defined $f_1, f_2 \in M_{*,+}$ such that

$$f = f_1 - f_2, \quad \|f\|_1 = \|f_1\|_1 + \|f_2\|_1.$$

This decomposition is called the *Jordan decomposition* (see [246, Theorem 4.2, p. 140]).

Let $A$ and $B$ be Banach spaces and $A \odot B$ be their algebraic tensor product. A norm $\alpha$ on $A \odot B$ is said to be a *cross-norm* if $\alpha(x \otimes y) = \alpha(x)\alpha(y)$ for all $x \in A$, $y \in B$. The greatest cross-norm $\gamma$ on $A \odot B$ is defined by

$$\gamma(z) = \inf\left\{ \sum_{j=1}^{n} \|x_j\| \|y_j\| \right\}, \quad z \in A \odot B,$$

where the inf is taken over all representations of $z = \sum_{j=1}^{n} x_j \otimes y_j$. The completion of $A \odot B$ with respect to $\gamma$ is denoted by $A \otimes_\gamma B$. For the theory of operator algebras and tensor products of Banach spaces, we refer to [226, 242, 246].

Let $M_n(\mathbb{C})$ be the algebra of $n \times n$ matrices over the complex field $\mathbb{C}$. Let $A$ and $B$ be two $C^*$-algebras with unit. Recall that a linear mapping $\Phi : A \to B$ is called

(i)   a *morphism* if $\Phi(x^*) = \Phi(x)^*$ for all $x \in A$;
(ii)  *positive* if $\Phi(x) \geq 0$ whenever $x \geq 0$;
(iii) *unital* if $\Phi(\mathbb{1}) = \mathbb{1}$;
(iv)  *n-positive* if the mapping $\Phi_n : M_n(A) \to M_n(B)$ defined by $\Phi_n(a_{ij}) = (\Phi(a_{ij}))$ is positive. Here $M_n(A)$ denotes the algebra of $n \times n$ matrices with $A$-valued entries;
(v)   *completely positive* if it is *n*-positive for all $n \in \mathbb{N}$;
(vi)  a *Kadison–Schwarz operator (KS-operator)*, if one has

$$\Phi(x)^* \Phi(x) \leq \Phi(x^*x) \quad \text{for all } x \in A. \tag{5.1}$$

It is well known [246] that complete positivity can be formulated as follows: for any two collections $a_1, \cdots, a_n \in A$ and $b_1, \cdots, b_n \in B$ the following relation holds

$$\sum_{i,j=1}^{n} b_i^* \Phi(a_i^* a_j) b_j \geq 0. \tag{5.2}$$

It is well known (cf. [203]) that $\sup_n \|\Phi_n\| = T(\mathbb{1})$ for completely positive maps. It is clear that the complete positivity of $T$ implies its positivity. The converse is not true in general.

Note that every unital completely positive map is a KS-operator, and a famous result of Kadison states that any positive unital map satisfies (5.1) for all self-adjoint elements.

There are several connections between CP and KS-operators. Namely, let $\Phi : A \to B$ be a given unital mapping. For a positive invertible $a \in A$ let us define

$$\Phi_{(a)}(x) = \Phi(a^2)^{-1/2}\Phi(axa)\Phi(a^2)^{-1/2}.$$

**Theorem 5.1.1 ([208])** *Let $\Phi : A \to B$ be a positive unital map. Then $\Phi$ is n-positive if and only if $(\Phi_{(a)})_n$ is a KS-operator for all positive invertible $a \in A$.*

By $\mathcal{KS}(A, B)$ we denote the set of all KS-operators mapping from $A$ to $B$.

**Theorem 5.1.2 ([179])** *Let $A$, $B$ and $C$ be $C^*$-algebras. The following statements hold:*

(i) *let $\Phi, \Psi \in \mathcal{KS}(A, B)$, then for any $\lambda \in [0, 1]$ the mapping $\Gamma_\lambda = \lambda\Phi + (1 - \lambda)\Psi$ belongs to $\mathcal{KS}(A, B)$. This means $\mathcal{KS}(A, B)$ is convex;*

(ii) *let $U, V$ be unitaries in $A$ and $B$, respectively, then for any $\Phi \in \mathcal{KS}(A, B)$ the mapping $\Psi_{U,V}(x) = U\Phi(VxV^*)U^*$ belongs to $\mathcal{KS}(A, B)$;*

(iii) *let $\Phi \in \mathcal{KS}(A, B)$ and $\Psi \in \mathcal{KS}(B, C)$, then $\Psi \circ \Phi \in \mathcal{KS}(A, C)$.*

*Proof*

(i). Let us show that $\Gamma_\lambda$ satisfies (5.1). Let $x \in M$, then one can see that

$$\Gamma_\lambda(x^*x) = \lambda\Phi(x^*x) + (1 - \lambda)\Psi(x^*x)$$
$$\geq \lambda\Phi(x)^*\Phi(x) + (1 - \lambda)\Psi(x)^*\Psi(x) \tag{5.3}$$

and

$$\Gamma_\lambda(x)^*\Gamma_\lambda(x) = \lambda^2\Phi(x)^*\Phi(x) + \lambda(1 - \lambda)\Phi(x)^*\Psi(x)$$
$$+ \lambda(1 - \lambda)\Psi(x)^*\Phi(x) + (1 - \lambda)^2\Psi(x)^*\Psi(x). \tag{5.4}$$

Hence, from (5.3)–(5.4) one gets

$$\Gamma_\lambda(x^*x) - \Gamma_\lambda(x)^*\Gamma_\lambda(x) \geq \lambda(1 - \lambda)\big(\Phi(x) - \Psi(x)\big)^*\big(\Phi(x) - \Psi(x)\big) \geq 0,$$

which proves the assertion.

(ii) For any $x \in A$ one has

$$\Psi_{U,V}(x^*x) = U\Phi\big((VxV^*)^*VxV^*\big)U^*$$
$$\geq U\Phi(VxV^*)^*\Phi(VxV^*)U^*$$
$$= U\Phi(VxV^*)^*U^*U\Phi(VxV^*)U^*$$
$$= \Psi_{U,V}(x)^*\Psi_{U,V}(x).$$

The statement (iii) is evident. This completes the proof.

In what follows, a unital positive (resp. completely positive) linear mapping $T$ : $A \to A$ is called a *Markov operator* (resp. *unital completely positive (ucp) map*).

From now, we restrict ourselves to von Neumann algebras, since such algebras have a very rich structure.

Let $M$ be a von Neumann algebra. The set of all linear continuous (in norm) maps of $M$ into itself is denoted by $B(M)$. On $B(M)$ we define a *weak topology* by seminorms of the following form

$$p_{\varphi,x}(T) = |\varphi(Tx)|, \quad T \in B(M), \ x \in M, \varphi \in M_*. \tag{5.5}$$

Now consider the tensor product $M_* \otimes_\gamma M$, where $\gamma$ is the greatest cross-norm on $M_* \odot M$.

**Theorem 5.1.3 ([208])** *Let $M$ be a von Neumann algebra. Then the conjugate space of $M_* \otimes_\gamma M$ is isomorphic to $B(M)$, i.e.*

$$(M_* \otimes_\gamma M)^* \cong B(M).$$

From this theorem it follows that the weak topology on $B(M)$ is the $\sigma(B(M), M_* \otimes_\gamma M)$-topology.

We note that any state on a von Neumann algebra $M$ can be considered as a unital ucp map. Namely, each state $\omega$ on $M$ defines a ucp map $T_\omega$ by $T_\omega(x) = \omega(x)\mathbb{1}$, $x \in M$. Thus the set of all states $S_1(M)$ can be seen as a subset of $B(M)$. It is known that $S_1(M)$ is a $\sigma(M^*, M)$-weakly compact set. Therefore, we are interested in similar results for the set of Markov operators.

By $\Sigma(M)$ (resp. $UCP(M)$) we denote the set of all Markov operators (resp. ucp maps) on $M$. It is clear that $UCP(M) \subset \Sigma(M) \subset B(M)_1$, where $B(M)_1$ is the unit ball in $B(M)$.

From (5.5) and (5.2) we may easily obtain that $\Sigma(M)$ and $UCP(M)$ are $\sigma(B(M), M_* \otimes_\gamma M)$-weakly closed. Consequently, we have the following.

**Theorem 5.1.4** *The spaces $\Sigma(M)$ and $UCP(M)$ are $*$-weakly compact.*

The proof immediately follows from Theorem 5.1.3 and the Banach–Alaoglu compactness theorem.

*Remark 5.1.1* The proved Theorem 5.1.4 generalizes a result obtained in [201], in which an analogous theorem was proved for a commutative von Neumann algebra.

**Corollary 5.1.5** *The set of all states $S_1(M)$ on $M$ is a $\sigma(B(M), M_* \otimes_\gamma M)$-weak closed subset of $\Sigma(M)$.*

*Proof* Let a net of states $\omega_\nu$ converge in the $\sigma(B(M), M_* \otimes_\gamma M)$-topology to $\omega$. Hence, for every $\varphi \otimes x \in M_* \otimes_\gamma M$ we have

$$\varphi \otimes x(\omega_\nu \mathbb{1}) = \varphi(\omega_\nu(x)\mathbb{1}) = \varphi(\mathbb{1})\omega_\nu(x) \to \varphi(\epsilon)\omega(x) \quad \text{as } \nu \to \infty.$$

This implies that $\omega_\nu \to \omega$, in the $\sigma(M^*, M)$-topology. Consequently, $\omega \in S_1(M)$.

**Corollary 5.1.6** *The induced $\sigma(B(M), M_* \otimes_\gamma M)$-topology on $S_1(M)$ coincides with the $\sigma(M^*, M)$-topology. Moreover, $S_1(M)$ is a $\sigma(B(M), M_* \otimes_\gamma M)$-weakly compact set.*

A $*$-morphism $T$ is called *ultraweakly continuous* if it is $(\sigma(M, M_*), \sigma(M, M_*))$-continuous.

**Lemma 5.1.7** *A Markov operator $T$ on a von Neumann algebra is ultraweakly continuous if and only if $\varphi \circ T \in S(M)$ for every $\varphi \in S(M)$.*

The proof is evident.

**Theorem 5.1.8** *Let $\mathscr{E}(M)$ be the set of all ultraweakly continuous Markov operators on a von Neumann algebra $M$, and $\mathscr{F}$ be a subset of $\mathscr{E}(M)$. The following assertions are equivalent:*

(i) *$\mathscr{F}$ is $\sigma(B(M), M_* \otimes_\gamma M)$-relatively compact.*
(ii) *For each state $\varphi \in S(M)$ and for any countable family of $\{e_i\}_{i \in I}$ of mutually orthogonal projections in $M$ one has*

$$\varphi(Te_i) \to 0 \text{ uniformly for } T \in \mathscr{F}. \tag{5.6}$$

*Proof* (i)$\Rightarrow$(ii). For each state $\varphi \in S(M)$, consider the set $S_{\mathscr{F}}^\varphi = \{\varphi \circ T | t \in \mathscr{F}\}$. It is clear that $S_{\mathscr{F}}^\varphi$ is $\sigma(M_*, M)$-relatively compact. By virtue of [242, Theorem 5.14], for any countable family of $\{e_i\}_{i \in I}$ of mutually orthogonal projection in $M$ one has (5.6).

(ii) $\Rightarrow$ (i). Since $\mathscr{F} \subset \Sigma(M)$ it follows that its $\sigma(B(M), M_* \otimes_\gamma M)$-closure $\overline{\mathscr{F}}$ in $\Sigma(M)$ is $\sigma(B(M), M_* \otimes_\gamma M)$-compact, because, by Theorem 5.1.4, $\Sigma(M)$ is $\sigma(B(M), M_* \otimes_\gamma M)$-compact. Hence, we have to show that $\overline{\mathscr{F}} \subset \mathscr{E}(M)$. Let a net $\{T_\nu\}_{\nu \in J}$ converge to $T$ in the $\sigma(B(M), M_* \otimes_\gamma M)$-topology. In particular, for each state $\varphi \in S(M)$ the following convergence holds

$$\varphi(T_\nu x) \to \varphi(Tx) \text{ as } \nu \to \infty, \; x \in M.$$

On the other hand, from condition (ii) of [242, Theorem 5.14] we obtain the $\sigma(M_*, M)$-relatively compactness of $S_{\mathscr{F}}^\varphi$. Consequently, we conclude that $\varphi \circ T$ is an ultraweakly continuous state. By virtue of Lemma 5.1.7, $T$ is ultraweakly continuous.

## 5.2  Quantum Quadratic Stochastic Operators

Let $M$ be a von Neumann algebra. Recall that the weak (operator) closure of the algebraic tensor product $M \odot M$ in $B(H \otimes H)$ is denoted by $M \otimes M$, and is called the *tensor product* of $M$ into itself. For details, we refer the reader to [25, 242, 246].

By $S(M \otimes M)$ we denote the set of all normal states on $M \otimes M$. Let $U : M \otimes M \to M \otimes M$ be a linear operator such that $U(x \otimes y) = y \otimes x$ for all $x, y \in M$.

**Definition 5.2.1** A linear operator $\Delta : M \to M \otimes M$ is said to be a *quantum quadratic stochastic operator (q.q.s.o.)* if it satisfies the following conditions:

(i) $\Delta$ is positive;
(ii) symmetric, i.e. $U\Delta = \Delta$;
(iii) unital, i.e. $\Delta \mathbb{1}_M = \mathbb{1}_{M \otimes M}$, where $\mathbb{1}_M$ and $\mathbb{1}_{M \otimes M}$ are units of the algebras $M$ and $M \otimes M$ respectively.

*Remark 5.2.1* We note that if one replaces a von Neumann algebra with a $C^*$-algebra, then in the same way, we can define quantum quadratic stochastic operators defined on $C^*$-algebras.

If a q.q.s.o. satisfies the following co-associativity condition

$$(\Delta \otimes id) \circ \Delta = (id \otimes \Delta) \circ \Delta,$$

where $id$ is the identity operator of $M$, then the q.q.s.o. is called a *co-associative co-multiplication*.

*Remark 5.2.2* We should stress that a q.q.s.o. is a more general notion than co-associative co-multiplication. In what follows, we do not require the co-associativity condition for q.q.s.o.s in our investigations.

A state $h \in S_1(M)$ is called *a Haar state* for a q.q.s.o. $\Delta$ if for every $x \in M$ one has

$$(h \otimes id) \circ \Delta(x) = (id \otimes h) \circ \Delta(x) = h(x)\mathbb{1}. \tag{5.7}$$

*Remark 5.2.3* Note that if a co-associative co-multiplication $\Delta$ on $M$ becomes a $*$-homomorphic map with the condition

$$\overline{\mathrm{Lin}}((\mathbb{1} \otimes M)\Delta(M)) = \overline{\mathrm{Lin}}((M \otimes \mathbb{1})\Delta(M)) = M \otimes M$$

then the pair $(M, P)$ is called a *compact quantum group* [240, 253]. It is known [253] that for any given compact quantum group, there exists a unique Haar state w.r.t. $\Delta$. There is a huge literature on quantum groups. However, in our investigations, we do not touch on quantum group aspects of q.q.s.o.s. The interested reader is referred to [111, 136, 205] for the general theory of quantum groups.

By $Q\Sigma(M)$ we denote the set of all q.q.s.o.s on $M$. Let us equip this set with a *weak topology* by the following seminorms

$$p_{\varphi,x}(\Delta) = |\varphi(\Delta x)|, \quad \Delta \in Q\Sigma(M), \ \varphi \in M_* \otimes_{\alpha_0^*} M_*, \ x \in M,$$

where $\alpha_0^*$ is the dual norm to the smallest $C^*$-cross-norm $\alpha_0$ on $M\otimes M$ (see [226, Sect. 1.22]).

Let $\varphi \in S(M)$ be a fixed state. We define the *conditional expectation* operator $E_\varphi : M \otimes M \to M$ on elements $a \otimes b, a, b \in M$ by

$$E_\varphi(a \otimes b) = \varphi(a)b \tag{5.8}$$

and extend it by linearity and continuity to $M \otimes M$. Clearly, such an operator is completely positive and $E_\varphi \mathbb{1}_{M\otimes M} = \mathbb{1}_M$ (see [246]).

Now let $\Delta \in Q\Sigma(M)$ be a q.q.s.o. Then we define a map $Q_\Delta : M \otimes M \to M \otimes M$ as follows

$$Q_\Delta(x) = \Delta(E_\varphi(x)), \quad x \in M \otimes M. \tag{5.9}$$

From Definition 5.2.1 one can see that $Q_\Delta$ is a Markov operator, (i.e. $Q_\Delta \in \Sigma(M \otimes M)$) with $UQ_\Delta = Q_\Delta$. Define

$$\Sigma_U(M \otimes M) = \{Q \in \Sigma(M \otimes M) : UQ = Q\}.$$

The defined set is $*$-weakly closed in $B(M \otimes M)$, hence in $\Sigma(M \otimes M)$. Indeed, let $Q_\nu \to Q$ in the $*$-weak topology in $B(M \otimes M)$, i.e. for any $f \in (M \otimes M)_*$ and $x \in M \otimes M$ one has $f(Q_\nu x) \to f(Qx)$. Replacing $f$ with $f \circ U$ and using $UQ_\nu = Q_\nu$ we obtain $UQ = Q$. Hence, thanks to Theorem 5.1.4, the set $\Sigma_U(M\otimes M)$ is $*$-weakly compact. Now define a mapping $\pi : \Sigma_U(M \otimes M) \mapsto Q\Sigma(M)$ by $\pi(Q)(x) = Q(x \otimes \mathbb{1})$. It is easy to check that $\pi$ is weakly continuous. Therefore, $Q\Sigma(M)$ is weakly compact too. So, we have proved the following

**Theorem 5.2.1** *The set $Q\Sigma(M)$ is weakly compact.*

By $Q\Sigma_u(M)$ we denote the set of all ultraweakly continuous q.q.s.o.s. Then each q.q.s.o. $\Delta \in Q\Sigma_u(M)$ defines a conjugate operator $\tilde{V}_\Delta : M_* \otimes_{\alpha_0^*} M_* \to M_*$ by

$$\tilde{V}_\Delta(f)(x) = f(\Delta x), \quad f \in M_* \otimes_{\alpha_0^*} M_*, \ x \in M. \tag{5.10}$$

Thanks to conditions (i) and (ii) of Definition 5.2.1 the operator $\tilde{V}_\Delta$ maps $S(M \otimes M)$ to $S(M)$. The operator $\tilde{V}_P$ is called the *conjugate quadratic operator (c.q.o.)*. In what follows, for the sake of brevity, instead of $\tilde{V}_\Delta(\varphi \otimes \psi)$ we will write $\tilde{V}_\Delta(\varphi, \psi)$, where $\varphi, \psi \in S(M)$. Note that the relation (iii) in Definition 5.2.1 implies that

$$\tilde{V}_\Delta(\varphi, \psi) = \tilde{V}_\Delta(\psi, \varphi). \tag{5.11}$$

By means of $\tilde{V}_\Delta$ one can define an operator $V_\Delta : S(M) \to S(M)$ by

$$V_\Delta(\varphi) = \tilde{V}_\Delta(\varphi, \varphi), \quad \varphi \in S(M), \tag{5.12}$$

which is called the *quadratic operator (q.o.)*. In some of the literature the operator $V_\Delta$ is called the quadratic convolution (see for example [45]).

*Remark 5.2.4* Now we would like to show that the defined quadratic operator reduces to a q.s.o. if $M$ is a commutative von Neumann algebra (in general, $M$ could be an algebra of bounded measurable functions on $(E, \Im)$), i.e. $M = L^\infty(X, \Im, \mu)$. Then $M_* = L^1(X, \Im, \mu)$. Assume that $\Delta$ is a q.q.s.o. on $L^\infty(X, \Im, \mu)$, i.e. $\Delta$ is a mapping from $L^\infty(X, \Im, \mu)$ to $L^\infty(X \times X, \Im \otimes \Im, \mu \otimes \mu)$. Then we define

$$P(x, y, A) = \Delta(\chi_A)(x, y), \quad x \in X,$$

where $\chi_A$ is the indicator function of a set $A \in \Im$. From Definition 5.2.1 we immediately find that $P(x, y, A)$ satisfies the following conditions:

 (i) for fixed $x, y \in X$ one has $P(x, y, \cdot)$ is a probability measure (which is absolutely continuous w.r.t. $\mu$);
(ii) for each fixed $B \in \Im$, $P(x, y, B)$ is a measurable function on $(X \times X, \Im \otimes \Im)$. Moreover, $P(x, y, B) = P(y, x, B)$ for any $x, y \in X$ and $B \in \Im$.

Then the corresponding quadratic operator (5.12) is defined by

$$(V_\Delta)(m)(B) = \int_X \int_X P(x, y, B) dm(x) dm(y),$$

where $m \in L^1(X, \Im, \mu)$. This means that any q.q.s.o. defines a q.s.o.

*Remark 5.2.5* We note that there is another approach to nonlinear quantum operators on $C^*$-algebras. In this approach a nonlinear mapping is defined on a $C^*$-algebra, which is a non-commutative variant of Koopman's construction. Such a construction may lead to quantum chaos (see [142]).

**Open problem 5.2.1** *Let $\Delta$ be an extremal point of $Q\Sigma(M)$. Would the corresponding q.o. $V_\Delta$ be a bijection of $S_1(M)$?*

**Open problem 5.2.2** *Describe the set of all bijective q.o. $V_\Delta$ of $S_1(M)$.*

*Example 5.2.1* Here we describe how linear operators and q.q.s.o.s are related to each other. Let $T : M \to M$ be a Markov operator. Define a linear operator $P : M \to M \otimes M$ as follows

$$P_T x = \frac{Tx \otimes \mathbb{1} + \mathbb{1} \otimes Tx}{2}, \quad x \in M. \tag{5.13}$$

It is clear that $P_T$ is a q.q.s.o. Then the associated c.q.o. and q.o. have the following forms, respectively:

$$\tilde{V}_{P_T}(\varphi, \psi)(x) = \frac{1}{2}(\varphi + \psi)(Tx),$$

$$V_{P_T}(\varphi)(x) = \varphi(Tx), \quad x \in M, \tag{5.14}$$

for every $\varphi, \psi \in S(M)$. Thus a linear operator can be viewed as a particular case of a q.q.s.o. If $T$ is the identity operator, then from (5.14) we can see that the associated q.o. would also be the identity operator of $S(M)$. The set of all q.q.s.o.s associated with linear operators we denote by $\mathscr{QL}(M)$.

We have the following

**Proposition 5.2.2** *To each q.q.s.o.* $\Delta \in Q\Sigma_u(M)$ *corresponds a linear operator* $T : M_* \to \Sigma(M)$ *defined by*

$$T(\varphi)(x) = E_\varphi(\Delta x), \quad \varphi \in M_*, \quad x \in M. \tag{5.15}$$

*Moreover, for every* $\varphi, \psi \in M_*$ *we have*

$$T_*(\varphi)\psi = T_*(\psi)\varphi, \quad \|T(\varphi)\| \le 2\|\varphi\|_1, \tag{5.16}$$

*where* $T_*(\varphi)\psi(a) = \psi(T(\varphi)(a))$. *In addition,*

$$\tilde{V}_\Delta(\varphi \otimes \psi) = T_*(\varphi)\psi, \quad \forall \varphi, \psi \in M_*. \tag{5.17}$$

The proof immediately follows from Definition 5.2.1 and equalities (5.10) and (5.15).

**Definition 5.2.2** A linear map $V : M_* \otimes_{\alpha_0^*} M_* \to M_*$ is called a *conjugate quadratic operator* if the following conditions hold

(i) $V(S(M \otimes M)) \subset S(M)$;
(ii) $V(\varphi \otimes \psi) = V(\psi \otimes \varphi), \quad \forall \varphi, \psi \in M_*.$

By $Q\Sigma^V(M)$ we denote the set of all conjugate quadratic operators.

**Proposition 5.2.3** *Every* $V \in Q\Sigma^V(M)$ *uniquely defines a q.q.s.o. belonging to* $Q\Sigma_u(M)$.

*Proof* Define a linear map $\Delta$ by setting $\Delta = V^*$. According to the equality $(M_* \otimes_{\alpha_0^*} M_*)^* \cong M \overline{\otimes} M$ (see [226, Definition 1.22.10]) it is clear that $\Delta : M \to M \overline{\otimes} M$. Now we show that $\Delta$ is indeed a q.q.s.o. Note that every element $x$ of $M$ can be viewed as a linear functional on $M_*$ via $\langle x, f \rangle = f(x)$. Therefore, we identify $\mathbb{1}$ with the functional $\langle \mathbb{1}, \varphi \rangle$, $\varphi \in M_*$ and

$$\langle (\Delta \mathbb{1}), \psi \rangle = \langle \mathbb{1}, V(\psi) \rangle = V(\psi)(\mathbb{1}) = 1,$$

where $\psi \in S(M \otimes M)$, whence $\Delta \mathbb{1} = \mathbb{1} \otimes \mathbb{1}$.

Now let $x \geq 0$, then $\langle x, \varphi \rangle \geq 0$ for every $\varphi \in S(M)$, consequently, it follows from (i) Definition 5.2.2 that

$$\langle \Delta x, \psi \rangle = \langle x, V(\psi) \rangle \geq 0, \quad \text{for every} \ \psi \in M_{*,+} \otimes_{\alpha_0^*} M_{*,+}.$$

Hence, $\Delta x \geq 0$. From (ii) we find $\langle \Delta x, \varphi \otimes \psi \rangle = \langle \Delta x, \psi \otimes \varphi \rangle$ which means $U\Delta = \Delta$. This proves the assertion.

Let $\Delta \in Q\Sigma(M)$, and consider the corresponding q.o. $V_\Delta$ on $S_1(M)$.

**Definition 5.2.3** A q.o. $V_\Delta$ is called

(i) *asymptotically stable* or *regular* if there exists a state $\mu \in S_1(M)$ such that for any $\varphi \in S_1(M)$ one has

$$\lim_{n \to \infty} \|V_\Delta^n(\varphi) - \mu\|_1 = 0, \tag{5.18}$$

where by $\| \cdot \|_1$ we denote the norm on $M^*$;

(ii) *weak asymptotically stable* if there exists a state $\mu \in S_1(M)$ such that for any $\varphi \in S_1(M)$ and $a \in M$ one has

$$\lim_{n \to \infty} V_\Delta^n(\varphi)(a) = \mu(a). \tag{5.19}$$

It is clear that asymptotical stability implies weak asymptotical stability, but in general, the converse is not true. Note that if we consider a q.o. $V_{P_T}$ associated with a Markov operator (see (5.14)), then the introduced notions, i.e. asymptotical stability and weak asymptotical stability, coincide with complete mixing and weak mixing, respectively, of the Markov operator $T$ (see [1, 128]). Therefore, one can find many examples of such operators [128].

**Open problem 5.2.3** *Find a weak asymptotically stable q.o. which is not asymptotically stable and not generated by a Markov operator.*

By $\hat{S}$ (resp. $\check{S}$) we denote the set of all functionals $g : M_+ \to \mathbb{R}_+$ such that

$$g(x + y) \leq g(x) + g(y), \quad \left( \text{resp.} \ g(x + y) \geq g(x) + g(y) \right), \ \forall x, y \in M_+,$$

$$g(\lambda x) = \lambda g(x), \quad \text{for all} \ \lambda \in \mathbb{R}_+, \ x \in M_+,$$

$$g(\mathbb{1}) = 1.$$

Let us put $\tilde{S} = \hat{S} \cup \check{S}$. We endow the set $\tilde{S}$ with the topology of pointwise convergence. By $C(S_1(M), \tilde{S})$ we denote the set of all strong continuous operators from $S_1(M)$ to $\tilde{S}$, i.e. $f \in C(S_1(M), \tilde{S})$ if whenever a sequence $x_n$ norm converges to $x$ in $S_1(M)$, then $f(x_\alpha)$ converges in $\tilde{S}$.

**Definition 5.2.4** A q.o. $V_\Delta$ is called $\eta$-*ergodic* if for $f \in C(S_1(M), \tilde{S})$ the equality $f(V_P(\varphi))(x) = f(\varphi)(x)$ for every $\varphi \in S_1(M)$ and $x \in M_+$ implies that $f(\varphi)$ does not depend on $\varphi$, i.e. there is a $\delta_f \in \tilde{S}$ such that $f(\varphi) = \delta_f$ for all $\varphi \in S_1(M)$.

One has the following

**Theorem 5.2.4** *Let* $\Delta \in Q\Sigma(M)$ *and* $V_\Delta$ *be the associated q.o. Then for the assertions*

(i) *the q.o.* $V_P$ *is asymptotically stable;*
(ii) *the q.o.* $V_P$ *is* $\eta$-*ergodic;*
(iii) *the q.o.* $V_P$ *is weak asymptotically stable,*

*the following implications holds true:* (i)$\Rightarrow$(ii)$\Rightarrow$(iii).

*Proof* (i) $\Rightarrow$ (ii). Let $V$ be regular and assume that for $f \in C(S_1(M), \tilde{S})$ one has $f(V(\varphi))(a) = f(\varphi)(a)$, for every $\varphi \in S_1(M)$ and $a \in M_+$. Due to the regularity of $V$ there exists a $\mu \in S_1(M)$ such that

$$\left\| V^n(\varphi) - \mu \right\|_1 \to 0 \quad \text{as} \quad n \to \infty.$$

From the continuity of $f$ one finds

$$f(\varphi) = f\big(V^n(\varphi)\big) \to f(\mu_1) \quad \text{as} \quad n \to \infty.$$

Hence, $f(\varphi) = f(\mu)$, which means $\eta$-ergodicity.

(ii) $\Rightarrow$ (iii). Let $V$ be $\eta$-ergodic. Then define functionals as follows:

$$\hat{f}(\varphi)(a) = \limsup_{n \to \infty} V^n(\varphi)(a), \quad \check{f}(\varphi)(a) = \liminf_{n \to \infty} V^n(\varphi)(a), \quad a \in M_+.$$

One can see that $\check{f}, \hat{f} \in C(S_1(M), \tilde{S})$. On the other hand, we have

$$\hat{f}(V(\varphi))(a) = \limsup_{n \to \infty} V^n(V(\varphi))(a) = \limsup_{n \to \infty} V^{n+1}(\varphi)(a) = \hat{f}(\varphi)(a), \quad a \in M_+.$$

Similarly, one gets

$$\check{f}(V(\varphi))(a) = \check{f}(\varphi)(a), \quad a \in M_+.$$

According to the $\eta$-ergodicity of $V$, there are $\delta(\hat{f})$, $\delta(\check{f}) \in \tilde{S}$ such that

$$\hat{f}(\varphi) = \delta(\hat{f}), \quad \check{f}(\varphi) = \delta(\check{f}), \quad \forall \varphi \in S.$$

Since $S_1(M)$ is $*$-weakly compact in $M^*$ and $*$-weak continuity of $V$ with $V(S_1(M)) \subset S_1(M)$ allows us to apply Schauder's theorem, one can find a fixed point $\mu$ of $V$, i.e. $V(\mu) = \mu$. Therefore, one has $\hat{f}(\mu) = \check{f}(\mu)$, which yields

$\hat{f}(\varphi)(a) = \check{f}(\varphi)(a)$, $a \in M_+$. The last equality implies the existence of the limit:

$$\lim_{n \to \infty} V^n(\varphi)(a) = \mu(a).$$

So, $V$ is weak asymptotically stable.

**Corollary 5.2.5** *Let $M$ be a finite-dimensional $C^*$-algebra. Then all the conditions of Theorem 5.2.4 are equivalent.*

We note that the defined $\eta$-ergodicity is similar to the ergodicity of dynamical systems [33].

**Open problem 5.2.4** *Investigate the reverse implications in Theorem 5.2.4.*

## 5.3  Quantum Markov Chains and q.q.s.o.s

In this section, we recall the general definitions and properties of quantum Markov chains taken from [9]. Moreover, using a given q.q.s.o. we construct a quantum Markov chain.

In the following, by $M$ we denote the algebra of $d \times d$ complex matrices (for a fixed arbitrary integer $d$) and $\mathbb{N}_0 = \{0\} \cup \mathbb{N}$. Consider the infinite $C^*$-algebra tensor product $\mathfrak{A} = \otimes_{n \in \mathbb{N}_0} M_n$, where $M_n = M$ for all $n \in \mathbb{N}_0$. For any subset $I \subseteq \mathbb{N}_0$ we denote by $\mathfrak{A}_I$ the $C^*$-subalgebra $\otimes_{i \in I} M_i$ of $\mathfrak{A}$. We will simply write $\mathfrak{A}_n$ for $\mathfrak{A}_{\{n\}}$. A *localization* on $\mathfrak{A}$ is a family $\{\mathfrak{A}_I : I \in \mathfrak{F}\}$ of subalgebras of $\mathfrak{A}$, where $\mathfrak{F}$ is an increasing net of subsets of $\mathbb{N}_0$ whose union coincides with $\mathbb{N}_0$. In what follows, we restrict our attention to the localization $\left(\mathfrak{A}_{[0,n]}\right)_{n \in \mathbb{N}}$.

A *quasi-conditional expectation* with respect to the triple $\mathfrak{A}_{[0,n-1]} \subseteq \mathfrak{A}_{[0,n]} \subseteq \mathfrak{A}_{[0,n+1]}$ is a completely positive identity preserving map $E_{n+1,n} : \mathfrak{A}_{[0,n+1]} \to \mathfrak{A}_{[0,n]}$ such that

$$E_{n+1,n}(ba) = bE_{n+1,n}(a); \quad \text{for all } b \in \mathfrak{A}_{[0,n-1]}, a \in \mathfrak{A}_{[0,n+1]}. \tag{5.20}$$

Equivalently, $E_{n+1,n}$ can be characterized as a completely positive identity preserving map $\mathfrak{A}_{[0,n+1]} \to \mathfrak{A}_{[0,n]}$ whose fixed point algebra contains $\mathfrak{A}_{[0,n-1]}$ (see [4]).

Condition (5.20) implies that

$$E_{n+1,n}(\mathfrak{A}_{[n,n+1]}) \subseteq \mathfrak{A}_n. \tag{5.21}$$

This relation is called *the quantum Markov property*. Note that if the $\mathfrak{A}'_i$s are abelian algebras and $E_{n+1,n}$ is a conditional expectation in the usual sense, then (5.21) is an equivalent formulation of the classical Markov property (see [3]).

**Definition 5.3.1** A state $\phi$ on $\mathfrak{A}$ is called a *Markov state* with respect to the localization $\{\mathfrak{A}_{[0,n]}\}$ if for each $n \in \mathbb{N}_0$ there exists a quasi-conditional expectation with respect to the triple $\mathfrak{A}_{[0,n-1]} \subseteq \mathfrak{A}_{[0,n]} \subseteq \mathfrak{A}_{[0,n+1]}$ such that

$$\phi_{[0,n+1]}(a) = \phi_{[0,n]}(E_{n+1,n}(a)) \quad \text{for all} \quad a \in \mathfrak{A}_{[0,n+1]}. \tag{5.22}$$

Here $\phi_{[0,k]}$ stands for the restriction of the state $\phi$ to $\mathfrak{A}_{[0,k]}$.

In this case, the quasi-conditional expectation $E_{n+1,n}$ is said to be *compatible* with the state $\phi$.

In what follows, we simply say that $\phi$ is a Markov state on $\mathfrak{A}$ without explicitly mentioning the localization $\{\mathfrak{A}_{[0,n]}\}$.

Define, for each $n \in \mathbb{N}$,

$$\mathscr{E}_n(a_{[n,n+1]}) = E_{n+1,n}(a_{[n,n+1]}).$$

Equivalently, $\mathscr{E}_n$ is the restriction of $E_{n+1,n}$ to the algebra $\mathfrak{A}_{[n,n+1]}$. It is clear that due to the Markov property, $\mathscr{E}_n$ maps $M_n \otimes M_{n+1}$ into $M_n$, and is completely positive and normalized, i.e.

$$\mathscr{E}_n(\mathbb{1} \otimes \mathbb{1}) = \mathbb{1}. \tag{5.23}$$

Moreover, because of (5.20) and (5.22) one has

$$\phi(a_{[0,n-1]} \otimes a_n \otimes a_{n+1}) = \phi(a_{[0,n-1]} E_{n+1,n}(a_n \otimes a_{n+1}))$$
$$= \phi(a_{[0,n-1]} E_{n+1,n}(a_n \otimes E_{n+2,n+1}(a_{n+1} \otimes \mathbb{1})))$$

for each $a_{[0,n-1]} \in \mathfrak{A}_{[0,n-1]}, a_n, a_{n+1} \in M$. This means

$$\phi_{[0,n]}(a_{[0,n-1]} \otimes \mathscr{E}_n(a_n \otimes a_{n+1})) = \phi_{[0,n]}(a_{[0,n-1]} \otimes \mathscr{E}_n(a_n \otimes \mathscr{E}_{n+1}(a_{n+1} \otimes \mathbb{1}))).$$

The last expression holds for all $a_{[n-1]} \in \mathfrak{A}_{[n-1]}$. Therefore, we simply write it as follows

$$\mathscr{E}_n(a \otimes b) = \mathscr{E}_n(a \otimes \mathscr{E}_{n+1}(b \otimes \mathbb{1})), \quad \text{mod} \; \{\phi_0, (\mathscr{E}_k)\} \tag{5.24}$$

for each $a, b \in M$.

Denote by $\phi_0$ the restriction of $\phi$ to $\mathfrak{A}_0$ (in the following, when no confusion is possible, this state will be identified with the state $\phi_0$ on $M$).

The state $\phi$ is completely determined by the pair $\{\phi_0; (\mathscr{E}_n)\}$ through the relation:

$$\phi(a_0 \otimes \cdots \otimes a_n) = \phi_0(\mathscr{E}_0(a_0 \otimes \mathscr{E}_1(a_1 \otimes \cdots \otimes \mathscr{E}_n(a_{n-1} \otimes a_n) \ldots)))$$
$$= \phi_0(\mathscr{E}_0(a_0 \otimes \mathscr{E}_1(a_1 \otimes \cdots \otimes \mathscr{E}_n(a_{n-1} \otimes \mathbb{1}) \ldots) \tag{5.25}$$

for all $a_0, \ldots, a_n \in M, n \in \mathbb{N}_0$.

Conversely, let $\{\phi_0; (\mathscr{E}_n)\}$ be a pair satisfying (5.23) and (5.24), where $\mathscr{E}_n : M_n \otimes M_{n+1} \rightarrow M_n$ is completely positive and $\phi_0$ is a state on $M$.

Then for each $n \in \mathbb{N}_0$, the right-hand side of (5.25) defines a state $\phi_{[0,n]}$ on $\mathfrak{A}_{[0,n]}$. One can see that the family of states $\phi_{[0,n]}$ is projective in the sense that

$$\phi_{[0,n+1]} \lceil \mathfrak{A}_{[0,n]} = \phi_{[0,n]}.$$

Therefore there exists a unique state $\phi$ on $\mathfrak{A}$ whose restriction on $\mathfrak{A}_{[0,n]}$ is $\phi_{[0,n]}$ ($n \in \mathbb{N}_0$). Let us prove that $\phi$ is a Markov state.

The map $E_{n+1,n} \mathfrak{A}_{[0,n+1]} \rightarrow \mathfrak{A}_{[0,n]}$ defined (for each $n$) by extension of

$$a_0 \otimes a_1 \otimes \cdots \otimes a_{n+1} \rightarrow a_0 \otimes a_1 \otimes \cdots a_{n-1} \otimes \mathscr{E}_n(a_n \otimes a_{n+1})$$

is clearly a quasi-conditional expectation with respect to the triple $\mathfrak{A}_{[0,n-1]} \subseteq \mathfrak{A}_{[0,n]} \subseteq \mathfrak{A}_{[0,n+1]}$.

Using (5.24) the state $\phi$ defined above satisfies the equalities

$$\phi(E_{n+1,n}(a_0 \otimes \cdots \otimes a_{n+1}))$$
$$= \phi_0(\mathscr{E}_0(a_0 \otimes \cdots \otimes \mathscr{E}_{n-1}(a_{n-1} \otimes \mathscr{E}_n(\mathscr{E}_n(a_n \otimes a_{n+1}) \otimes \mathbb{1}))) \ldots)$$
$$= \phi_0(\mathscr{E}_0(a_0 \otimes \cdots \otimes \mathscr{E}_{n-1}(a_{n-1} \otimes \mathscr{E}_n(a_n \otimes a_{n+1})) \ldots))$$
$$= \phi_0(\mathscr{E}_0(a_0 \otimes \cdots \otimes \mathscr{E}_{n-1}(a_{n-1} \otimes \mathscr{E}_n(a_n \otimes \mathscr{E}_{n+1}(a_{n+1} \otimes \mathbb{1})))) \qquad (5.26)$$

for all $(a_0, \ldots, a_n \in M)$, where in the last equality we also used (5.24).

Because of the definition of $\phi$ from (5.26) one finds

$$\phi_{[0,n]}(E_{n+1,n}(a)) = \phi_{[0,n+1]}(a); \forall a \in \mathfrak{A}_{[0,n+1]}.$$

Hence $\phi$ is a Markov state on $\mathfrak{A}$.

**Definition 5.3.2** Let $\phi_0$ be a state on $M$. A family of linear maps $\mathscr{E}_n : M \otimes M \rightarrow M$ such that for each $n \in \mathbb{N}_0$

1. $\mathscr{E}_n$ is completely positive,
2. $\mathscr{E}_n(\mathbb{1} \otimes \mathbb{1}) = \mathbb{1}$,
3. $\mathscr{E}_n(a \otimes b) = \mathscr{E}_n(a \otimes \mathscr{E}_{n+1}(b \otimes \mathbb{1}))$, $a, b \in M$, $\mod \{\phi_n, (\mathscr{E}_k)\}$

will be called *a family of transition expectations* with initial distribution $\phi_0$.

Hence, we get the following result.

**Theorem 5.3.1** *Every Markov state $\phi$ on $\mathfrak{A}$ is determined by a pair $\{\phi_0; (\mathscr{E}_n)\}$ such that $\phi$ is a state on $M$ and $(\mathscr{E}_n)$ is a family of transition expectations with initial distribution $\phi_0$. Conversely, every such family defines a unique Markov state on $\mathfrak{A}$.*

*Remark 5.3.1* From the proof of Theorem 5.3.1 it is clear that any family $\mathscr{E}_n : M \otimes M \rightarrow M$ of completely positive unital maps defines, through (5.25), a unique state

$\varphi$ on $\mathfrak{A}$. In this case, however, property (5.22) might fail, so that $\varphi$ might not be a Markov state. Since the structure of joint expectations for such states is very similar to that of Markov states, these states are called *quantum Markov chains* (see [3]).

Now we are going to provide a construction of a quantum Markov chain associated with given a q.q.s.o.

Let $\Delta : M \to M \otimes M$ be a completely positive q.q.s.o. defined on $M$. Here $M$ is, as before, the algebra of $d \times d$ matrices. Recall that by $V_\Delta$ we denote the associated quadratic operator.

Take an arbitrary state $\psi \in S(M)$ and define

$$\psi_n = V_\Delta^n(\psi), \quad n \in \mathbb{N}.$$

For each $n \in \mathbb{N}$ let us define a completely positive mapping $\mathscr{E}_{n,\Delta} : M \otimes M \to M$ by

$$\mathscr{E}_{n,\Delta,\psi} = E_{\psi_n} \circ \Delta \circ E_{\psi_{n+1}}. \tag{5.27}$$

Here $E_{\psi_n}$ is the conditional expectation given by (5.8).

According to Remark 5.3.1 the defined family $\{\mathscr{E}_{n,\Delta,\psi}\}$ defines a quantum Markov chain on $\mathfrak{A} = \otimes_{\mathbb{N}_0} M$, which is denoted by $\varphi_{\Delta,\psi}$. Using (5.25) let us explicitly find its values on tensor monomials, i.e.

$\varphi_{\Delta,\psi}(a_0 \otimes \cdots \otimes a_n)$

$= \psi(\mathscr{E}_{0,\Delta,\psi}(a_0 \otimes \mathscr{E}_{1,\Delta,\psi}(a_1 \otimes \cdots \otimes \mathscr{E}_{n,\Delta,\psi}(a_n \otimes \mathbb{1}) \ldots)$

$= \psi(\mathscr{E}_{0,\Delta,\psi}(a_0 \otimes \mathscr{E}_{1,\Delta,\psi}(a_1 \otimes \cdots \otimes \mathscr{E}_{n-1,\Delta,\psi}(a_{n-1} \otimes E_{\psi_n} \Delta(a_n))$

$= \psi(\mathscr{E}_{0,\Delta,\psi}(a_0 \otimes \mathscr{E}_{1,\Delta,\psi}(a_1 \otimes \cdots \otimes \mathscr{E}_{n-2,\Delta,\psi}(a_{n-2} \otimes E_{\psi_{n-1}} \Delta(a_{n-1}))\psi_{n+1}(a_n)$

$\vdots$

$= \psi_1(a_0)\psi_2(a_1) \cdots \psi_{n+1}(a_n)$

for all $a_0 \ldots, a_n \in M$, $n \in \mathbb{N}_0$.

Hence, the quantum Markov chain $\varphi_{\Delta,\psi}$ is a product state associated with $\Delta$.

One can see that if $\psi$ is a fixed point of the quadratic operator $V_\Delta$, then $\varphi_{\Delta,\psi}$ is a product state of the form $\otimes_{\mathbb{N}_0} \psi$.

**Open problem 5.3.1** *Let $\psi$, $\psi_1$ be given states. Investigate the quasi-equivalence or disjointness (see [25] for definitions) of the quantum Markov chains $\varphi_{\Delta,\psi}$ and $\varphi_{\Delta,\psi_1}$ with respect to the q.q.s.o. $\Delta$.*

**Open problem 5.3.2** *Let $\pi_{\varphi_{\Delta,\psi}}$ be the GNS-representation associated with $\varphi_{\Delta,\psi}$. Then investigate the von Neumann algebra $\pi_{\varphi_{\Delta,\psi}}(\mathfrak{A})''$ with respect to $\Delta$ and $\psi$.*

## 5.4   Comments and References

Entanglement is one of the essential features of quantum physics and is fundamental
in modern quantum technologies [199]. One of the central problems in the theory of
entanglement is the discrimination between separable and entangled states. There
are several tools which can be used for this purpose. The most general consists in
applying the theory of linear positive maps [203]. In these studies, one of the goals
is to construct a map from the state space of a system to the state space of another
system. In the literature on quantum information and communication systems, such
a map is called a channel [199]. The concept of a state in a physical system is a
powerful weapon in the study of the dynamical behavior of that system.

There are many systems which are described by nonlinear operators, one of the
simplest being the quadratic one. In the previous chapters, we have considered
classical (i.e. commutative) quadratic operators. These operators were defined
over commutative algebras. However, such operators do not cover the case of
quantum systems. Therefore, in [57–59] quantum quadratic operators acting on a
von Neumann algebra were defined and studied. In Sect. 5.1, we have collected some
well-known facts on positive and completely positive mappings (see [203, 246]),
but some of the results are taken from [153]. Results concerning the compactness
of the set of q.q.s.o.s first appeared in [153, 164]. Theorem 5.2.4 was proved in
[158, 159]. Several ergodic type theorems for q.q.s.o.s have been investigated in
[161, 162, 166, 167, 170]. In those papers, the trajectory of quadratic operators were
essentially defined according to some recurrence rule, which made it possible to
study the asymptotic behavior of the dynamics of these operators. However, with a
given quadratic operator one can also define a non-linear operator whose dynamics
(in the non-commutative setting) has not yet been well studied.

In [142] another construction of nonlinear quantum maps was suggested and
some physical explanations of such nonlinear quantum dynamics were discussed.
In this type of approach, a nonlinear mapping is defined on a $C^*$-algebra which
is a non-commutative variant of Koopman's construction. Certain applications to
quantum chaos are also indicated.

On the other hand, [45] considered the ergodic averages of the dynamics
of $\varphi^{[n]}$, where $\varphi^{[n]} = \underbrace{\tilde{V}(\varphi, \tilde{V}(\varphi, \ldots, \tilde{V}(\varphi, \varphi)) \ldots)}_{n}$, of the state $\varphi$ generated by
the quadratic operator $\tilde{V}$ which is associated with coassociative co-multiplication
in quantum groups [253]. The investigation of convergence of the averages is
reduced to the convergence of the ergodic averages of absolute contractions of von
Neumann algebras [115]. Actually, due to this reduction, the nonlinear dynamics
of convolution operators were not investigated. Therefore, a complete analysis of
the dynamics of quadratic operators associated with quantum groups is still not
well studied. A certain class of quantum groups on $\mathbb{M}_2(\mathbb{C})$ was investigated in
[240]. There are many books and papers on quantum groups. We only refer to
[111, 136, 205] for the general theory of quantum groups.

Classical Markov chains are defined through an intrinsic statistical property (the Markov property) which allows the explicit form of their finite-dimensional joint expectation (correlation functions) to be determined. This explicit structure of the correlation function can in turn be generalized and gives rise to a strictly larger class of stochastic processes (generalized classical Markov chains). Moreover, it is well known that any classical Markov chain defines, uniquely up to the boundary terms, a *potential function* which completely determines the conditional probability matrices of the chain [116]. Therefore, quantum Markov chains were introduced in [2] through an intrinsic definition which allowed the explicit structure of their correlation function to be determined, and an extension of the resulting construction led to the introduction of the strictly larger class of 'generalized quantum Markov chains'. In [3] connections are given between quantum mechanics and quantum Markov chains. Namely, it is shown that one could 'potentially' construct explicit examples of quantum (and generalized quantum) Markov chains, e.g. the Ising (resp. Heisenberg) potential gives rise to quantum (resp. generalized quantum) Markov chains. For the recent development of the theory of quantum Markov states and chains, we refer the reader to [6–8, 10–12]. We stress that there are also different approaches to the definition of quantum Markov chains which could be found in [41, 94, 95, 129].

# Chapter 6
# Quantum Quadratic Stochastic Operators on $\mathbb{M}_2(\mathbb{C})$

In this chapter, we are going to study the nonlinear dynamics of quantum quadratic stochastic operators (q.q.s.o.s) acting on the algebra of $2 \times 2$ matrices $\mathbb{M}_2(\mathbb{C})$. Since positive trace-preserving maps arise naturally in quantum information theory (see e.g. [199]) and in other settings in which one wishes to restrict attention to a quantum system that should properly be considered a subsystem of a larger system which it interacts with, we describe quadratic operators with a Haar state (invariant with respect to the trace). Then we characterize q.q.s.o.s with the Kadison–Schwarz property (which is a stronger condition than positivity). By means of such a description we provide an example of a positive q.q.s.o. which is not a Kadison–Schwarz operator. Note that such a characterization is related to the separability condition, which plays an important role in quantum information. We also study the stability of the dynamics of quadratic operators associated with q.q.s.o.s.

## 6.1 Description of Quantum Quadratic Stochastic Operators on $\mathbb{M}_2(\mathbb{C})$

In what follows, by $\mathbb{M}_2(\mathbb{C})$ we denote an algebra of $2 \times 2$ matrices over the complex field $\mathbb{C}$. By $\mathbb{M}_2(\mathbb{C}) \otimes \mathbb{M}_2(\mathbb{C})$ we mean the tensor product of $\mathbb{M}_2(\mathbb{C})$ into itself. We note that such a product can be considered as an algebra of $4 \times 4$ matrices $\mathbb{M}_4(\mathbb{C})$ over $\mathbb{C}$. In the sequel $\mathbb{1}$ means an identity matrix, i.e. $\mathbb{1} = \begin{pmatrix} 1 & 0 \\ 0 & 1 \end{pmatrix}$. By $S(\mathbb{M}_2(\mathbb{C}))$ we denote the set of all states (i.e. linear positive functionals which take value 1 at $\mathbb{1}$) defined on $\mathbb{M}_2(\mathbb{C})$.

In this section we are going to describe q.q.s.o.s on $\mathbb{M}_2(\mathbb{C})$ and find necessary conditions for such operators to satisfy the Kadison–Schwarz property.

© Springer International Publishing Switzerland 2015
F. Mukhamedov, N. Ganikhodjaev, *Quantum Quadratic Operators and Processes*,
Lecture Notes in Mathematics 2133, DOI 10.1007/978-3-319-22837-2_6

Recall [25] that the identity and Pauli matrices $\{\mathbb{1}, \sigma_1, \sigma_2, \sigma_3\}$ form a basis for $M_2(\mathbb{C})$, where

$$\sigma_1 = \begin{pmatrix} 0 & 1 \\ 1 & 0 \end{pmatrix} \quad \sigma_2 = \begin{pmatrix} 0 & -i \\ i & 0 \end{pmatrix} \quad \sigma_3 = \begin{pmatrix} 1 & 0 \\ 0 & -1 \end{pmatrix}.$$

In this basis every matrix $x \in M_2(\mathbb{C})$ can be written as $x = w_0 \mathbb{1} + \mathbf{w}\sigma$ with $w_0 \in \mathbb{C}$, $\mathbf{w} = (w_1, w_2, w_3) \in \mathbb{C}^3$. Here $\mathbf{w}\sigma = w_1\sigma_1 + w_2\sigma_2 + w_3\sigma_3$. In what follows, we frequently use the notation $\overline{\mathbf{w}} = (\overline{w_1}, \overline{w_2}, \overline{w_3})$.

**Lemma 6.1.1 ([219])** *The following assertions hold true:*

*(a) $x$ is self-adjoint iff $w_0$, $\mathbf{w}$ are reals;*
*(b) $\mathrm{Tr}(x) = 1$ iff $w_0 = 0.5$, here $\mathrm{Tr}$ is the trace of a matrix $x$;*
*(c) $x \geq 0$ iff $\|\mathbf{w}\| \leq w_0$, where $\|\mathbf{w}\| = \sqrt{|w_1|^2 + |w_2|^2 + |w_3|^2}$;*
*(d) A linear functional $\varphi$ on $M_2(\mathbb{C})$ is a state iff it can be represented by*

$$\varphi(w_0 \mathbb{1} + \mathbf{w}\sigma) = w_0 + \langle \mathbf{w}, \mathbf{f} \rangle, \tag{6.1}$$

*where $\mathbf{f} = (f_1, f_2, f_3) \in \mathbb{R}^3$ such that $\|\mathbf{f}\| \leq 1$. Here, as before, $\langle \cdot, \cdot \rangle$ stands for the scalar product in $\mathbb{C}^3$.*

In the sequel we shall identify a state with a vector $\mathbf{f} \in \mathbb{R}^3$ with $\|\mathbf{f}\| \leq 1$. By $\tau$ we denote a normalized trace, i.e.

$$\tau \begin{pmatrix} x_{11} & x_{12} \\ x_{21} & x_{22} \end{pmatrix} = \frac{x_{11} + x_{22}}{2},$$

i.e. $\tau(x) = \frac{1}{2}\mathrm{Tr}(x)$, $x \in M_2(\mathbb{C})$. Clearly, $\tau$ corresponds to the vector $(0, 0, 0)$.

It is clear that the system $\{\mathbb{1} \otimes \mathbb{1}, \mathbb{1} \otimes \sigma_i, \sigma_j \otimes \mathbb{1}, \sigma_i \otimes \sigma_j\}_{i,j=1}^3$ forms a basis in $M_2(\mathbb{C}) \otimes M_2(\mathbb{C})$.

Let $\Delta : M_2(\mathbb{C}) \to M_2(\mathbb{C}) \otimes M_2(\mathbb{C})$ be a q.q.s.o. Then we write the operator $\Delta$ in terms of a basis of $M_2(\mathbb{C}) \otimes M_2(\mathbb{C})$ formed by the Pauli matrices, as follows

$$\Delta\mathbb{1} = \mathbb{1} \otimes \mathbb{1};$$

$$\Delta(\sigma_i) = b_i(\mathbb{1} \otimes \mathbb{1}) + \sum_{j=1}^3 b_{ji}^{(1)}(\mathbb{1} \otimes \sigma_j) + \sum_{j=1}^3 b_{ji}^{(2)}(\sigma_j \otimes \mathbb{1}) + \sum_{m,l=1}^3 b_{ml,i}(\sigma_m \otimes \sigma_l),$$

$$\tag{6.2}$$

where $i = 1, 2, 3$.

Hence, using Definition 5.2.1 we obtain $b_{uv,k} = b_{vu,k}$ and $b_{u,k} := b_{u,k}^{(1)} = b_{u,k}^{(2)}$. Therefore we have

$$\Delta(\sigma_k) = b_k \mathbb{1} + \sum_{u,v=1}^3 b_{uv,k}\sigma_u \otimes \sigma_u + \sum_{u=1}^3 b_{u,k}(\sigma_u \otimes \mathbb{1} + \mathbb{1} \otimes \sigma_u). \tag{6.3}$$

Now we consider a conjugate quadratic operator $\tilde{V}_\Delta$ related to $\Delta$. Taking into account (6.1) from (5.10), Chap. 5 and (6.3) we infer that

$$\tilde{V}_\Delta(\mathbf{f}, \mathbf{p})(\sigma_k) = b_k + \sum_{u,v=1}^{3} b_{uv,k} f_u p_v + \sum_{u=1}^{3} b_{u,k}(f_u + p_u), \tag{6.4}$$

where $\mathbf{f} = (f_1, f_2, f_3)$, $\mathbf{p} = (p_1, p_2, p_3)$.

Therefore, the associated quadratic operator has the form

$$V_\Delta(\mathbf{f})(\sigma_k) = b_k + \sum_{u,v=1}^{3} b_{uv,k} f_u f_v + 2 \sum_{u=1}^{3} b_{u,k} f_u, \tag{6.5}$$

where $\mathbf{f} = (f_1, f_2, f_3)$.

**Observation 6.1.1** *Consider the q.q.s.o. $P_T$ defined by (5.13). From (6.3) one gets*

$$P_T(\sigma_k) = b_k^{(T)} \mathbb{1} + \sum_{u=1}^{3} b_{u,k}^{(T)}(\sigma_u \otimes \mathbb{1} + \mathbb{1} \otimes \sigma_u)$$

*and the corresponding q.o. has the form*

$$V_T(\mathbf{f})(\sigma_k) = b_k^{(T)} + 2 \sum_{u=1}^{3} b_{u,k}^{(T)} f_u.$$

**Observation 6.1.2** *Let us consider the commutative quadratic stochastic operator (q.s.o.) defined by the cubic matrix $\{p_{ij,k}\}$ with properties*

$$p_{ij,k} \geq 0, \quad p_{ij,k} = p_{ji,k}, \quad p_{ij,1} + p_{ij,2} = 1, \quad \forall i,j,k \in \{1,2\}. \tag{6.6}$$

*Define an operator $\mathbb{P} : \mathbb{C}^2 \to \mathbb{C}^2 \otimes \mathbb{C}^2$ by*

$$(\mathbb{P}(\mathbf{x}))_{i,j} = \sum_{k=1}^{2} p_{ij,k} x_k, \quad i,j \in \{1,2\}, \tag{6.7}$$

*where $\mathbf{x} = (x_1, x_2)$. Here $\mathbb{C}^2$ is considered with a norm $\|\mathbf{x}\| = \max\{|x_1|, |x_2|\}$.*

*By $DM_2(\mathbb{C})$ we denote the commutative subalgebra of $\mathbb{M}_2(\mathbb{C})$ generated by $\mathbb{1}$ and $\sigma_3$. It is obvious that $DM_2(\mathbb{C})$ can be identified with $\mathbb{C}^2$, and further we will use this identification. Let $E : \mathbb{M}_2(\mathbb{C}) \to DM_2(\mathbb{C})$ be the canonical conditional expectation. Now define another operator $P_\mathbb{P} : \mathbb{M}_2(\mathbb{C}) \to DM_2(\mathbb{C}) \otimes DM_2(\mathbb{C})$ by*

$$\Delta_\mathbb{P}(x) = \mathbb{P}(E(x)), \quad x \in \mathbb{M}_2(\mathbb{C}). \tag{6.8}$$

*From (6.7) and the properties of the conditional expectation one concludes that $P_\mathbb{P}$ is a q.q.s.o. Now we rewrite it in the form (6.3). From (6.8) and (6.7) we get $\Delta_\mathbb{P}(\sigma_1) = \Delta_\mathbb{P}(\sigma_2) = 0$ and*

$$\Delta_\mathbb{P}(\sigma_3) = (p_{11,1} + 2p_{12,1} + p_{22,1} - 2)\mathbb{1}$$
$$+(p_{11,1} - p_{22,2})(\sigma_3 \otimes \mathbb{1} + \mathbb{1} \otimes \sigma_3)$$
$$+(p_{11,1} - 2p_{12,1} + p_{22,1})\sigma_3 \otimes \sigma_3.$$

*Hence for the corresponding q.o. we have $V_\mathbb{P}(\mathbf{f})(\sigma_1) = V_\mathbb{P}(\mathbf{f})(\sigma_2) = 0$ and*

$$V_\mathbb{P}(\mathbf{f})(\sigma_3) = (p_{11,1} + 2p_{12,1} + p_{22,1} - 2) + (p_{11,1} - p_{22,2})f_3$$
$$+(p_{11,1} - 2p_{12,1} + p_{22,1})f_3^2. \tag{6.9}$$

*As before, $\mathbf{f} = (f_1, f_2, f_3) \in \mathbb{R}^3$.*

In general, the description of positive operators is one of the main problems of quantum information. In the literature most tractable maps are positive and trace-preserving, since such maps arise naturally in quantum information theory (see [118, 125, 199, 219]). Therefore, in the sequel we shall restrict ourselves to the trace preserving q.q.s.o.s, i.e. $\tau \otimes \tau \circ \Delta = \tau$. So, we would like to describe all such kind of maps.

**Proposition 6.1.2** *Let $\Delta : \mathbb{M}_2(\mathbb{C}) \to \mathbb{M}_2(\mathbb{C}) \otimes \mathbb{M}_2(\mathbb{C})$ be a trace preserving q.q.s.o., then in (6.2) one has $b_j = 0$, and $b_{ij}$ and $b_{ij,k} = b_{ji,k}$ are reals for every $i, j, k \in \{1, 2, 3\}$. Moreover, $\Delta$ has the following form:*

$$\Delta(x) = w_0 \mathbb{1} \otimes \mathbb{1} + \mathbf{Bw} \cdot \sigma \otimes \mathbb{1} + \mathbb{1} \otimes \mathbf{Bw} \cdot \sigma + \sum_{m,l=1}^{3} \langle \mathbf{b}_{ml}, \overline{\mathbf{w}} \rangle \sigma_m \otimes \sigma_l, \tag{6.10}$$

*where $x = w_0 + \mathbf{w}\sigma$, $\mathbf{b}_{ml} = (b_{ml,1}, b_{ml,2}, b_{ml,3})$, and $\mathbf{B} = (b_{ij})_{i,j=1}^{3}$. Here, as before, $\langle \cdot, \cdot \rangle$ stands for the standard scalar product in $\mathbb{C}^3$.*

*Proof* From the positivity of $\Delta$ we get that $\Delta x^* = (\Delta x)^*$. Therefore

$$\Delta(\sigma_i^*) = \overline{b_i}(\mathbb{1} \otimes \mathbb{1}) + \sum_{j=1}^{3} \overline{b_{ji}}\big((\mathbb{1} \otimes \sigma_j) + (\sigma_j \otimes \mathbb{1})\big) + \sum_{m,l=1}^{3} \overline{b_{ml,i}}(\sigma_m \otimes \sigma_l).$$

This yields that $b_i = \overline{b_i}$, $b_{ji} = \overline{b_{ji}}$ and $b_{ml,i} = \overline{b_{ml,i}}$, i.e. all coefficients are real numbers.

From the trace preserving condition we get $\tau \otimes \tau(\Delta(\sigma_i)) = \tau(\sigma_i) = 0$, which yields $b_j = 0, j = 1, 2, 3$.

Hence, $\Delta$ has the following form

$$\Delta(\sigma_i) = \sum_{j=1}^{3} b_{ji}\big((\mathbb{1} \otimes \sigma_j) + (\sigma_j \otimes \mathbb{1})\big) + \sum_{m,l=1}^{3} b_{ml,i}(\sigma_m \otimes \sigma_l). \qquad (6.11)$$

Defining

$$\mathbf{B} = (b_{ij})_{i,j=1}^{3}, \quad \mathbf{b}_{ml} = (b_{ml,1}, b_{ml,2}, b_{ml,3}) \qquad (6.12)$$

and taking any $x = w_0 \mathbb{1} + \mathbf{w}\sigma \in \mathbb{M}_2(\mathbb{C})$, from (6.11) we immediately find (6.10). This completes the proof.

One can rewrite (6.10) as follows

$$\Delta(x) = \lambda \Delta_1(x) + (1 - \lambda)\Delta_2(x), \qquad (6.13)$$

where

$$\Delta_1(x) = w_0 \mathbb{1} \otimes \mathbb{1} + \frac{1}{\lambda} \sum_{m,l=1}^{3} \langle \mathbf{b}_{ml}, \overline{\mathbf{w}} \rangle \sigma_m \otimes \sigma_l, \qquad (6.14)$$

$$\Delta_2(x) = w_0 \mathbb{1} \otimes \mathbb{1} + \frac{1}{1 - \lambda}\left(\mathbf{B}\mathbf{w} \cdot \sigma \otimes \mathbb{1} + \mathbb{1} \otimes \mathbf{B}\mathbf{w} \cdot \sigma\right). \qquad (6.15)$$

A q.q.s.o. of the form (6.14) (resp. (6.15)) is called *simple* (resp. *non-simple*). So, any q.q.s.o. is a convex combination of simple and non-simple q.q.s.o.s. In the sequel we are going to investigate simple and non-simple q.q.s.o.s one by one.

*Remark 6.1.1* Note that if $\tau$ is a Haar state for $\Delta$ (see (5.7)), then $\Delta$ can be written as follows

$$\Delta(x) = w_0 \mathbb{1} \otimes \mathbb{1} + \sum_{m,l=1}^{3} \langle \mathbf{b}_{ml}, \overline{\mathbf{w}} \rangle \sigma_m \otimes \sigma_l, \qquad (6.16)$$

which means that $\Delta$ is non-simple.

Let us turn to the positivity of $\Delta$. Given a vector $\mathbf{f} = (f_1, f_2, f_3) \in \mathbb{R}^3$ put

$$\beta(\mathbf{f})_{ij} = \sum_{k=1}^{3} b_{ki,j} f_k. \qquad (6.17)$$

Define a matrix $\mathbb{B}(\mathbf{f}) = (\beta(\mathbf{f})_{ij})_{ij=1}^{3}$, and by $\|\mathbb{B}(\mathbf{f})\|$ denote its norm associated with the Euclidean norm in $\mathbb{R}^3$.

Given a state $\varphi$, by $E_\varphi$ we denote the canonical conditional expectation defined by $E_\varphi(x \otimes y) = \varphi(x)y$, where $x, y \in \mathbb{M}_2(\mathbb{C})$.

In the sequel, we denote the unit ball in $\mathbb{R}^3$ by $\mathbf{S}$, i.e.

$$\mathbf{S} = \{\mathbf{p} = (p_1, p_2, p_3) \in \mathbb{R}^3 : p_1^2 + p_2^2 + p_3^2 \le 1\}.$$

Let us define

$$\||\mathbb{B}\|| = \sup_{\mathbf{f} \in S} \|\mathbb{B}(\mathbf{f})\|.$$

**Proposition 6.1.3** *Let $\Delta$ be a trace preserving q.q.s.o., then one has*

$$|\langle \mathbf{Bw}, \mathbf{f} \rangle| \le 1, \quad \|(\mathbf{B} + \mathbb{B}(\mathbf{f}))\mathbf{w}\| \le 1 + |\langle \mathbf{Bw}, \mathbf{f} \rangle|, \tag{6.18}$$

*for all* $\mathbf{w}, \mathbf{f} \in \mathbf{S}$.

*Proof* Let $x \in \mathbb{M}_2(\mathbb{C})$ (i.e. $x = w_0 \mathbb{1} + \mathbf{w}\sigma$) be a positive element. Without loss of generality we may assume that $w_0 = 1$. The positivity of $x$ implies $\|\mathbf{w}\| \le 1$. Then for any state $\varphi(x) = w_0 + \langle \mathbf{f}, \mathbf{w} \rangle$ (here $\mathbf{f} = (f_1, f_2, f_3) \in \mathbf{S}$) from (6.10) and (6.17) one finds

$$E_\varphi(\Delta(x)) = (1 + \langle \mathbf{Bw}, \mathbf{f} \rangle)\mathbb{1} + \mathbf{Bw} \cdot \sigma + \sum_{i,j=1}^{3} \langle \mathbf{b}_{ij}, \overline{\mathbf{w}} \rangle f_i \sigma_j$$

$$= (1 + \langle \mathbf{Bw}, \mathbf{f} \rangle)\mathbb{1} + (\mathbf{B} + \mathbb{B}(\mathbf{f}))\mathbf{w} \cdot \sigma$$

where we have used $\varphi(\sigma_i) = f_i$ and

$$\sum_{i=1}^{3} \langle \mathbf{b}_{ij}, \overline{\mathbf{w}} \rangle f_i = \sum_{l=1}^{3} \sum_{i=1}^{3} b_{ij,l} f_i w_l$$

$$= \sum_{l=1}^{3} \beta_{jl}(\mathbf{f}) w_l$$

$$= (\mathbb{B}(\mathbf{f})\mathbf{w})_j.$$

We know that $E_\varphi$ is a positive mapping, therefore, the positivity of $x$ yields that $E_\varphi(\Delta(x))$ is positive, for all states $\varphi$. Hence, according to Lemma 6.1.1 the positivity of $E_\varphi(\Delta(x))$ is equivalent to (6.18). This completes the proof.

**Corollary 6.1.4** *Let $\Delta$ be a q.q.s.o. with Haar state $\tau$, then one has $\||\mathbb{B}\|| \le 1$.*

*Proof* In this case, $\mathbf{B} = 0$, therefore, from (6.18) one finds $\|\mathbb{B}(\mathbf{f})\mathbf{w}\| \le 1$ for all $\mathbf{f}, \mathbf{w} \in \mathbf{S}$. Consequently, $\|\mathbb{B}(\mathbf{f})\| = \sup_{\|\mathbf{w}\| \le 1} \|\mathbb{B}(\mathbf{f})\mathbf{w}\| \le 1$, which yields the assertion.

*Remark 6.1.2* Similar characterizations of positive maps defined on $M_2(\mathbb{C})$ were considered in [144]. The characterization of completely positive mappings from $M_2(\mathbb{C})$ into itself with invariant state $\tau$ was established in [219].

## 6.2 Simple Kadison–Schwarz Type q.q.s.o.s

In this section we are going to study simple q.q.s.o.s which satisfy the Kadison–Schwarz condition (5.1) and complete positivity.

First we need the following auxiliary

**Lemma 6.2.1** *Let* $x = w_0\mathbb{1}\otimes\mathbb{1}+\mathbf{w}\cdot\sigma\otimes\mathbb{1}+\mathbb{1}\otimes\mathbf{r}\cdot\sigma$. *Then the following statements hold true:*

*(i)* $x$ *is self-adjoint if and only if* $w_0 \in \mathbb{R}$ *and* $\mathbf{w}, \mathbf{r} \in \mathbb{R}^3$;
*(ii)* $x$ *is positive if and only if* $w_0 > 0$ *and* $\|\mathbf{w}\| + \|\mathbf{r}\| \leq w_0$.

*Proof*

(i). One can see that

$$x^* = \overline{w_0}\mathbb{1}\otimes\mathbb{1}+\overline{\mathbf{w}}\cdot\sigma\otimes\mathbb{1}+\mathbb{1}\otimes\overline{\mathbf{r}}\cdot\sigma.$$

So, the self-adjointness of $x$ implies $\overline{w_0} = w_0$, $\overline{\mathbf{w}} = \mathbf{w}$, $\overline{\mathbf{r}} = \mathbf{r}$.
(ii). Let $x$ be self-adjoint. Then from the definition of Pauli matrices one finds

$$x = \begin{pmatrix} w_0+w_3+r_3 & w_1-iw_2 & r_1-ir_2 & 0 \\ w_1+iw_2 & w_0-w_3+r_3 & 0 & r_1-ir_2 \\ r_1+ir_2 & 0 & w_0+w_3-r_3 & w_1-iw_2 \\ 0 & r_1+ir_2 & w_1+iw_2 & w_0-w_3-r_3 \end{pmatrix}.$$

It is easy to calculate that the eigenvalues of the last matrix are the following

$$\lambda_1 = w_0 - \|\mathbf{r}\| + \|\mathbf{w}\|, \quad \lambda_2 = w_0 - \|\mathbf{r}\| - \|\mathbf{w}\|,$$

$$\lambda_3 = w_0 + \|\mathbf{r}\| + \|\mathbf{w}\|, \quad \lambda_4 = w_0 + \|\mathbf{r}\| - \|\mathbf{w}\|.$$

So, one concludes that $x$ is positive if and only if the smallest eigenvalue is positive. This means $w_0 - \|\mathbf{r}\| - \|\mathbf{w}\| \geq 0$, which completes the proof.

Let $\Delta : M_2(\mathbb{C}) \to M_2(\mathbb{C}) \otimes M_2(\mathbb{C})$ be a linear operator which is given by

$$\Delta(w_0\mathbb{1} + \mathbf{w}\cdot\sigma) = w_0\mathbb{1}\otimes\mathbb{1} + \mathbf{Bw}\cdot\sigma\otimes\mathbb{1} + \mathbb{1}\otimes\mathbf{Bw}\cdot\sigma, \qquad (6.19)$$

where $\mathbf{B}$ is a linear operator on $\mathbb{C}^3$.

Clearly, $\Delta$ is unital. Now let us first find conditions when $\Delta$ is positive, i.e. a q.q.s.o. This is given by the following

**Theorem 6.2.2** *The mapping $\Delta$ given by (6.19) is a q.q.s.o. if and only if $\|\mathbf{B}\| \leq 1/2$.*

*Proof* Let $x = w_0 \mathbb{1} + \mathbf{w} \cdot \sigma$ be positive, i.e. $w_0 > 0$, $\|\mathbf{w}\| \leq w_0$. Without loss of generality we may assume $w_0 = 1$. Now Lemma 6.2.1 yields that $\Delta(x)$ is positive if and only if $2\|\mathbf{B}\mathbf{w}\| \leq 1$. This yields the assertion.

Now let us turn to the Kadison–Schwarz property.
Define the following mapping

$$\Phi(x) = w_0 \mathbb{1} + 2\mathbf{B}\mathbf{w} \cdot \sigma. \tag{6.20}$$

Then from (6.19) and (6.20) one finds

$$\Delta(x) = \frac{1}{2}\Big( \Phi(x) \otimes \mathbb{1} + \mathbb{1} \otimes \Phi(x) \Big). \tag{6.21}$$

From Theorem 5.1.2 we immediately have

**Corollary 6.2.3** *If the mapping $\Phi$ given by (6.20) is a KS-operator, then $\Delta$ given by (6.21) is also a KS-operator.*

We are interested in finding a more general condition than the formulated one.

**Theorem 6.2.4** *Let $\Delta$ be a simple q.q.s.o. given by (6.21). If one has*

$$\|\mathbf{B}\| \leq \frac{1}{2} \tag{6.22}$$

$$2\|\mathbf{B}[\mathbf{w}, \overline{\mathbf{w}}] - 2[\mathbf{B}\mathbf{w}, \mathbf{B}\overline{\mathbf{w}}]\| \| \leq \|\mathbf{w}\|^2 - 4\|\mathbf{B}\mathbf{w}\|^2. \tag{6.23}$$

*Then $\Delta$ is a Kadison–Schwarz operator.*

*Proof* From (6.21) one finds that

$$\Delta(x^*x) - \Delta(x)^*\Delta(x) = \frac{1}{2}\Big( \big(\Phi(x^*x) - \Phi(x)^*\Phi(x)\big) \otimes \mathbb{1}$$

$$+ \mathbb{1} \otimes \big(\Phi(x^*x) - \Phi(x)^*\Phi(x)\big) \Big)$$

$$+ \frac{1}{4}\Big( \mathbb{1} \otimes \Phi(x) - \Phi(x) \otimes \mathbb{1} \Big)^* \Big( \mathbb{1} \otimes \Phi(x) - \Phi(x) \otimes \mathbb{1} \Big). \tag{6.24}$$

Now taking into account the following formula

$$x^*x = \left(|w_0|^2 + \|\mathbf{w}\|^2\right)\mathbb{1} + \left(w_0\overline{\mathbf{w}} + \overline{w_0}\mathbf{w} - i[\mathbf{w}, \overline{\mathbf{w}}]\right) \cdot \sigma, \tag{6.25}$$

from (6.20) we have

$$\Phi(x^*x) - \Phi(x)^*\Phi(x) = \left(\|\mathbf{w}\|^2 - \|2\mathbf{Bw}\|^2\right)\mathbb{1} - 2i\left(\mathbf{B}[\mathbf{w}, \overline{\mathbf{w}}] - 2[\mathbf{Bw}, \mathbf{B}\overline{\mathbf{w}}]\right)\sigma.$$

Therefore, one gets

$$\left(\Phi(x^*x) - \Phi(x)^*\Phi(x)\right) \otimes \mathbb{1} + \mathbb{1} \otimes \left(\Phi(x^*x) - \Phi(x)^*\Phi(x)\right)$$

$$= \left(\left(\|\mathbf{w}\|^2 - 4\|\mathbf{Bw}\|^2\right)\mathbb{1} - 2i\left(\mathbf{B}[\mathbf{w}, \overline{\mathbf{w}}] - 2[\mathbf{Bw}, \mathbf{B}\overline{\mathbf{w}}]\right)\sigma\right) \otimes \mathbb{1}$$

$$+ \mathbb{1} \otimes \left(\left(\|\mathbf{w}\|^2 - 4\|\mathbf{Bw}\|^2\right)\mathbb{1} - 2i\left(\mathbf{B}[\mathbf{w}, \overline{\mathbf{w}}] - 2[\mathbf{Bw}, \mathbf{B}\overline{\mathbf{w}}]\right)\sigma\right)$$

$$= \left(2\|\mathbf{w}\|^2 - 8\|\mathbf{Bw}\|^2\right)\mathbb{1} \otimes \mathbb{1}$$

$$- 2i\left(\mathbf{B}[\mathbf{w}, \overline{\mathbf{w}}] - 2[\mathbf{Bw}, \mathbf{B}\overline{\mathbf{w}}]\right)\sigma \otimes \mathbb{1} - \mathbb{1} \otimes 2i\left(\mathbf{B}[\mathbf{w}, \overline{\mathbf{w}}] - 2[\mathbf{Bw}, \mathbf{B}\overline{\mathbf{w}}]\right)\sigma.$$

According to Lemma 6.2.1 we conclude that the last expression is positive if and only if (6.22) and (6.23) are satisfied. Consequently, from (6.24) we infer that under the last conditions the mapping $\Delta$ is a KS-operator. This completes the proof.

We should stress that the conditions (6.22) and (6.23) are sufficient for $\Delta$ to be a KS-operator.

*Remark 6.2.1* We have to stress that if $\Delta$ is a KS-operator, then the mapping $\Phi$, in general, need not be a KS-operator.

Let $D$ be a simple q.q.s.o. given by (6.19). Then following [118] let us decompose the matrix $\mathbf{B}$ as $\mathbf{B} = \mathbf{RA}$, where $\mathbf{R}$ is a rotation and $\mathbf{A}$ is a self-adjoint matrix (see [118]). Define a mapping $\Delta_{\mathbf{A}}$ as follows

$$\Delta_{\mathbf{A}}(w_0\mathbb{1} + \mathbf{w} \cdot \sigma) = w_0\mathbb{1} \otimes \mathbb{1} + \mathbf{Aw} \cdot \sigma \otimes \mathbb{1} + \mathbb{1} \otimes \mathbf{Aw} \cdot \sigma.$$

Every rotation is implemented by a unitary matrix in $M_2(\mathbb{C})$. Therefore there is a unitary $U \in M_2(\mathbb{C})$ such that

$$\Delta(x) = U\Delta_{\mathbf{A}}(x)U^*, \quad x \in M_2(\mathbb{C}). \tag{6.26}$$

On the other hand, every self-adjoint operator $\mathbf{A}$ can be diagonalized by some unitary operator, i.e. there is a unitary $V \in M_2(\mathbb{C})$ such that $\mathbf{A} = V D_{\lambda_1,\lambda_2,\lambda_3} V^*$, where

$$
D_{\lambda_1,\lambda_2,\lambda_3} = \begin{pmatrix} \lambda_1 & 0 & 0 \\ 0 & \lambda_2 & 0 \\ 0 & 0 & \lambda_3 \end{pmatrix}, \tag{6.27}
$$

where $\lambda_1, \lambda_2, \lambda_3 \in \mathbb{R}$.

Consequently, the mapping $\Delta$ can be represented by

$$
\Delta(x) = \tilde{U} \Delta_{D_{\lambda_1,\lambda_2,\lambda_3}}(x) \tilde{U}^*, \quad x \in M_2(\mathbb{C}) \tag{6.28}
$$

for some unitary $\tilde{U}$. Due to Theorem 5.1.2, the mapping $\Delta_{D_{\lambda_1,\lambda_2,\lambda_3}}$ is also a KS-operator. Hence, all simple q.q.s.o.s with the KS-property can be characterized by $\Delta_{D_{\lambda_1,\lambda_2,\lambda_3}}$ and unitaries. In what follows, for the sake of brevity, by $\Delta_{(\lambda_1,\lambda_2,\lambda_3)}$ we denote the mapping $\Delta_{D_{\lambda_1,\lambda_2,\lambda_3}}$. From Theorem 6.2.2 one finds that $|\lambda_k| \le 1/2$, for $k = 1, 2, 3$.

Next, we want to characterize KS-operators of the form $\Delta_{(\lambda_1,\lambda_2,\lambda_3)}$.

**Theorem 6.2.5** *If*

$$
(1 + 4\lambda_1^2)(3 + 4\lambda_2^2 + 4\lambda_3^2 - 4\lambda_1^2) \le 4(1 + 8\lambda_1\lambda_2\lambda_3),
$$
$$
(1 + 4\lambda_2^2)(3 + 4\lambda_1^2 + 4\lambda_3^2 - 4\lambda_2^2) \le 4(1 + 8\lambda_1\lambda_2\lambda_3),
$$
$$
(1 + 4\lambda_3^2)(3 + 4\lambda_1^2 + 4\lambda_2^2 - 4\lambda_3^2) \le 4(1 + 8\lambda_1\lambda_2\lambda_3)
$$

*are satisfied, then $\Delta_{(\lambda_1,\lambda_2,\lambda_3)}$ is a KS-operator.*

*Proof* Taking $\mathbf{B} = D_{\lambda_1,\lambda_2,\lambda_3}$ in (6.23), we obtain

$$
4A_1|w_2\overline{w_3} - \overline{w_2}w_3|^2 + 4A_2|\overline{w_1}w_3 - w_1\overline{w_3}|^2 + 4A_3|w_1\overline{w_2} - \overline{w_1}w_2|^2
$$
$$
\le \left( B_1|w_1|^2 + B_2|w_2|^2 + B_3|w_3|^2 \right)^2, \tag{6.29}
$$

where $\mathbf{w} = (w_1, w_2, w_3) \in \mathbb{C}^3$ and

$$
A_1 = |\lambda_1 - 2\lambda_2\lambda_3|^2, \ A_2 = |\lambda_2 - 2\lambda_1\lambda_3|^2, \ A_3 = |\lambda_3 - 2\lambda_1\lambda_2|^2, \tag{6.30}
$$
$$
B_1 = (1 - 4\lambda_1^2), \ B_2 = (1 - 4\lambda_2^2), \ B_3 = (1 - 4\lambda_3^2). \tag{6.31}
$$

Due to the inequality $|2\Im(uv)| \le |u|^2 + |v|^2$, one has

$$
|w_i\overline{w_j} - w_j\overline{w_i}|^2 = |2\Im(w_iw_j)|^2 \le |w_i|^4 + 2|w_i|^2|w_j|^2 + |w_j|^4 \ (i \ne j). \tag{6.32}
$$

Note that equality is attainable by an appropriate choice of values $w_i$ and $w_j$.

Hence the LHS of (6.29) can be evaluated as follows

$$4A_1\Big(|w_2|^4 + 2|w_2|^2|w_3|^2 + |w_3|^4\Big) + 4A_2\Big(|w_1|^4 + 2|w_1|^2|w_3|^2 + |w_3|^4\Big)$$
$$+4A_3\Big(|w_1|^4 + 2|w_1|^2|w_2|^2 + |w_2|^4\Big).$$

Therefore, from (6.29) one gets

$$\Big(B_1^2 - 4A_2 - 4A_3\Big)|w_1|^4 + \Big(B_2^2 - 4A_1 - 4A_3\Big)|w_2|^4 + \Big(B_3^2 - 4A_1 - 4A_2\Big)|w_3|^4$$
$$+2|w_2|^2|w_3|^2(B_2B_3 - 4A_1) + 2|w_1|^2|w_3|^2(B_1B_3 - 4A_2)$$
$$+2|w_1|^2|w_2|^2(B_1B_2 - 4A_3) \geq 0.$$

It is obvious that the inequality given above is satisfied if one has

$$B_1^2 \geq 4A_2 + 4A_3, \quad B_2^2 \geq 4A_1 + 4A_3, \quad B_3^2 \geq 4A_1 + 4A_2,$$
$$B_2B_3 \geq 4A_1, \quad B_1B_3 \geq 4A_2, \quad B_1B_2 \geq 4A_3.$$

Substituting the above (6.30) and (6.31) into the last inequalities, and doing some calculations, one derives

$$4(1 + 8\lambda_1\lambda_2\lambda_3) \geq (1 + 4\lambda_1^2)(3 + 4\lambda_2^2 + 4\lambda_3^2 - 4\lambda_1^2), \qquad (6.33)$$
$$4(1 + 8\lambda_1\lambda_2\lambda_3) \geq (1 + 4\lambda_2^2)(3 + 4\lambda_1^2 + 4\lambda_3^2 - 4\lambda_2^2), \qquad (6.34)$$
$$4(1 + 8\lambda_1\lambda_2\lambda_3) \geq (1 + 4\lambda_3^2)(3 + 4\lambda_1^2 + 4\lambda_2^2 - 4\lambda_3^2), \qquad (6.35)$$
$$1 + 16\lambda_1\lambda_2\lambda_3 \geq 4\lambda_1^2 + 4\lambda_2^2 + 4\lambda_3^2. \qquad (6.36)$$

These inequalities imply that $\lambda_1, \lambda_2, \lambda_3 \in \Big[-\frac{1}{2}, \frac{1}{2}\Big]$.

Now let us show that (6.36) is a redundant condition, i.e. it is always satisfied when (6.33), (6.34) and (6.35) are true.

Suppose that

$$1 + 16\lambda_1\lambda_2\lambda_3 = 4\lambda_1^2 + 4\lambda_2^2 + 4\lambda_3^2 \qquad (6.37)$$

is true. We will show that the elements of the surface do not satisfy the inequalities (6.33)–(6.35) except for $(0, 0, 0)$, $(\pm 1/2, \pm 1/2, \pm 1/2)$. Using simple algebra, from (6.33)–(6.35) with (6.36), we obtain

$$(1 - 4\lambda_1^2)(\lambda_1^2 - 2\lambda_1\lambda_2\lambda_3) \leq 0;$$
$$(1 - 4\lambda_2^2)(\lambda_2^2 - 2\lambda_1\lambda_2\lambda_3) \leq 0;$$
$$(1 - 4\lambda_3^2)(\lambda_3^2 - 2\lambda_1\lambda_2\lambda_3) \leq 0,$$

where $\lambda_1, \lambda_2, \lambda_3 \in [-1/2, 1/2]$.

Due to our assumption $\lambda_1 \neq \pm 1/2, \lambda_2 \neq \pm 1/2, \lambda_3 \neq \pm 1/2$ from the last inequalities one finds

$$\lambda_1(\lambda_1 - 2\lambda_2\lambda_3) \leq 0, \tag{6.38}$$

$$\lambda_2(\lambda_2 - 2\lambda_1\lambda_3) \leq 0, \tag{6.39}$$

$$\lambda_3(\lambda_3 - 2\lambda_1\lambda_2) \leq 0, \tag{6.40}$$

where $\lambda_1, \lambda_2, \lambda_3 \in (-1/2, 1/2)$.

Assume that $\lambda_1 > 0$, then from (6.38) one gets $\lambda_1 \leq 2\lambda_2\lambda_3$. This yields that $\lambda_2 > 0, \lambda_3 > 0$ or $\lambda_2 < 0, \lambda_3 < 0$. First suppose that $\lambda_2 > 0, \lambda_3 > 0$, then from (6.39) and (6.40) we obtain

$$\lambda_2 \leq 2\lambda_1\lambda_3, \quad \lambda_3 \leq 2\lambda_1\lambda_2.$$

From $\lambda_1 \leq 2\lambda_2\lambda_3$ and the last inequalities one finds $\lambda_2 \leq 4\lambda_2\lambda_3^2$. This means $|\lambda_3| \geq 1/2$. This contradicts our assumption.

Now let $\lambda_2 < 0$ and $\lambda_3 < 0$, then from (6.39) and (6.40) one finds

$$\lambda_2 \geq 2\lambda_1\lambda_3, \quad \lambda_3 \geq 2\lambda_1\lambda_2. \tag{6.41}$$

From (6.41), one finds $\lambda_3 \geq 4\lambda_1^2\lambda_3$, which implies that $|\lambda_1| \geq 1/2$. This is again a contradiction. In case $\lambda_1 < 0$, using a similar argument, we again get a contradiction.

The reverse implication is obvious, and this completes the proof.

It is interesting to study when the operator $\Delta_{(\lambda_1,\lambda_2,\lambda_3)}$ is complete positive. The next result characterizes the complete positivity of $\Delta_{(\lambda_1,\lambda_2,\lambda_3)}$.

**Theorem 6.2.6** *A map $\Delta_{(\lambda_1,\lambda_2,\lambda_3)}$ is completely positive if and only if one of the following inequalities is satisfied*

*(1)* $|\lambda_3| < \frac{1}{2}$,
$$4\lambda_1^2 + 4\lambda_2^2 + 4\lambda_3^2 \leq 1 + 16\lambda_1\lambda_2\lambda_3,$$
$$\lambda_1^2 + \lambda_2^2 + \sqrt{\left(\lambda_1^2 + \lambda_2^2\right)^2 - 4\lambda_1\lambda_2\lambda_3 + \lambda_3^2} \leq \frac{1}{2};$$
*(2)* $\lambda_3 = \frac{1}{2}, |\lambda_1| \leq \frac{1}{2}, |\lambda_2| \leq \frac{1}{2}$;
*(3)* $\lambda_3 = -\frac{1}{2}, \lambda_1 = \pm\frac{1}{2}, \lambda_2 = \mp\frac{1}{2}$.

*Proof* From [26], we know that the complete positivity of $\Delta_{(\lambda_1,\lambda_2,\lambda_3)}$ is equivalent to the positivity of the following matrix

$$\hat{\Delta}_{(\lambda_1,\lambda_2,\lambda_3)} = \begin{pmatrix} \Delta_{(\lambda_1,\lambda_2,\lambda_3)}(e_{11}) & \Delta_{(\lambda_1,\lambda_2,\lambda_3)}(e_{12}) \\ \Delta_{(\lambda_1,\lambda_2,\lambda_3)}(e_{21}) & \Delta_{(\lambda_1,\lambda_2,\lambda_3)}(e_{22}) \end{pmatrix}.$$

It is clear that

$$\Delta_{(\lambda_1,\lambda_2,\lambda_3)}(e_{11}) = \frac{1}{2}\begin{pmatrix} 1+2\lambda_3 & 0 & 0 & 0 \\ 0 & 1 & 0 & 0 \\ 0 & 0 & 1 & 0 \\ 0 & 0 & 0 & 1-2\lambda_3 \end{pmatrix},$$

$$\Delta_{(\lambda_1,\lambda_2,\lambda_3)}(e_{12}) = \frac{1}{2}\begin{pmatrix} 0 & \lambda_1+\lambda_2 & \lambda_1+\lambda_2 & 0 \\ \lambda_1-\lambda_2 & 0 & 0 & \lambda_1+\lambda_2 \\ \lambda_1-\lambda_2 & 0 & 0 & \lambda_1+\lambda_2 \\ 0 & \lambda_1-\lambda_2 & \lambda_1-\lambda_2 & 0 \end{pmatrix}$$

and $\Delta_{(\lambda_1,\lambda_2,\lambda_3)}(e_{22}) = \mathbb{1}\otimes\mathbb{1} - T_{(\lambda_1,\lambda_2,\lambda_3)}(e_{11})$, $\Delta_{(\lambda_1,\lambda_2,\lambda_3)}(e_{21}) = T_{(\lambda_1,\lambda_2,\lambda_3)}(e_{12})^*$.

(1). According to Theorem 1.3.3 in [20], the matrix $\hat{\Delta}_{(\lambda_1,\lambda_2,\lambda_3)}$ is positive if and only if

$$\Delta_{(\lambda_1,\lambda_2,\lambda_3)}(e_{11}) - \Delta_{(\lambda_1,\lambda_2,\lambda_3)}(e_{12})\Delta_{(\lambda_1,\lambda_2,\lambda_3)}(e_{22})^{-1}\Delta_{(\lambda_1,\lambda_2,\lambda_3)}(e_{21}) \geq 0,$$

$$(6.42)$$

where $\Delta_{(\lambda_1,\lambda_2,\lambda_3)}(e_{11})$ and $\Delta_{(\lambda_1,\lambda_2,\lambda_3)}(e_{22})$ are positive matrices.

It is easy to see that $\Delta_{(\lambda_1,\lambda_2,\lambda_3)}(e_{11})$ and $\Delta_{(\lambda_1,\lambda_2,\lambda_3)}(e_{22})$ are positive if and only if

$$|\lambda_3| \leq \frac{1}{2}. \qquad (6.43)$$

One can calculate that (6.42) is equivalent to

$$\begin{pmatrix} \alpha_1 & 0 & 0 & \alpha_4 \\ 0 & 1+\alpha_3 & \alpha_3 & 0 \\ 0 & \alpha_3 & 1+\alpha_3 & 0 \\ \alpha_4 & 0 & 0 & \alpha_2 \end{pmatrix} \geq 0$$

where

$$\alpha_1 = 1 + 2\lambda_3 - 2(\lambda_1+\lambda_2)^2, \quad \alpha_2 = 1 - 2\lambda_3 - 2(\lambda_1-\lambda_2)^2,$$

$$\alpha_3 = \frac{(\lambda_1-\lambda_2)^2}{2\lambda_3-1} - \frac{(\lambda_1+\lambda_2)^2}{2\lambda_3+1}, \quad \alpha_4 = -2\left(\lambda_1^2-\lambda_2^2\right).$$

A matrix is positive if and only if its eigenvalues are positive. The eigenvalues of the last matrix can be calculated as follows

$$s_1 = 1, \quad s_2 = \frac{4\lambda_1^2 + 4\lambda_2^2 + 4\lambda_3^2 - 16\lambda_1\lambda_2\lambda_3 - 1}{4\lambda_3^2 - 1},$$

$$s_3 = 1 - 2\lambda_1^2 - 2\lambda_2^2 + 2\sqrt{\left(\lambda_1^2 + \lambda_2^2\right)^2 - 4\lambda_1\lambda_2\lambda_3 + \lambda_3^2},$$

$$s_4 = 1 - 2\lambda_1^2 - 2\lambda_2^2 - 2\sqrt{\left(\lambda_1^2 + \lambda_2^2\right)^2 - 4\lambda_1\lambda_2\lambda_3 + \lambda_3^2}.$$

To check their positivity, it is enough to have $s_2 \geq 0$ and $s_4 \geq 0$. This means that

$$\lambda_3 \neq \frac{1}{2}; \tag{6.44}$$

$$4\lambda_1^2 + 4\lambda_2^2 + 4\lambda_3^2 \leq 1 + 16\lambda_1\lambda_2\lambda_3; \tag{6.45}$$

$$\lambda_1^2 + \lambda_2^2 + \sqrt{\left(\lambda_1^2 + \lambda_2^2\right)^2 - 4\lambda_1\lambda_2\lambda_3 + \lambda_3^2} \leq \frac{1}{2}. \tag{6.46}$$

Note that the expression standing inside the square root is always positive. Indeed, we have

$$\left(\lambda_1^2 + \lambda_2^2\right)^2 + \lambda_3^2 \geq 2\left(\lambda_1^2 + \lambda_2^2\right)\lambda_3 \geq 2(2\lambda_1\lambda_2)\lambda_3 = 4\lambda_1\lambda_2\lambda_3.$$

Therefore, from (6.43), (6.44), (6.45) and (6.46) one has

$$|\lambda_3| < \frac{1}{2};$$

$$4\lambda_1^2 + 4\lambda_2^2 + 4\lambda_3^2 \leq 1 + 16\lambda_1\lambda_2\lambda_3;$$

$$\lambda_1^2 + \lambda_2^2 + \sqrt{\left(\lambda_1^2 + \lambda_2^2\right)^2 - 4\lambda_1\lambda_2\lambda_3 + \lambda_3^2} \leq \frac{1}{2}.$$

(2). Let $\lambda_3 = \frac{1}{2}$. Then $\hat{\Delta}_{(\lambda_1,\lambda_2,\lambda_3)}$ has the following form

$$\hat{\Delta}_{(\lambda_1,\lambda_2,\frac{1}{2})} = \begin{pmatrix} 2 & 0 & 0 & 0 & 0 & \beta_1 & \beta_1 & 0 \\ 0 & 1 & 0 & 0 & \beta_2 & 0 & 0 & \beta_1 \\ 0 & 0 & 1 & 0 & \beta_2 & 0 & 0 & \beta_1 \\ 0 & 0 & 0 & 0 & 0 & \beta_2 & \beta_2 & 0 \\ 0 & \beta_2 & \beta_2 & 0 & 0 & 0 & 0 & 0 \\ \beta_1 & 0 & 0 & \beta_2 & 0 & 1 & 0 & 0 \\ \beta_1 & 0 & 0 & \beta_2 & 0 & 0 & 1 & 0 \\ 0 & \beta_1 & \beta_1 & 0 & 0 & 0 & 0 & 2 \end{pmatrix},$$

where $\beta_1 = \lambda_1 + \lambda_2$, $\beta_2 = \lambda_1 - \lambda_2$, $\lambda_1, \lambda_2 \in \left[-\frac{1}{2}, \frac{1}{2}\right]$. According to Silvester's criterion, the matrix given above is positive if and only if the leading principal minors are positive. Let $D_n (n = \overline{1, 8})$ be the leading principal minor of $\hat{\Delta}_{(\lambda_1, \lambda_2, \frac{1}{2})}$. One can see that for each $n \in \{1, \ldots, 8\}$, the minor $D_n$ is positive. Hence, if $\lambda_3 = \frac{1}{2}$ then $\hat{\Delta}_{(\lambda_1, \lambda_2, \frac{1}{2})}$ is positive.

(3). Now assume $\lambda_3 = -\frac{1}{2}$. Then one finds

$$\hat{\Delta}_{(\lambda_1, \lambda_2, -\frac{1}{2})} = \begin{pmatrix} 0 & 0 & 0 & 0 & 0 & \beta_1 & \beta_1 & 0 \\ 0 & 1 & 0 & 0 & \beta_2 & 0 & 0 & \beta_2 \\ 0 & 0 & 1 & 0 & \beta_2 & 0 & 0 & \beta_1 \\ 0 & 0 & 0 & 2 & 0 & \beta_2 & \beta_2 & 0 \\ 0 & \beta_2 & \beta_1 & 0 & 2 & 0 & 0 & 0 \\ \beta_1 & 0 & 0 & \beta_2 & 0 & 1 & 0 & 0 \\ \beta_1 & 0 & 0 & \beta_1 & 0 & 0 & 1 & 0 \\ 0 & \beta_1 & \beta_1 & 0 & 0 & 0 & 0 & 0 \end{pmatrix},$$

where as before $\beta_1 = \lambda_1 + \lambda_2$, $\beta_2 = \lambda_1 - \lambda_2$, $\lambda_1, \lambda_2 \in \left[-\frac{1}{2}, \frac{1}{2}\right]$. One can calculate that the principal minors of the last matrix are

$$D_n = 0, \ (n = \overline{1, 5}),$$

$$D_6 = (\lambda_1 + \lambda_2)^2 \left(4(\lambda_1 - \lambda_2)^2 - 4\right),$$

$$D_7 = (\lambda_1 + \lambda_2)^2 \left(8(\lambda_1 - \lambda_2)^2 - 8\right),$$

$$D_8 = 16(\lambda_1 + \lambda_2)^4.$$

It is easy to see that $\hat{\Delta}_{(\lambda_1, \lambda_2, -\frac{1}{2})}$ is positive if $D_6 \geq 0$ and $D_7 \geq 0$. This means that $\lambda_1 = \pm \frac{1}{2}$, $\lambda_2 = \mp \frac{1}{2}$. This completes the proof.

## 6.3 Non-Simple Kadison–Schwarz Type q.q.s.o.s

In this section, we are going to find some conditions for non-simple q.q.s.o.s to be Kadison–Schwarz operators, meaning we consider a q.q.s.o. $\Delta$ with a Haar state $\tau$, i.e. $\Delta$ has the following form

$$\Delta(x) = w_0 \mathbb{1} \otimes \mathbb{1} + \sum_{m,l=1}^{3} \langle \mathbf{b}_{ml}, \overline{\mathbf{w}} \rangle \sigma_m \otimes \sigma_l. \tag{6.47}$$

Before formulating the result, we need the following auxiliary fact.

**Lemma 6.3.1** *Let* $\mathbf{a}, \mathbf{c} \in \mathbb{C}^3$. *Then one has*

$$(\mathbf{a}\sigma) \cdot (\overline{\mathbf{c}}\sigma) - (\mathbf{c}\sigma) \cdot (\overline{\mathbf{a}}\sigma) = \big(\langle \mathbf{a}, \mathbf{c} \rangle - \langle \mathbf{c}, \mathbf{a} \rangle\big)\mathbb{1} + i\big([\mathbf{a}, \overline{\mathbf{c}}] + [\overline{\mathbf{a}}, \mathbf{c}]\big)\sigma, \tag{6.48}$$

$$(\mathbf{a}\sigma) \cdot (\overline{\mathbf{a}}\sigma) = \|\mathbf{a}\|^2 \mathbb{1} + i[\mathbf{a}, \overline{\mathbf{a}}]\sigma. \tag{6.49}$$

The proof is straightforward.

Let us introduce some notation. Given $x = w_0 + \mathbf{w}\sigma$ and a vector $\mathbf{f} \in S$ we define

$$x_{ml} = \langle \mathbf{b}_{ml}, \mathbf{w} \rangle, \quad \mathbf{x}_m = \big(\langle \mathbf{b}_{m1}, \mathbf{w} \rangle, \langle \mathbf{b}_{m2}, \mathbf{w} \rangle, \langle \mathbf{b}_{m3}, \mathbf{w} \rangle\big), \tag{6.50}$$

$$\alpha_{ml} = \langle \mathbf{x}_m, \mathbf{x}_l \rangle - \langle \mathbf{x}_l, \mathbf{x}_m \rangle, \quad \gamma_{ml} = [\mathbf{x}_m, \overline{\mathbf{x}_l}] + [\overline{\mathbf{x}_m}, \mathbf{x}_l], \tag{6.51}$$

$$\mathbf{q}(\mathbf{f}, \mathbf{w}) = \big(\langle \beta(\mathbf{f})_1, [\mathbf{w}, \overline{\mathbf{w}}] \rangle, \langle \beta(\mathbf{f})_2, [\mathbf{w}, \overline{\mathbf{w}}] \rangle, \langle \beta(\mathbf{f})_3, [\mathbf{w}, \overline{\mathbf{w}}] \rangle\big), \tag{6.52}$$

where $\beta(\mathbf{f})_m = \big(\beta(\mathbf{f})_{m1}, \beta(\mathbf{f})_{m2}, \beta(\mathbf{f})_{m3}\big)$ and as before $\mathbf{b}_{ml} = (b_{ml,1}, b_{ml,2}, b_{ml,3})$.

By $\pi$ we denote the mapping $\{1, 2, 3, 4\}$ to $\{1, 2, 3\}$ defined by $\pi(1) = 2$, $\pi(2) = 3$, $\pi(3) = 1$, $\pi(4) = \pi(1)$.

**Theorem 6.3.2** *Let* $\Delta : \mathbb{M}_2(\mathbb{C}) \to \mathbb{M}_2(\mathbb{C}) \otimes \mathbb{M}_2(\mathbb{C})$ *be a q.q.s.o. with a Haar state* $\tau$ *(see (6.47)). Assume that* $\Delta$ *is a Kadison–Schwarz operator. Then the coefficients* $\{b_{ml,k}\}$ *satisfy the following conditions*

$$\|\mathbf{w}\|^2 \geq i\sum_{m=1}^{3} f_m \alpha_{\pi(m), \pi(m+1)} + \sum_{m=1}^{3} \|\mathbf{x}_m\|^2 \tag{6.53}$$

$$\left\| \mathbf{q}(\mathbf{f}, \mathbf{w}) - i\sum_{m=1}^{3} f_m \gamma_{\pi(m), \pi(m+1)} - [\mathbf{x}_m, \overline{\mathbf{x}}_m] \right\| \leq \|\mathbf{w}\|^2 - i\sum_{k=1}^{3} f_k \alpha_{\pi(k), \pi(k+1)}$$

$$- \sum_{m=1}^{3} \|\mathbf{x}_m\|^2 \tag{6.54}$$

*for all* $\mathbf{f} \in S, \mathbf{w} \in \mathbb{C}^3$.

*Proof* Let $x \in \mathbb{M}_2(\mathbb{C})$ be an arbitrary element, i.e. $x = w_0 \mathbb{1} + \mathbf{w} \cdot \sigma$. According to Remark 6.1.1, the operator $\Delta$ has the form (6.47). Therefore, taking into account

(6.25) with (6.50), one finds

$$\Delta(x^*x) = (|w_0|^2 + \|\mathbf{w}\|^2)\mathbb{1} + \sum_{m,l=1}^{3} (\overline{w}_0 \overline{x}_{m,l} + w_0 x_{m,l})\sigma_m \otimes \sigma_l$$

$$+i \sum_{m,l=1}^{3} \langle \mathbf{b}_{m,l}, [\mathbf{w}, \overline{\mathbf{w}}] \rangle \sigma_m \otimes \sigma_l, \qquad (6.55)$$

$$\Delta(x)^*\Delta(x) = |w_0|^2 \mathbb{1} + \sum_{m,l=1}^{3} (\overline{w}_0 \overline{x}_{m,l} + w_0 x_{m,l})\sigma_m \otimes \sigma_l$$

$$+\left( \sum_{m,l=1}^{3} x_{m,l}\sigma_m \otimes \sigma_l \right)\left( \sum_{m,l=1}^{3} \overline{x_{m,l}}\sigma_m \otimes \sigma_l \right). \qquad (6.56)$$

Noting that $\mathbf{x}_m = (x_{m1}, x_{m2}, x_{m3})$, $m = 1, 2, 3$ we rewrite the last term of the equality (6.56) as follows

$$\left( \sum_{m,l=1}^{3} x_{m,l}\sigma_m \otimes \sigma_l \right)\left( \sum_{m,l=1}^{3} \overline{x_{m,l}}\sigma_m \otimes \sigma_l \right) = \left( \sum_{m=1}^{3} \sigma_m \otimes (\mathbf{x}_m\sigma) \right)\left( \sum_{m=1}^{3} \sigma_m \otimes (\overline{\mathbf{x}}_m\sigma) \right)$$

$$= \mathbb{1} \otimes \sum_{k=1}^{3} (\mathbf{x}_k\sigma) \cdot (\overline{\mathbf{x}}_k\sigma)$$

$$+i\sigma_1 \otimes \left( (\mathbf{x}_2\sigma) \cdot (\overline{\mathbf{x}}_3\sigma) - (\mathbf{x}_3\sigma) \cdot (\overline{\mathbf{x}}_2\sigma) \right)$$

$$+i\sigma_2 \otimes \left( (\mathbf{x}_3\sigma) \cdot (\overline{\mathbf{x}}_1\sigma) - (\mathbf{x}_1\sigma) \cdot (\overline{\mathbf{x}}_3\sigma) \right)$$

$$+i\sigma_3 \otimes \left( (\mathbf{x}_1\sigma) \cdot (\overline{\mathbf{x}}_2\sigma) - (\mathbf{x}_2\sigma) \cdot (\overline{\mathbf{x}}_1\sigma) \right).$$

According to Lemma 6.3.1 and (6.51), the last equality can be rewritten as

$$X := \mathbb{1} \otimes \left( \sum_{j=1}^{3} \|\mathbf{x}_j\|^2 \mathbb{1} + i \sum_{j=1}^{3} [\mathbf{x}_j, \overline{\mathbf{x}}_j]\sigma) \right)$$

$$+i \sum_{m=1}^{3} \sigma_m \otimes \left( \alpha_{\pi(m),\pi(m+1)} \mathbb{1} + i\gamma_{\pi(m),\pi(m+1)}\sigma \right). \qquad (6.57)$$

Then from (6.55) and (6.56) one gets

$$\Delta(x^*x) - \Delta(x)^*\Delta(x) = \|\mathbf{w}\|^2 \mathbb{1} + \sum_{m,l=1}^{3} \langle \mathbf{b}_{ml}, [\mathbf{w}, \overline{\mathbf{w}}] \rangle \sigma_m \otimes \sigma_l - X. \qquad (6.58)$$

Now taking an arbitrary state $\varphi \in S(\mathbb{M}_2(\mathbb{C}))$ and applying $E_\varphi$ to (6.58) we have

$$E_\varphi(\Delta(x^*x) - \Delta(x)^*\Delta(x)) = \|w\|^2 \mathbb{1} + i \sum_{m,l=1}^{3} \langle \mathbf{b}_{ml}, [\mathbf{w}, \overline{\mathbf{w}}]\rangle f_m \sigma_l$$
$$-E_\varphi(X), \tag{6.59}$$

where $\varphi(\sigma_m) = f_m$.

From (6.57), one immediately finds

$$E_\varphi(X) = \sum_{m=1}^{3} \|\mathbf{x}_m\|^2 \mathbb{1} + i \sum_{m=1}^{3} [\mathbf{x}_m, \overline{\mathbf{x}}_m]\sigma$$
$$+ i \sum_{m=1}^{3} f_m(\alpha_{\pi(m),\pi(m+1)} \mathbb{1} + i\gamma_{\pi(m),\pi(m+1)}\sigma). \tag{6.60}$$

Now substituting the last equality into (6.59) together with (6.52), we obtain

$$E_\varphi(\Delta(x^*x) - \Delta(x)^*\Delta(x)) = \left( \|\mathbf{w}\|^2 - i \sum_{m=1}^{3} f_m \alpha_{\pi(m),\pi(m+1)} - \sum_{m=1}^{3} \|\mathbf{x}_m\|^2 \right) \mathbb{1}$$
$$+ i \left( \mathbf{q}(\mathbf{f}, \mathbf{w}) - i \sum_{m=1}^{3} f_m \gamma_{\pi(m),\pi(m+1)} - [\mathbf{x}_m, \overline{\mathbf{x}}_m] \right) \sigma.$$

So, thanks to Lemma 6.1.1, the right-hand side of (6.61) is positive if and only if (6.53) and (6.54) are satisfied for all $\mathbf{f} \in S, \mathbf{w} \in \mathbb{C}^3$. Note that the numbers $\alpha_{ml}$ are skew-symmetric, i.e. $\overline{\alpha_{ml}} = -\alpha_{ml}$, therefore, the equality (6.53) has a meaning.

Let us define

$$\mathbf{h}(\mathbf{w}) = \big(\langle \mathbf{b}_{11}, [\mathbf{w}, \overline{\mathbf{w}}]\rangle, \langle \mathbf{b}_{12}, [\mathbf{w}, \overline{\mathbf{w}}]\rangle, \langle \mathbf{b}_{13}, [\mathbf{w}, \overline{\mathbf{w}}]\rangle\big).$$

Then one has the following

**Corollary 6.3.3** *Let* $\Delta : \mathbb{M}_2(\mathbb{C}) \to \mathbb{M}_2(\mathbb{C}) \otimes \mathbb{M}_2(\mathbb{C})$ *be a Kadison–Schwarz operator given by (6.47), then the coefficients* $\{b_{ml,k}\}$ *satisfy the following conditions*

$$\sum_{m=1}^{3} \|\mathbf{x}_m\|^2 + i\alpha_{2,3} \leq \|\mathbf{w}\|^2, \tag{6.61}$$

$$\left\| h(\mathbf{w}) - i\gamma_{2,3} + i\sum_{m=1}^{3} [\mathbf{x}_m, \overline{\mathbf{x}}_m] \right\| \leq \|\mathbf{w}\|^2 - i\alpha_{2,3} - \sum_{m=1}^{3} \|\mathbf{x}_m\|^2, \tag{6.62}$$

*for all* $\mathbf{w} \in \mathbb{C}^3$.

The proof immediately follows from the previous Theorem 6.3.2 by taking $\mathbf{f} = (1, 0, 0)$ in (6.53) and (6.54).

*Remark 6.3.1* The provided characterization with [145] allows us to construct examples of positive or Kadison–Schwarz operators which are not completely positive (see Sect. 4.3).

Now we are going to give a general characterization of KS-operators. But let us first give some notation. For a given mapping $\Delta : M_2(\mathbb{C}) \to M_2(\mathbb{C}) \otimes M_2(\mathbb{C})$, by $\Delta(\sigma)$ we denote the vector $(\Delta(\sigma_1), \Delta(\sigma_2), \Delta(\sigma_3))$, and by $\mathbf{w}\Delta(\sigma)$ we mean the following

$$\mathbf{w}\Delta(\sigma) = w_1\Delta(\sigma_1) + w_2\Delta(\sigma_2) + w_3\Delta(\sigma_3), \tag{6.63}$$

where $\mathbf{w} \in \mathbb{C}^3$. Note that the last equality (6.63), due to the linearity of $\Delta$, can also be written as $\mathbf{w}\Delta(\sigma) = \Delta(\mathbf{w}\sigma)$.

**Theorem 6.3.4** *Let $\Delta : M_2(\mathbb{C}) \to M_2(\mathbb{C}) \otimes M_2(\mathbb{C})$ be a unital $*$-preserving linear mapping. Then $\Delta$ is a KS-operator if and only if one has*

$$i[\mathbf{w}, \overline{\mathbf{w}}]\Delta(\sigma) + (\mathbf{w}\Delta(\sigma))(\overline{\mathbf{w}}\Delta(\sigma)) \leq \mathbb{1} \otimes \mathbb{1}, \tag{6.64}$$

*for all $\mathbf{w} \in \mathbb{C}^3$ with $\|\mathbf{w}\| = 1$.*

*Proof* Let $x \in M_2(\mathbb{C})$ be an arbitrary element, i.e. $x = w_0\mathbb{1} + \mathbf{w}\sigma$. Then from (6.25), we have

$$\Delta(x) = w_0\mathbb{1} \otimes \mathbb{1} + \mathbf{w}\Delta(\sigma), \quad \Delta(x^*) = \overline{w_0}\mathbb{1} \otimes \mathbb{1} + \overline{\mathbf{w}}\Delta(\sigma), \tag{6.65}$$

$$\Delta(x^*x) = \left(|w_0|^2 + \|\mathbf{w}\|^2\right)\mathbb{1} \otimes \mathbb{1} + \left(w_0\overline{\mathbf{w}} + \overline{w_0}\mathbf{w} - i[\mathbf{w}, \overline{\mathbf{w}}]\right)\Delta(\sigma), \tag{6.66}$$

$$\Delta(x)^*\Delta(x) = |w_0|^2\mathbb{1} \otimes \mathbb{1} + \left(w_0\overline{\mathbf{w}} + \overline{w_0}\mathbf{w}\right)\Delta(\sigma) + (\mathbf{w}\Delta(\sigma))(\overline{\mathbf{w}}\Delta(\sigma)). \tag{6.67}$$

From (6.65)–(6.67) one gets

$$\Delta(x^*x) - \Delta(x)^*\Delta(x) = \|\mathbf{w}\|^2\mathbb{1} \otimes \mathbb{1} - i[\mathbf{w}, \overline{\mathbf{w}}]\Delta(\sigma) - (\mathbf{w}\Delta(\sigma))(\overline{\mathbf{w}}\Delta(\sigma)).$$

So, the positivity of the last equality implies that

$$\|\mathbf{w}\|^2\mathbb{1} \otimes \mathbb{1} - i[\mathbf{w}, \overline{\mathbf{w}}]\Delta(\sigma) - (\mathbf{w}\Delta(\sigma))(\overline{\mathbf{w}}\Delta(\sigma)) \geq 0.$$

Now dividing both sides by $\|\mathbf{w}\|^2$ we get the required inequality. Hence, this completes the proof.

## 6.4   An Example of a Non-Simple q.q.s.o. Which Is Not Kadison–Schwarz

In this section we are going to provide an example of a q.q.s.o. which is not a Kadison–Schwarz operator. Let us consider the matrix $\{b_{ij,k}\}$ given by:

$$b_{11,1} = \varepsilon; \quad b_{11,2} = 0; \quad b_{11,3} = 0;$$

$$b_{12,1} = 0; \quad b_{12,2} = 0; \quad b_{12,3} = \varepsilon;$$

$$b_{13,1} = 0; \quad b_{13,2} = \varepsilon; \quad b_{13,3} = 0;$$

$$b_{22,1} = 0; \quad b_{22,2} = \varepsilon; \quad b_{22,3} = 0;$$

$$b_{23,1} = \varepsilon; \quad b_{23,2} = 0; \quad b_{23,3} = 0;$$

$$b_{33,1} = 0; \quad b_{33,2} = 0; \quad b_{33,3} = \varepsilon;$$

and $b_{ij,k} = b_{ji,k}$.

By (6.16) we define a linear operator $\Delta_\varepsilon$, for which $\tau$ is a Haar state. In the sequel, we would like to find some conditions on $\varepsilon$ which ensures the positivity of $\Delta_\varepsilon$.

For the given $\{b_{ijk}\}$ one can easily find the form of $\Delta_\varepsilon$ as follows

$$\Delta_\varepsilon(x) = w_0 \mathbb{1} \otimes \mathbb{1} + \varepsilon \omega_1 \sigma_1 \otimes \sigma_1 + \varepsilon \omega_3 \sigma_1 \otimes \sigma_2 + \varepsilon \omega_2 \sigma_1 \otimes \sigma_3$$

$$+ \varepsilon \omega_3 \sigma_2 \otimes \sigma_1 + \varepsilon \omega_2 \sigma_2 \otimes \sigma_2 + \varepsilon \omega_1 \sigma_2 \otimes \sigma_3$$

$$+ \varepsilon \omega_2 \sigma_3 \otimes \sigma_1 + \varepsilon \omega_1 \sigma_3 \otimes \sigma_2 + \varepsilon \omega_3 \sigma_3 \otimes \sigma_3, \tag{6.68}$$

where as before $x = w_0 \mathbb{1} + \mathbf{w}\sigma$.

**Theorem 6.4.1** *A linear operator $\Delta_\varepsilon$ given by (6.68) is a q.q.s.o. if and only if $|\varepsilon| \le \frac{1}{3}$.*

*Proof* Let $x = w_0 \mathbb{1} + \mathbf{w}\sigma$ be a positive element from $\mathbb{M}_2(\mathbb{C})$. Let us show the positivity of the matrix $\Delta_\varepsilon(x)$. To do this, we rewrite (6.68) as $\Delta_\varepsilon(x) = w_0 \mathbb{1} + \varepsilon \mathbf{D}$. Here

$$\mathbf{D} = \begin{pmatrix} \omega_3 & \omega_2 - i\omega_1 & \omega_2 - i\omega_1 & \omega_1 - 2i\omega_3 - \omega_2 \\ \omega_2 + i\omega_1 & -\omega_3 & \omega_1 + \omega_2 & -\omega_2 + i\omega_1 \\ \omega_2 + i\omega_1 & \omega_1 + \omega_2 & -\omega_3 & -\omega_2 + i\omega_1 \\ \omega_1 + 2i\omega_3 - \omega_2 & -\omega_2 - i\omega_1 & -\omega_2 - i\omega_1 & \omega_3 \end{pmatrix},$$

where the positivity of $x$ yields that $w_0, \omega_1, \omega_2, \omega_3$ are real numbers. In what follows, without loss of generality, we may assume that $w_0 = 1$, and therefore $\|\mathbf{w}\| \le 1$. It is known that the positivity of $\Delta_\varepsilon(x)$ is equivalent to the positivity of the eigenvalues of $\Delta_\varepsilon(x)$.

Let us first examine the eigenvalues of $\mathbf{D}$. Simple algebra shows us that all eigenvalues of $\mathbf{D}$ can be written as follows

$$\lambda_1(\mathbf{w}) = \omega_1 + \omega_2 + \omega_3 + 2\sqrt{\omega_1^2 + \omega_2^2 + \omega_3^2 - \omega_1\omega_2 - \omega_1\omega_3 - \omega_2\omega_3},$$

$$\lambda_2(\mathbf{w}) = \omega_1 + \omega_2 + \omega_3 - 2\sqrt{\omega_1^2 + \omega_2^2 + \omega_3^2 - \omega_1\omega_2 - \omega_1\omega_3 - \omega_2\omega_3},$$

$$\lambda_3(\mathbf{w}) = \lambda_4(\mathbf{w}) = -\omega_1 - \omega_2 - \omega_3.$$

Now we examine the maximum and minimum values of the functions $\lambda_1(\mathbf{w}), \lambda_2(\mathbf{w}), \lambda_3(\mathbf{w}), \lambda_4(\mathbf{w})$ on the ball $\|\mathbf{w}\| \le 1$.

One can see that

$$|\lambda_3(\mathbf{w})| = |\lambda_4(\mathbf{w})| \le \sum_{k=1}^{3} |\omega_k| \le \sqrt{3} \sum_{k=1}^{3} |\omega_k|^2 \le \sqrt{3}. \tag{6.69}$$

Note that the functions $\lambda_3, \lambda_4$ attain the values $\pm\sqrt{3}$ at $\pm(1/\sqrt{3}, 1/\sqrt{3}, 1/\sqrt{3})$. Now let us rewrite $\lambda_1(\mathbf{w})$ and $\lambda_2(\mathbf{w})$ as follows

$$\lambda_1(\mathbf{w}) = \omega_1 + \omega_2 + \omega_3 + \frac{2}{\sqrt{2}}\sqrt{3(\omega_1^2 + \omega_2^2 + \omega_3^2) - (\omega_1 + \omega_2 + \omega_3)^2},$$
$$\tag{6.70}$$

$$\lambda_2(\mathbf{w}) = \omega_1 + \omega_2 + \omega_3 - \frac{2}{\sqrt{2}}\sqrt{3(\omega_1^2 + \omega_2^2 + \omega_3^2) - (\omega_1 + \omega_2 + \omega_3)^2}. \tag{6.71}$$

One can see that

$$\lambda_k(h\omega_1, h\omega_2, h\omega_3) = h\lambda_k(\omega_1, \omega_2, w_3), \quad \text{if } h \ge 0, \tag{6.72}$$

$$\lambda_1(h\omega_1, h\omega_2, h\omega_3) = h\lambda_2(\omega_1, \omega_2, w_3), \quad \text{if } h \le 0, \tag{6.73}$$

where $k = 1, 2$. Therefore, the functions $\lambda_k(\mathbf{w})$, $k = 1, 2$, reach their maximum and minimum on the sphere $\omega_1^2 + \omega_2^2 + \omega_3^2 = 1$ (i.e. $\|\mathbf{w}\| = 1$). Hence, with $t = \omega_1 + \omega_2 + \omega_3$, from (6.72) and (6.71) we introduce the following functions

$$g_1(t) = t + \frac{2}{\sqrt{2}}\sqrt{3 - t^2}, \quad g_2(t) = t - \frac{2}{\sqrt{2}}\sqrt{3 - t^2}$$

where $|t| \le \sqrt{3}$.

One can find that the critical values of $g_1$ are $t = \pm 1$, and the critical value of $g_2$ is $t = -1$. Consequently, the extremal values of $g_1$ and $g_2$ on $|t| \le \sqrt{3}$ are the following:

$$\min_{|t| \leq \sqrt{3}} g_1(t) = -\sqrt{3}, \quad \max_{|t| \leq \sqrt{3}} g_1(t) = 3,$$

$$\min_{|t| \leq \sqrt{3}} g_2(t) = -3, \quad \max_{|t| \leq \sqrt{3}} g_2(t) = \sqrt{3}.$$

Therefore, from (6.72) and (6.73) we conclude that

$$-3 \leq \lambda_k(\mathbf{w}) \leq 3, \quad \text{for any } \|\mathbf{w}\| \leq 1, \ k = 1, 2. \tag{6.74}$$

It is known that for the spectrum of $\mathbb{1} + \varepsilon \mathbf{D}$ one has

$$Sp(\mathbb{1} + \varepsilon \mathbf{D}) = 1 + \varepsilon Sp(\mathbf{D}).$$

Therefore,

$$Sp(\mathbb{1} + \varepsilon \mathbf{D}) = \{1 + \varepsilon \lambda_k(\mathbf{w}) : k = \overline{1, 4}\}.$$

So, if

$$|\varepsilon| \leq \frac{1}{\max\limits_{\|\mathbf{w}\| \leq 1} |\lambda_k(\mathbf{w})|}, \quad k = \overline{1, 4}$$

then one can see that $1 + \varepsilon \lambda_k(\mathbf{w}) \geq 0$ for all $\|\mathbf{w}\| \leq 1$ $k = \overline{1, 4}$. This implies that the matrix $\mathbb{1} + \varepsilon \mathbf{D}$ is positive for all $\mathbf{w}$ with $\|\mathbf{w}\| \leq 1$.

Now assume that $\Delta_\varepsilon$ is positive. Then $\Delta_\varepsilon(x)$ is positive whenever $x$ is positive. This means that $1 + \varepsilon \lambda_k(\mathbf{w}) \geq 0$ for all $\|\mathbf{w}\| \leq 1$ ($k = \overline{1, 4}$). From (6.69) and (6.74) we conclude that $|\varepsilon| \leq 1/3$. This completes the proof.

**Theorem 6.4.2** *Let* $\varepsilon = \frac{1}{3}$. *Then the corresponding q.q.s.o.* $\Delta_\varepsilon$ *is not a KS-operator.*

*Proof* It is enough to show the dissatisfaction of (6.54) at some values of $\mathbf{w}$ ($\|\mathbf{w}\| \leq 1$) and $\mathbf{f} = (f_1, f_1, f_2)$.

Assume that $\mathbf{f} = (1, 0, 0)$, then a little algebra shows that (6.54) reduces to the following

$$\sqrt{A + B + C} \leq D, \tag{6.75}$$

where

$$A = |\varepsilon(\overline{\omega}_2 \omega_3 - \overline{\omega}_3 \omega_2) - i\varepsilon^2(2\overline{\omega}_2 \omega_3 - 2|\omega_1|^2 - \overline{\omega}_2 \omega_1 + \overline{\omega}_1 \omega_2 - \overline{\omega}_1 \omega_3 + \overline{\omega}_3 \omega_1)|^2,$$

$$B = |\varepsilon(\overline{\omega}_1 \omega_2 - \overline{\omega}_2 \omega_1) - i\varepsilon^2(2\overline{\omega}_1 \omega_2 - 2|\omega_3|^2 - \overline{\omega}_1 \omega_3 + \overline{\omega}_3 \omega_1 - \overline{\omega}_3 \omega_2 + \overline{\omega}_2 \omega_3)|^2,$$

$$C = |\varepsilon(\overline{\omega}_3\omega_1 - \overline{\omega}_1\omega_3) - i\varepsilon^2(2\overline{\omega}_3\omega_1 - 2|\omega_2|^2 - \overline{\omega}_3\omega_2 + \overline{\omega}_2\omega_3 - \overline{\omega}_2\omega_1 + \overline{\omega}_1\omega_2)|^2,$$

$$D = (1 - 3|\varepsilon|^2)(|\omega_1|^2 + |\omega_2|^2 + |\omega_3|^2)$$
$$-i\varepsilon^2(\overline{\omega}_3 w_2 - \overline{\omega}_2\omega_3 + \overline{\omega}_2\omega_1 - \overline{\omega}_1\omega_2 + \overline{\omega}_1\omega_3 - \overline{\omega}_3\omega_1).$$

Now choose **w** as follows:

$$\omega_1 = -\frac{1}{9}; \qquad \omega_2 = \frac{5}{36}; \qquad \omega_3 = \frac{5i}{27}.$$

Then calculations show that

$$A = \frac{9594}{19131876}; \qquad B = \frac{19625}{86093442};$$

$$C = \frac{1625}{3779136}; \qquad D = \frac{589}{17496}.$$

Hence, we find

$$\sqrt{\frac{9594}{19131876} + \frac{19625}{86093442} + \frac{1625}{3779136}} > \frac{589}{17496},$$

which means that (6.75) is not satisfied. Hence, $\Delta_\varepsilon$ is not a KS-operator at $\varepsilon = 1/3$.

**Theorem 6.4.3** *Let* $\Delta_\varepsilon : M_2(\mathbb{C}) \to M_2(\mathbb{C}) \otimes M_2(\mathbb{C})$ *be given by (6.68). Then* $\Delta_\varepsilon$ *is completely positive if and only if* $|\varepsilon| \leq \frac{1}{3\sqrt{3}}$.

*Proof* To check the complete positivity of $\Delta_\varepsilon$ we need to show the positivity of the following matrix (see [26])

$$\hat{\Delta}_\varepsilon = \begin{pmatrix} \Delta_\varepsilon(e_{11}) & \Delta_\varepsilon(e_{12}) \\ \Delta_\varepsilon(e_{21}) & \Delta_\varepsilon(e_{22}) \end{pmatrix}.$$

Here, as before, the $e_{ij}$ are the matrix units in $M_2(\mathbb{C})$.
   From (6.68), one can calculate that

$$\Delta_\varepsilon(e_{11}) = \frac{1}{2}\mathbb{1} \otimes \mathbb{1} + \varepsilon D_{11}, \quad \Delta_\varepsilon(e_{22}) = \frac{1}{2}\mathbb{1} \otimes \mathbb{1} - \varepsilon D_{11},$$

$$\Delta_\varepsilon(e_{12}) = \varepsilon D_{12}, \quad \Delta_\varepsilon(e_{21}) = \varepsilon D_{12}^*,$$

where

$$D_{11} = \begin{pmatrix} \frac{1}{2} & 0 & 0 & -i \\ 0 & -\frac{1}{2} & 0 & 0 \\ 0 & 0 & -\frac{1}{2} & 0 \\ i & 0 & 0 & \frac{1}{2} \end{pmatrix}, \quad D_{12} = \begin{pmatrix} 0 & 0 & 0 & \frac{1-i}{2} \\ i & 0 & \frac{1+i}{2} & 0 \\ i & \frac{1+i}{2} & 0 & 0 \\ \frac{1-i}{2} & -i & -i & 0 \end{pmatrix}.$$

Hence, we find

$$2\hat{\Delta}_\varepsilon = \mathbb{1}_8 + \varepsilon\mathbb{D},$$

where $\mathbb{1}_8$ is the unit matrix in $\mathbb{M}_8(\mathbb{C})$ and

$$\mathbb{D} = \begin{pmatrix} 1 & 0 & 0 & -2i & 0 & 0 & 0 & 1-i \\ 0 & -1 & 0 & 0 & 2i & 0 & 1+i & 0 \\ 0 & 0 & -1 & 0 & 2i & 1+i & 0 & 0 \\ 2i & 0 & 0 & 1 & 1-i & -2i & -2i & 0 \\ 0 & -2i & -2i & 1+i & -1 & 0 & 0 & 2i \\ 0 & 0 & 1-i & 2i & 0 & 1 & 0 & 0 \\ 0 & 1-i & 0 & 2i & 0 & 0 & 1 & 0 \\ 1+i & 0 & 0 & 0 & -2i & 0 & 0 & -1 \end{pmatrix}.$$

So, the matrix $\hat{\Delta}_\varepsilon$ is positive if and only if

$$|\varepsilon| \le \frac{1}{\lambda_{\max}(\mathbb{D})},$$

where $\lambda_{\max}(\mathbb{D}) = \max\limits_{\lambda \in Sp(\mathbb{D})} |\lambda|$. One can easily calculate that $\lambda_{\max}(\mathbb{D}) = 3\sqrt{3}$. This completes the proof.

## 6.5  The Dynamics of the Quadratic Operator Associated with $\Delta_\varepsilon$

In this section, we are going to study the dynamics of the quadratic operator $V_\Delta$ associated with a q.q.s.o. $\Delta$ defined on $\mathbb{M}_2(\mathbb{C})$.

**Proposition 6.5.1** *Let* $\Delta : \mathbb{M}_2(\mathbb{C}) \to \mathbb{M}_2(\mathbb{C}) \otimes \mathbb{M}_2(\mathbb{C})$ *be a linear operator with a Haar state* $\tau$. *Then* $\Delta^*(\varphi \otimes \psi) \in S(\mathbb{M}_2(\mathbb{C}))$ *for any* $\varphi, \psi \in S(\mathbb{M}_2(\mathbb{C}))$ *if and only if*

$$\sum_{k=1}^{3} \left| \sum_{i,j=1}^{3} b_{ij,k} f_i p_j \right|^2 \le 1 \quad \text{for all } \mathbf{f}, \mathbf{p} \in \mathbf{S}. \tag{6.76}$$

*Proof* Take arbitrary states $\varphi, \psi \in S(\mathbb{M}_2(\mathbb{C}))$ and let $\mathbf{f}, \mathbf{p} \in \mathbf{S}$ be the corresponding vectors (see (6.1)). Then from (6.10), one finds

$$\Delta^*(\varphi \otimes \psi)(\sigma_k) = \sum_{i,j=1}^{3} b_{ij,k} f_i p_j, \quad k = 1, 2, 3.$$

Due to Lemma 6.1.1(d), the functional $\Delta^*(\varphi \otimes \psi)$ is a state if and only if the vector

$$\mathbf{f}_{\Delta^*(\varphi,\psi)} = \left( \sum_{i,j=1}^{3} b_{ij,1} f_i p_j, \sum_{i,j=1}^{3} b_{ij,2} f_i p_j, \sum_{i,j=1}^{3} b_{ij,3} f_i p_j \right)$$

satisfies $\|\mathbf{f}_{\Delta^*(\varphi,\psi)}\| \leq 1$, which is the required assertion.

*Remark 6.5.1* The positivity of $\Delta^*(\varphi \otimes \psi)$ for all states $\varphi$ and $\psi$ is called the *block positivity* of the operator $\Delta$. It is clear that the positivity of $\Delta$ implies the block positivity, but the reverse is not true (see Remark 6.5.2).

From the proof of Corollary 6.1.4 and 6.5.1 we get the following

**Corollary 6.5.2** *Let $\mathbb{B}(\mathbf{f})$ be the matrix corresponding to an operator given by (6.10). Then $\|\|\mathbb{B}\|\| \leq 1$ if and only if (6.76) is satisfied.*

Let us find a sufficient condition for the coefficients $\{b_{ij,k}\}$ to satisfy (6.76).

**Corollary 6.5.3** *Let*

$$\sum_{i,j,k=1}^{3} |b_{ij,k}|^2 \leq 1 \tag{6.77}$$

*be satisfied, then (6.76) holds.*

*Proof* Let (6.77) be satisfied. Take any $\mathbf{f}, \mathbf{p} \in S$, then

$$\left| \sum_{i,j=1}^{3} b_{ij,k} f_i p_j \right|^2 \leq \left( \sum_{i,j=1}^{3} |b_{ij,k}| |f_i p_j| \right)^2$$

$$\leq \sum_{i,j=1}^{3} |b_{ij,k}|^2 \sum_{i=1}^{3} |f_i|^2 \sum_{j=1}^{3} |p_j|^2$$

$$\leq \sum_{i,j=1}^{3} |b_{ij,k}|^2,$$

which implies the assertion.

Let us consider the quadratic operator defined by $V_\Delta(\varphi) = \Delta^*(\varphi \otimes \varphi)$, $\varphi \in S(\mathbb{M}_2(\mathbb{C}))$. According to Proposition 6.5.1 and Corollary 6.5.2 we conclude that the operator $V_\Delta$ maps $S(\mathbb{M}_2(\mathbb{C}))$ into itself if and only if $\|\|\mathbb{B}\|\| \leq 1$. To study the dynamics of $V_\Delta$ on $S(\mathbb{M}_2(\mathbb{C}))$ it is enough to investigate the behavior of the corresponding vector $\mathbf{f}_{V_\Delta(\varphi)}$ in $\mathbb{R}^3$. Therefore, from (6.10) we find that

$$V_\Delta(\varphi)(\sigma_k) = \sum_{i,j=1}^{3} b_{ij,k} f_i f_j, \quad \mathbf{f} \in \mathbf{S}.$$

This suggests that we should consider the nonlinear operator $V : \mathbf{S} \to \mathbf{S}$ defined by

$$V(\mathbf{f})_k = \sum_{i,j=1}^{3} b_{ij,k} f_i f_j, \quad k = 1, 2, 3, \tag{6.78}$$

where $\mathbf{f} = (f_1, f_2, f_3) \in \mathbf{S}$. Here, as before, $\mathbf{S} = \{\mathbf{f} = (f_1, f_2, f_3) \in \mathbb{R}^3 : f_1^2 + f_2^2 + f_3^2 \le 1\}$.

In this section, we are going to study the dynamics of the quadratic operator $V_\varepsilon$ corresponding to $\Delta_\varepsilon$ (see (6.68)), which has the following form

$$\begin{cases} V_\varepsilon(f)_1 = \varepsilon(f_1^2 + 2f_2 f_3), \\ V_\varepsilon(f)_2 = \varepsilon(f_2^2 + 2f_1 f_3), \\ V_\varepsilon(f)_3 = \varepsilon(f_3^2 + 2f_1 f_2). \end{cases} \tag{6.79}$$

Let us first find some condition on $\varepsilon$ which ensures (6.76).

**Proposition 6.5.4** *Let $V_\varepsilon$ be given by (6.79). Then $V_\varepsilon$ maps $\mathbf{S}$ into itself if and only if $|\varepsilon| \le \frac{1}{\sqrt{3}}$ is satisfied.*

*Proof* "If" part. Assume that $V_\varepsilon$ maps $\mathbf{S}$ into itself. Then (6.76) is satisfied. Take $\mathbf{f} = (1/\sqrt{3}, 1/\sqrt{3}, 1/\sqrt{3})$, $\mathbf{p} = \mathbf{f}$. Then from (6.76) one finds

$$\sum_{k=1}^{3} \left| \sum_{i,j=1}^{3} b_{ij,k} f_i p_j \right|^2 = 3\varepsilon^2 \le 1,$$

which yields $|\varepsilon| \le 1/\sqrt{3}$.

"Only if" part. Assume that $|\varepsilon| \le 1/\sqrt{3}$. Take any $\mathbf{f} = (f_1, f_2, f_3)$, $\mathbf{p} = (p_1, p_2, p_3) \in \mathbf{S}$. Then one finds

$$\sum_{k=1}^{3} \left| \sum_{i,j=1}^{3} b_{ij,k} f_i p_j \right|^2$$

$$= \varepsilon^2 (|f_1 p_1 + f_3 p_2 + f_2 p_3|^2 + |f_3 p_1 + f_2 p_2 + f_1 p_3|^2 + |f_2 p_1 + f_1 p_2 + f_3 p_3|^2)$$

$$\le \varepsilon^2 ((f_1^2 + f_2^2 + f_3^2)(p_1^2 + p_2^2 + p_3^2) + (f_3^2 + f_2^2 + f_1^2)(p_1^2 + p_2^2 + p_3^2)$$

$$+(p_1^2 + p_2^2 + p_3^2)(f_2^2 + f_1^2 + f_3^2))$$

$$\leq \varepsilon^2(1 + 1 + 1) = 3\varepsilon^2 \leq 1.$$

This completes the proof.

From this proposition, we immediately get the following important corollary.

**Corollary 6.5.5** *Let $\Delta_\varepsilon$ be given by (6.68). Then $\Delta_\varepsilon$ is block positive if and only if $|\varepsilon| \leq 1/\sqrt{3}$.*

*Remark 6.5.2* As previously mentioned, the condition (6.76) is necessary for $\Delta$ to be a positive operator. Indeed, from Theorem 6.4.1 and Corollary 6.5.5, we conclude that if $\varepsilon \in (\frac{1}{3}, \frac{1}{\sqrt{3}}]$ then the operator $\Delta_\varepsilon$ is not positive, while $\Delta$ is block positive. Other examples of block positive but not positive operators can be found in [30].

In what follows, to study the dynamics of $V_\varepsilon$ we assume $|\varepsilon| \leq \frac{1}{\sqrt{3}}$. Recall that a vector $\mathbf{f} \in \mathbf{S}$ is a fixed point of $V_\varepsilon$ if $V_\varepsilon(\mathbf{f}) = \mathbf{f}$. Clearly $(0, 0, 0)$ is a fixed point of $V_\varepsilon$. The next result describes other fixed points.

**Proposition 6.5.6** *If $|\varepsilon| < \frac{1}{\sqrt{3}}$ then $V_\varepsilon$ has a unique fixed point $(0, 0, 0)$ in $\mathbf{S}$. If $|\varepsilon| = \frac{1}{\sqrt{3}}$ then $V_\varepsilon$ has the fixed points $(0, 0, 0)$ and $(\pm\frac{1}{\sqrt{3}}, \pm\frac{1}{\sqrt{3}}, \pm\frac{1}{\sqrt{3}})$ in $\mathbf{S}$.*

*Proof* To find the fixed points of $V$, we need to solve the following equation

$$\begin{cases} \varepsilon(f_1^2 + 2f_2f_3) = f_1, \\ \varepsilon(f_2^2 + 2f_1f_3) = f_2, \\ \varepsilon(f_3^2 + 2f_1f_2) = f_3. \end{cases} \tag{6.80}$$

If $f_k = 0$ for some $k \in \{1, 2, 3\}$, then due to $|\varepsilon| \leq \frac{1}{\sqrt{3}}$, one can see that the only solution of (6.80) belonging to $\mathbf{S}$ is $f_1 = f_2 = f_3 = 0$. Therefore, we assume that $f_k \neq 0$ ($k = 1, 2, 3$). So, from (6.80), one finds

$$\begin{cases} \frac{f_1^2 + 2f_2f_3}{f_2^2 + 2f_1f_3} = \frac{f_1}{f_2}, \\ \frac{f_1^2 + 2f_2f_3}{f_3^2 + 2f_1f_2} = \frac{f_1}{f_3}, \\ \frac{f_2^2 + 2f_1f_3}{f_3^2 + 2f_1f_2} = \frac{f_2}{f_3}. \end{cases} \tag{6.81}$$

Defining

$$x = \frac{f_1}{f_2}, \ y = \frac{f_1}{f_3}, \ z = \frac{f_2}{f_3}, \tag{6.82}$$

from (6.81) it follows that

$$
\begin{cases}
x\left(\dfrac{x\left(1+\frac{2}{xy}\right)}{1+\frac{2x}{z}} - 1\right) = 0, \\[3ex]
y\left(\dfrac{y\left(1+\frac{2}{xy}\right)}{1+2yz} - 1\right) = 0, \\[3ex]
z\left(\dfrac{z\left(1+\frac{2x}{z}\right)}{1+2yz} - 1\right) = 0.
\end{cases}
\tag{6.83}
$$

According to our assumption $x, y, z$ are nonzero, so from (6.83) one gets

$$
\begin{cases}
\dfrac{x\left(1+\frac{2}{xy}\right)}{1+\frac{2x}{z}} = 1, \\[3ex]
\dfrac{y\left(1+\frac{2}{xy}\right)}{1+2yz} = 1, \\[3ex]
\dfrac{z\left(1+\frac{2x}{z}\right)}{1+2yz} = 1,
\end{cases}
\tag{6.84}
$$

where $2x \neq -z$ and $2yz \neq -1$.

Dividing the second equality of (6.84) by the first one of (6.84) we find

$$
\frac{y\left(1+\frac{2x}{z}\right)}{x(1+2yz)} = 1
$$

which with $xz = y$ yields

$$
y + 2x^2 = x + 2y^2.
$$

Simplifying the last equality one gets

$$
(y - x)(1 - 2(y + x)) = 0.
$$

This means that either $y = x$ or $x + y = \frac{1}{2}$.

Assume that $x = y$. Then from $xz = y$, one finds $z = 1$. Moreover, from the second equality of (6.84) we have $y + \frac{2}{y} = 1 + 2y$. So, $y^2 + y - 2 = 0$. The solutions of the last equation are $y_1 = 1, y_2 = -2$. Hence, $x_1 = 1, x_2 = -2$.

Now suppose that $x + y = \frac{1}{2}$. Then $x = \frac{1}{2} - y$. We note that $y \neq 1/2$, since $x \neq 0$. So, from the second equality of (6.84) we find

$$
y + \frac{4}{1 - 2y} = 1 + \frac{4y^2}{1 - 2y}.
$$

So, $2y^2 - y - 1 = 0$, which yields the solutions $y_3 = -\frac{1}{2}, y_4 = 1$. Therefore, we obtain $x_3 = 1, z_3 = -\frac{1}{2}$ and $x_4 = -\frac{1}{2}, z_4 = -2$.

Consequently, the solutions of (6.84) are the following

$$(1,1,1),\ \left(1,-\frac{1}{2},-\frac{1}{2}\right),\ \left(-\frac{1}{2},1,-2\right),\ (-2,-2,1).$$

Now owing to (6.82), we need to solve the following equations

$$\begin{cases} \frac{f_1}{f_2} = x_k, \\ \frac{f_2}{f_3} = z_k, \end{cases} \quad k = \overline{1,4}. \tag{6.85}$$

According to our assumption, $f_k \neq 0$. We therefore consider cases when $x_k z_k \neq 0$. Now let us consider several cases:

CASE 1.   Let $x_2 = 1,\ z_2 = 1$. Then from (6.85) one gets $f_1 = f_2 = f_3$. So, from (6.80) we find $3\varepsilon f_1^2 = f_1$, i.e. $f_1 = \frac{1}{3\varepsilon}$. Now taking into account that $f_1^2 + f_2^2 + f_3^2 \leq 1$, one gets $\frac{1}{3\varepsilon^2} \leq 1$. From the last inequality we have $|\varepsilon| \geq \frac{1}{\sqrt{3}}$. Due to Lemma 6.5.4, the operator $V_\varepsilon$ is well defined iff $|\varepsilon| \leq \frac{1}{\sqrt{3}}$. Therefore, one gets $|\varepsilon| = \frac{1}{\sqrt{3}}$. Hence, in this case a solution is $\left(\pm\frac{1}{\sqrt{3}};\pm\frac{1}{\sqrt{3}};\pm\frac{1}{\sqrt{3}}\right)$.

CASE 2.   Let $x_2 = 1,\ z_2 = -1/2$. Then from (6.85) one finds $f_1 = f_2, 2f_2 = -f_3$. Substituting the latter into (6.80) we get $f_1 + 3f_1^2\varepsilon = 0$. Then, we have $f_1 = -\frac{1}{3\varepsilon}, f_2 = -\frac{1}{3\varepsilon}, f_3 = \frac{2}{3\varepsilon}$. Taking into account $f_1^2 + f_2^2 + f_3^2 \leq 1$, we find $\frac{1}{9\varepsilon^2} + \frac{4}{9\varepsilon^2} + \frac{1}{9\varepsilon^2} \leq 1$. This means $|\varepsilon| \geq \sqrt{\frac{2}{3}}$. Due to Lemma 6.5.4, in this case the operator $V_\varepsilon$ is not well defined. Therefore, we conclude that there is no fixed point of $V_\varepsilon$ belonging to $\mathbf{S}$.

Using the same argument for the remaining cases we conclude the absence of solutions. This shows that if $|\varepsilon| < 1/\sqrt{3}$, the operator $V_\varepsilon$ has a unique fixed point in $S$. If $|\varepsilon| = 1/\sqrt{3}$, then $V_\varepsilon$ has three fixed points belonging to $S$. This completes the proof.

Now we are going to study the dynamics of $V_\varepsilon$.

**Theorem 6.5.7** *Let $V_\varepsilon$ be given by (6.79). Then the following assertions hold:*

*(i) if $|\varepsilon| < 1/\sqrt{3}$, then for any $\mathbf{f} \in \mathbf{S}$ one has $V_\varepsilon^n(\mathbf{f}) \to (0,0,0)$ as $n \to \infty$;*

*(ii) if $|\varepsilon| = 1/\sqrt{3}$, then for any $\mathbf{f} \in \mathbf{S}$ with $\mathbf{f} \notin \{(\pm\frac{1}{\sqrt{3}},\pm\frac{1}{\sqrt{3}},\pm\frac{1}{\sqrt{3}})\}$ one has $V_\varepsilon^n(\mathbf{f}) \to (0,0,0)$ as $n \to \infty$.*

*Proof* Let us consider the following function $\rho(\mathbf{f}) = f_1^2 + f_2^2 + f_3^2$. Then we have

$$\rho(V_\varepsilon(\mathbf{f})) = \varepsilon^2\left((f_1^2 + 2f_2f_3)^2 + (f_2^2 + 2f_1f_3)^2 + (f_3^2 + 2f_1f_2)^2\right)$$
$$\leq \varepsilon^2\left(f_1^4 + 2|f_2||f_3| + f_2^4 + 2|f_1||f_3| + f_3^4 + 2|f_1||f_2|\right)$$
$$\leq \varepsilon^2\left(f_1^2 + f_2^2 + f_3^2 + f_2^2 + f_1^2 + f_3^2 + f_3^2 + f_1^2 + f_2^2\right)$$
$$= 3\varepsilon^2(f_1^2 + f_2^2 + f_3^2) = 3\varepsilon^2\rho(\mathbf{f}).$$

This means

$$\rho(V_\varepsilon(\mathbf{f})) \leq 3\varepsilon^2\rho(\mathbf{f}). \tag{6.86}$$

Due to $\varepsilon^2 \leq \frac{1}{3}$, from (6.86) one finds

$$\rho(V_\varepsilon^{n+1}(\mathbf{f})) \leq \rho(V_\varepsilon^n(\mathbf{f})),$$

which yields that the sequence $\{\rho(V_\varepsilon^n(\mathbf{f}))\}$ is convergent. Next we would like to find the limit of $\{\rho(V_\varepsilon^n(\mathbf{f}))\}$.

(i). First we assume that $|\varepsilon| < \frac{1}{\sqrt{3}}$. Then from (6.86) we obtain

$$\rho(V_\varepsilon^n(\mathbf{f})) \leq 3\varepsilon^2\rho(V_\varepsilon^{n-1}(\mathbf{f})) \leq \cdots \leq (3\varepsilon^2)^n\rho(\mathbf{f}).$$

This yields that $\rho(V_\varepsilon^n(\mathbf{f})) \to 0$ as $n \to \infty$, for all $\mathbf{f} \in \mathbf{S}$.

(ii). Now let $|\varepsilon| = \frac{1}{\sqrt{3}}$. Then consider two distinct subcases.

CASE (A).   Let $f_1^2 + f_2^2 + f_3^2 < 1$ and define $d = f_1^2 + f_2^2 + f_3^2$. Then one gets

$$\rho(V_\varepsilon(\mathbf{f})) \leq \varepsilon^2((f_1^2 + 2|f_2||f_3|)^2 + (f_2^2 + 2|f_1||f_3|)^2 + (f_3^2 + 2|f_1||f_2|)^2)$$

$$\leq \varepsilon^2((f_1^2 + f_2^2 + f_3^2)^2 + (f_2^2 + f_1^2 + f_3^2)^2 + (f_3^2 + f_1^2 + f_2^2)^2)$$

$$= 3\varepsilon^2 d^2 = dd = d\rho(\mathbf{f}).$$

Hence, we have $\rho(V_\varepsilon(\mathbf{f})) \leq d\rho(\mathbf{f})$. This means $\rho(V_\varepsilon^n(\mathbf{f})) \leq d^n\rho(\mathbf{f}) \to 0$. Hence, $V_\varepsilon^n(\mathbf{f}) \to 0$ as $n \to \infty$.

CASE (B).   Now take $f_1^2 + f_2^2 + f_3^2 = 1$ and assume that $\mathbf{f}$ is not a fixed point. Therefore, we may assume that $f_i \neq f_j$ for some $i \neq j$, otherwise from Lemma 6.5.6 we conclude that $\mathbf{f}$ is a fixed point. Hence, from (6.79) one finds

$$V_\varepsilon(\mathbf{f})_1 = \varepsilon(f_1^2 + 2f_2f_3) = \varepsilon(1 - f_2^2 - f_3^2 + 2f_2f_3) = \varepsilon(1 - (f_2 - f_3)^2).$$

Similarly one gets

$$V_\varepsilon(\mathbf{f})_2 = \varepsilon(1 - (f_1 - f_3)^2),$$

$$V_\varepsilon(\mathbf{f})_3 = \varepsilon(1 - (f_1 - f_2)^2).$$

It is clear that $|V_\varepsilon(\mathbf{f})_k| \leq |\varepsilon|$ ($k = 1, 2, 3$), and according to our assumption $f_i \neq f_j$ ($i \neq j$), we conclude that one of $|V_\varepsilon(\mathbf{f})_k|$ is strictly less than $\frac{1}{\sqrt{3}}$. This yields that $V_\varepsilon(\mathbf{f})_1^2 + V_\varepsilon(\mathbf{f})_2^2 + V_\varepsilon(\mathbf{f})_3^2 < 1$. Therefore, from the case (a), one finds that $V_\varepsilon^n(\mathbf{f}) \to 0$ as $n \to \infty$.

## 6.6   Stability of the Dynamics of Non-Simple q.q.s.o.s

In this section we are going to study the stability of the dynamics of $V_\Delta$ associated with a q.q.s.o. $\Delta$ defined on $\mathbb{M}_2(\mathbb{C})$.

Let $V$ be an operator given by

$$V(\mathbf{f})_k = \sum_{i,j=1}^{3} b_{ij,k} f_i f_j, \quad k = 1, 2, 3, \tag{6.87}$$

where $\mathbf{f} = (f_1, f_2, f_3) \in S$. One can see that $V$ has a fixed point $(0, 0, 0)$. Furthermore, we will be interested in the uniqueness (stability) of this fixed point.

Define

$$\alpha_k = \sqrt{\sum_{j=1}^{3}\left(\sum_{i=1}^{3}|b_{ij,k}|\right)^2} + \sqrt{\sum_{i=1}^{3}\left(\sum_{j=1}^{3}|b_{ij,k}|\right)^2}, \quad \alpha = \sum_{k=1}^{3}\alpha_k^2. \tag{6.88}$$

**Theorem 6.6.1** *If $\alpha < 1$, then $V$ is a contraction, hence $(0, 0, 0)$ is a unique stable fixed point.*

*Proof* Let us take $\mathbf{f}, \mathbf{p} \in S$ and consider the difference

$$|V(\mathbf{f})_k - V(\mathbf{p})_k| \leq \sum_{i,j=1}^{3}|b_{ij,k}||f_i f_j - p_i p_j|$$

$$\leq \sum_{i,j=1}^{3}|b_{ij,k}||f_i||f_j - p_j| + \sum_{i,j=1}^{3}|b_{ij,k}||p_j||f_i - p_i|$$

$$\leq \sum_{i,j=1}^{3}|b_{ij,k}||f_j - p_j| + \sum_{i,j=1}^{3}|b_{ij,k}||f_i - p_i|$$

$$\leq \left(\sqrt{\sum_{j=1}^{3}\left(\sum_{i=1}^{3}|b_{ij,k}|\right)^2} + \sqrt{\sum_{i=1}^{3}\left(\sum_{j=1}^{3}|b_{ij,k}|\right)^2}\right)\|\mathbf{f} - \mathbf{p}\|$$

$$= \alpha_k\|\mathbf{f} - \mathbf{p}\|,$$

where $k = 1, 2, 3$. Hence, $V$ is a contraction, so it has a unique fixed point. This completes the proof.

Note that the condition $\alpha < 1$ in Theorem 6.6.1 is too strong, therefore it would be interesting to find weaker conditions than the provided one.

Put

$$\delta_k = \sum_{i,j=1}^{3} |b_{ij,k}|, \quad k = 1, 2, 3 \tag{6.89}$$

and let $\mathbf{d} = (\delta_1, \delta_2, \delta_3)$.

Given a quadratic operator $V$ by (6.78), define a new operator $\tilde{V} : \mathbb{R}^3 \to \mathbb{R}^3$ by

$$\tilde{V}(\mathbf{p})_k = \sum_{i,j=1}^{3} |b_{ij,k}| p_i p_j, \quad \mathbf{p} \in \mathbb{R}^3, \ k = 1, 2, 3. \tag{6.90}$$

For any given $\mathbf{f} \in S$, we let $\gamma_{\mathbf{f}} = \max\{|f_1|, |f_2|, |f_3|\}$. It is clear that $\gamma_{\mathbf{f}} \leq 1$.

**Proposition 6.6.2** *If the sequence $\{\tilde{V}^n(\mathbf{d})\}$ is bounded, then for any $\mathbf{f} \in S$ with $\gamma_{\mathbf{f}} < 1$ one has $V^n(\mathbf{f}) \to (0, 0, 0)$ as $n \to \infty$.*

*Proof* From (6.78) we immediately find

$$|V(\mathbf{f})_k| \leq \gamma_{\mathbf{f}}^2 \sum_{i,j=1}^{3} |b_{ij,k}| = \gamma_{\mathbf{f}}^2 \delta_k, \quad k = 1, 2, 3.$$

Hence, the last inequality implies that

$$|V^2(\mathbf{f})_k| \leq \sum_{i,j=1}^{3} |b_{ij,k}| |V(\mathbf{f})_i| |V(\mathbf{f})_j| \leq \gamma_{\mathbf{f}}^{2^2} \tilde{V}(\mathbf{d})_k, \quad k = 1, 2, 3. \tag{6.91}$$

Here, as before, $\mathbf{d} = (\delta_1, \delta_2, \delta_3)$.

Hence, using mathematical induction one can get

$$|V^n(\mathbf{f})_k| \leq \gamma_{\mathbf{f}}^{2^n} \tilde{V}^{n-1}(\mathbf{d})_k, \quad \text{for any } n \in \mathbb{N}, \ k = 1, 2, 3. \tag{6.92}$$

Due to $\gamma_{\mathbf{f}} < 1$ and the boundedness of $\{\tilde{V}^n(\mathbf{d})_k\}$, from (6.92) we obtain the desired assertion.

The next lemma provides us with a sufficient condition for the boundedness of $\{\tilde{V}^n(\mathbf{d})_k\}$.

**Lemma 6.6.3** *Assume that one has*

$$\sum_{i,j=1}^{3} |b_{ij,k}| \leq 1, \quad k = 1, 2, 3. \tag{6.93}$$

*Then the sequence $\{\tilde{V}^n(\mathbf{d})_k\}$ is bounded.*

*Proof* From (6.93) we conclude that $\delta_k \le 1$ for $k = 1, 2, 3$. Therefore, it follows from (6.90) that

$$|\tilde{V}(\mathbf{d})_k| = \sum_{i,j=1}^{3} |b_{ij,k}|\delta_i\delta_j \le \delta_k \le 1.$$

Now assume that $|\tilde{V}^m(\mathbf{d})_k| \le \delta_k$ for $k = 1, 2, 3$. Then, by assumption, from (6.93) one gets

$$|\tilde{V}^{m+1}(\mathbf{d})_k| = \sum_{i,j=1}^{3} |b_{ij,k}||\tilde{V}^m(\mathbf{d})_i||\tilde{V}^m(\mathbf{d})_j|$$

$$\le \sum_{i,j=1}^{3} |b_{ij,k}|\delta_i\delta_j$$

$$\le \delta_k.$$

Hence, mathematical induction implies that $|\tilde{V}^n(\mathbf{d})_k| \le \delta_k$ for every $n \in \mathbb{N}$, $k = 1, 2, 3$. This completes the proof.

Now we are interested in when the sequence $\{\tilde{V}^n(\mathbf{d})\}$ converges to $(0, 0, 0)$.

**Lemma 6.6.4** *Assume that* (6.93) *is satisfied. If there is an* $n_0 \in \mathbb{N}$ *such that* $\tilde{V}^{n_0}(\mathbf{d})_k < 1$ *for* $k = 1, 2, 3$, *then* $\tilde{V}^n(\mathbf{d}) \to (0, 0, 0)$ *as* $n \to \infty$.

*Proof* Let $v = \max\{V^{n_0}(\mathbf{d})_1, V^{n_0}(\mathbf{d})_k, V^{n_0}(\mathbf{d})_3\}$. Then, by assumption, one has $0 < v < 1$. It then follows from (6.90) and (6.93) that

$$\tilde{V}^{n_0+1}(\mathbf{d})_k = \sum_{i,j=1}^{3} |b_{ij,k}|V^{n_0}(\mathbf{d})_i V^{n_0}(\mathbf{d})_j \le v^2\delta_k \le v^2.$$

Iterating this procedure we obtain $\tilde{V}^{n+n_0}(\mathbf{d})_k \le v^{2^n}$ for every $n \in \mathbb{N}$, $k = 1, 2, 3$. This yields the assertion.

Now we are ready to formulate the main result concerning the stability of the unique fixed point $(0, 0, 0)$ for $V$.

**Theorem 6.6.5** *Assume that* (6.93) *is satisfied. If there is a* $k_0 \in \{1, 2, 3\}$ *such that* $\delta_{k_0} < 1$ *and for each* $k = 1, 2, 3$ *one can find an* $i_0 \in \{1, 2, 3\}$ *with* $|b_{i_0,k_0,k}| + |b_{k_0,i_0,k}| \ne 0$, *then* $(0, 0, 0)$ *is a unique stable fixed point, i.e. for every* $\mathbf{f} \in \mathbf{S}$ *one has* $V^n(\mathbf{f}) \to (0, 0, 0)$ *as* $n \to \infty$.

*Proof* Take any $k \in \{1, 2, 3\}$, then due to the condition one can find $i_0$ such that $|b_{i_0,k_0,k}| + |b_{k_0,i_0,k}| \neq 0$. Then from (6.90) with (6.93) we have

$$\tilde{V}(\mathbf{d})_k = \sum_{i,j=1}^{3} |b_{ij,k}| \delta_j \delta_j$$

$$= \sum_{j=1}^{3} |b_{k_0j,k}| \delta_{k_0} \delta_j + \sum_{i=1}^{3} |b_{ik_0,k}| \delta_i \delta_{k_0} + \sum_{\substack{i,j=1 \\ i,j \neq k_0}}^{3} |b_{ij,k}| \delta_i \delta_j - |b_{k_0k_0,k}| \delta_{k_0}^2$$

$$\leq \sum_{j=1}^{3} |b_{k_0j,k}| \delta_{k_0} + \sum_{i=1}^{3} |b_{ik_0,k}| \delta_{k_0} + \sum_{\substack{i,j=1 \\ i,j \neq k_0}}^{3} |b_{ij,k}| - |b_{k_0k_0,k}| \delta_{k_0}^2$$

$$= \delta_k - (1 - \delta_{k_0}) \sum_{j=1}^{3} (|b_{k_0j,k}| + |b_{jk_0,k}|) + |b_{k_0k_0,k}|(1 - \delta_{k_0}^2)$$

$$= \delta_k - (1 - \delta_{k_0}) \left( \sum_{j=1}^{3} (|b_{k_0j,k}| + |b_{jk_0,k}|) - (1 + \delta_{k_0})|b_{k_0k_0,k}| \right)$$

$$= \delta_k - (1 - \delta_{k_0}) \left( \sum_{\substack{j=1 \\ j \neq k_0}}^{3} (|b_{k_0j,k}| + |b_{jk_0,k}|) + (1 - \delta_{k_0})|b_{k_0k_0,k}| \right)$$

$$= \delta_k - (1 - \delta_{k_0}) \sum_{\substack{j=1 \\ j \neq k_0}}^{3} (|b_{k_0j,k}| + |b_{jk_0,k}|) - (1 - \delta_{k_0})^2 |b_{k_0k_0,k}|$$

$$< \delta_k \leq 1.$$

Hence from Lemma 6.6.4 we find that $\tilde{V}^n(\mathbf{d}) \to (0, 0, 0)$ as $n \to \infty$. So, from (6.92) one gets the desired assertion.

We call a quadratic operator $V$ given by (6.78) *diagonal* if $b_{ij,k} = 0$ for all $i, j$ with $i \neq j$. In what follows, for the sake of brevity, we write $b_{ik}$ instead of $b_{ii,k}$. Hence from (6.78) we derive

$$(V(\mathbf{f}))_k = \sum_{i=1}^{3} b_{ik} f_i^2, \quad \mathbf{f} = (f_1, f_2, f_3) \in S. \tag{6.94}$$

First we are interested when $V$ maps $\mathbf{S}$ into itself, i.e. $V(\mathbf{S}) \subset \mathbf{S}$. If the coefficients $\{b_{ik}\}$ satisfy (6.76) then from Proposition 6.5.1 we conclude the desired inclusion. The next lemma provides us with a sufficient condition on $\{b_{ik}\}$ for the satisfaction of (6.76).

**Lemma 6.6.6** *Let V be a diagonal quadratic operator given by* (6.94). *Assume that*

$$\sum_{k=1}^{3} \max_{i}\{|b_{ik}|^2\} \le 1. \tag{6.95}$$

*Then* (6.76) *is satisfied.*

*Proof* Let us check (6.76). Take any $\mathbf{f}, \mathbf{p} \in \mathbf{S}$, then taking into account the definition of a diagonal operator and using our notation we get

$$\left| \sum_{i,j=1}^{3} b_{ij,k} f_i p_j \right| \le \sum_{i=1}^{3} |b_{ik}| |f_i| |p_i|$$

$$\le \max_{i}\{|b_{ik}|\} \sum_{i=1}^{3} |f_i| |p_i|$$

$$\le \max_{i}\{|b_{ik}|\} \|\mathbf{f}\| \|\mathbf{p}\|$$

$$\le \max_{i}\{|b_{ik}|\},$$

which implies the desired inequality.

*Remark 6.6.1* It is easy to see that the condition (6.95) is weaker than (6.77).

**Theorem 6.6.7** *Let V be a diagonal quadratic operator given by* (6.94). *Assume that*

$$\sum_{k=1}^{3} \max_{i}\{|b_{i,k}|^2\} < 1. \tag{6.96}$$

*Then the operator has a unique stable fixed point* $(0, 0, 0)$.

*Proof* First, from (6.96) with Lemma 6.6.6 we conclude that $V$ maps $\mathbf{S}$ into itself. Now, let $a_k := \max_{i}\{|b_{i,k}|\}$ and put

$$\gamma := \sum_{k=1}^{3} a_k^2. \tag{6.97}$$

Take any $\mathbf{f} = (f_1, f_2, f_3) \in \mathbf{S}$. Then from (6.94) we find

$$|V(\mathbf{f})_k| \le \sum_{i=1}^{3} |b_{ik}| f_i^2 \le a_k \sum_{i=1}^{3} f_i^2 \le a_k, \quad k = 1, 2, 3.$$

From the last inequality and (6.94) we have

$$|V^2(\mathbf{f})_k| \leq a_k \gamma, \quad k = 1, 2, 3.$$

Now iterating this procedure, we derive

$$|V^n(\mathbf{f})_k| \leq a_k \gamma^{n-1}, \quad k = 1, 2, 3, \tag{6.98}$$

for every $n \geq 2$. Due to (6.96) we have $\gamma < 1$, therefore (6.98) implies that $V^n(\mathbf{f}) \to 0$ as $n \to \infty$, and the arbitrariness of $\mathbf{f}$ proves the theorem.

*Remark 6.6.2* Note that if (6.96) is not satisfied, then the corresponding quadratic operator may have more than one fixed point. Indeed, let us consider the following diagonal operator defined by $V_0(\mathbf{f}) = (f_1^2, 0, 0)$, where $\mathbf{f} = (f_1, f_2, f_3)$. One can see that for this operator the condition (6.95) is satisfied, but (6.96) does not hold. It is clear that $V_0$ has two fixed points $(1, 0, 0)$ and $(0, 0, 0)$.

## 6.7  Example 1

Let us consider the diagonal quadratic operator defined by

$$\begin{cases} (V(\mathbf{f}))_1 = f_1^2, \\ (V(\mathbf{f}))_2 = a f_2^2 + b f_3^2, \quad \mathbf{f} = (f_1, f_2, f_3). \\ (V(\mathbf{f}))_3 = c f_3^2, \end{cases} \tag{6.99}$$

We can immediately observe that for the given operator the condition (6.95) is not satisfied, if one of the coefficients $a, b, c$ is non-zero, since $b_{11} = 1$.

By $\Delta_{a,b,c}$ we denote the linear operator from $M_2(\mathbb{C})$ to $M_2(\mathbb{C}) \otimes M_2(\mathbb{C})$ corresponding to (6.99).

**Lemma 6.7.1** *Let*

$$\max\{a^2, b^2\} + c^2 \leq 1 \tag{6.100}$$

*be satisfied. Then $\Delta_{a,b,c}$ is block positive, i.e. (6.76) is satisfied.*

*Proof* Take any $\mathbf{f}, \mathbf{p} \in \mathbf{S}$, and let

$$z = |f_2 p_2| + |f_3 p_3|.$$

Then using $|f_1p_1| + |f_2p_2| + |f_3p_3| \le 1$ we have

$$\sum_{k=1}^{3} \left| \sum_{m,l=1}^{3} b_{ml,k} f_m p_l \right|^2 - 1 = |f_1p_1|^2 + |af_2p_2 + bf_3p_3|^2 + |cf_3p_3|^2 - 1$$

$$\le |f_1p_1|^2 + \max\{a^2, b^2\}(|f_2p_2| + |f_3p_3|)^2 + c^2|f_3p_3| - 1$$

$$\le (f_1p_1)^2 + \max\{a^2, b^2\}(|f_2p_2| + |f_3p_3|)^2$$
$$+ c^2(|f_2p_2| + |f_3p_3|) - 1$$

$$\le (1 - |f_2p_2| - |f_3p_3|)^2 + \max\{a^2, b^2\}(|f_2p_2| + |f_3p_3|)^2$$
$$+ c^2(|f_2p_2| + |f_3p_3|) - 1$$

$$\le (1 - z)^2 + \max\{a^2, b^2\}z^2 + c^2z - 1$$

$$= z(z(1 + \max\{a^2, b^2\}) + c^2 - 2). \qquad (6.101)$$

Due to $0 \le z \le 1$, we conclude that (6.101) is less than zero, if one has

$$\max\{a^2, b^2\} + c^2 - 1 \le 0,$$

which implies the assertion.

The proved lemma implies that the operator (6.99) maps **S** into itself. Therefore, we examine the dynamics of (6.99) on **S**.

**Theorem 6.7.2** *Let V be a quadratic operator given by (6.99), and assume (6.100) is satisfied. Then the following assertions hold:*

*(i)* $(0, 0, 0)$ *and* $(1, 0, 0)$ *are fixed points of V;*

*(ii)* *if* $|f_1| = 1$, *then* $V^n(\mathbf{f}) = (1, 0, 0)$ *for all* $n \in \mathbb{N}$;

*(iii)* *Let* $|c| = 1$. *Then there is another fixed point* $(0, 0, c)$. *Moreover, if* $|f_3| = 1$, *then* $V^n(\mathbf{f}) = (0, 0, c)$ *for every* $n \ge 2$, *and if* $\max\{|f_1|, |f_3|\} < 1$, *then* $V^n(\mathbf{f}) \to (0, 0, 0)$ *as* $n \to \infty$;

*(iv)* *Let* $|a| = 1$. *Then there is another fixed point* $(0, a, 0)$. *If* $|af_2^2 + bf_3^2| = 1$, *then* $V^n(\mathbf{f}) = (0, a, 0)$ *for all* $n \ge 2$, *and if* $|af_2^2 + bf_3^2| < 1$ *and* $|f_1| < 1$, *then* $V^n(\mathbf{f}) = (0, 0, 0)$ *as* $n \to \infty$;

*(v)* *Let* $|b| = 1$, $|a| < 1$. *If* $|f_1| < 1$, *then* $V^n(\mathbf{f}) = (0, 0, 0)$ *as* $n \to \infty$;

*(vi)* *Let* $\max\{a^2, b^2\} + c^2 < 1$. *If* $|f_1| < 1$, *then* $V^n(\mathbf{f}) = (0, 0, 0)$ *as* $n \to \infty$.

*Proof* The statements (i) and (ii) are obvious. We further assume $|f_1| < 1$. Now let us consider (iii). If $|c| = 1$, then from (6.100) one gets that $a = b = 0$. Hence, in this case, we have another fixed point $(0, 0, c)$. One can see that $V(0, 0, -c) = (0, 0, c)$. So, if $|f_3| = 1$, then $V^n(\mathbf{f}) = (0, 0, c)$ for every $n \ge 2$. If $\max\{|f_1|, |f_3|\} < 1$, then from (6.99) we find $V^n(\mathbf{f}) = (f_1^{2^n}, 0, cf_3^{2^n}) \to (0, 0, 0)$ as $n \to \infty$.

(iv). Let $|a| = 1$. Then from (6.100) one finds $c = 0$, which implies the existence of another fixed point $(0, a, 0)$. From (6.99) we find

$$(V^n(\mathbf{f}))_2 = a(af_2^2 + bf_3^2)^{2^{n-1}}. \tag{6.102}$$

Hence, if $|af_2^2 + bf_3^2| = 1$ then $V^n(\mathbf{f}) = (0, a, 0)$ for every $n \geq 2$. If $|af_2^2 + bf_3^2| < 1$, $|f_1| < 1$ then $V^n(\mathbf{f}) \to (0, 0, 0)$ as $n \to \infty$.

(v). Let $|b| = 1$, $|a| < 1$. Then we have $c = 0$. In this case, one has $|af_2^2 + bf_3^2| < 1$ for every $\mathbf{f} \in S$. Therefore, (6.102) yields the desired assertion.

(vi). Let us assume that $\max\{a^2, b^2\} + c^2 < 1$. Then the modulus of all the coefficients are strictly less than one. For the sake of simplicity, we let $m = \max\{|a|, |b|\}$. From (6.99) we have

$$\begin{cases} |(V(\mathbf{f}))_2| \leq m, \\ |(V(\mathbf{f}))_3| \leq |c|, \end{cases} \tag{6.103}$$

for every $\mathbf{f} \in S$.

Then with $\kappa = m^2 + |c|^2$, from (6.99) with (6.103) one gets

$$\begin{cases} |(V^2(\mathbf{f}))_2| \leq m\kappa, \\ |(V^2(\mathbf{f}))_3| \leq |c|^3. \end{cases} \tag{6.104}$$

Assume that

$$\begin{cases} |(V^m(\mathbf{f}))_2| \leq m\kappa^{2^{m-1}-1}, \\ |(V^m(\mathbf{f}))_3| \leq |c|^{2^{m+1}-1}, \end{cases} \tag{6.105}$$

for some $m \geq 2$. Then from (6.99) with (6.105) we derive

$$\begin{aligned}
|(V^{m+1}(\mathbf{f}))_2| &\leq m\big(m^2\kappa^{2^m-2} + |c|^{2^{m+2}-2}\big) \\
&= m\big(m^2\kappa^{2^m-2} + (|c|^2)^{2^m-2}|c|^{2^{m+1}+2}\big), \\
&\leq m\big(m^2\kappa^{2^m-2} + \kappa^{2^m-2}|c|^2\big), \\
&= m\kappa^{2^m-1}.
\end{aligned}$$

Here we have used $|c^2| \leq \kappa$.

One can see that

$$|(V^{m+1}(\mathbf{f}))_3| \leq |c|^{2^{m+2}-1}.$$

Consequently, by induction, we conclude that (6.105) is valid for all $m \geq 2$.

According to our assumption, one has $\kappa < 1$, therefore (6.105) with (6.99) implies that $V^n(\mathbf{f}) \to (0, 0, 0)$ $(n \to \infty)$ when $|f_1| < 1$.

Now we would like to choose parameters $a, b, c$ so that $\Delta_{a,b,c}$ is not a KS-operator.

**Theorem 6.7.3** *Assume that* (6.100) *is satisfied. If* $|a| + |b| > 1$, *then* $\Delta_{a,b,c}$ *is not a KS-operator.*

*Proof* To prove the statement, it is enough to choose numbers $a, b, c$ so that the conditions of Corollary 6.3.3 are not satisfied. Let us begin with (6.61). A little calculation shows that

$$\mathbf{x}_1 = (\overline{w}_1, 0, 0), \mathbf{x}_2 = (0, a\overline{w}_2, 0), \mathbf{x}_3 = (0, 0, b\overline{w}_2 + c\overline{w}_3), \qquad (6.106)$$

where $(w_1, w_2, w_3) \in \mathbb{C}^3$. So, from (6.51) we immediately find

$$\alpha_{2,3} = \langle \mathbf{x}_2, \mathbf{x}_3 \rangle - \langle \mathbf{x}_3, \mathbf{x}_2 \rangle = 0.$$

Hence, from the last equality with (6.106) we infer that (6.61) is reduced to

$$|a|^2 |w_2|^2 + |b\overline{w}_2 + c\overline{w}_3|^2 \le |w_2|^2 + |w_3|^2. \qquad (6.107)$$

Now let us estimate the left-hand side of (6.107):

$$|a|^2 |w_2|^2 + |b\overline{w}_2 + c\overline{w}_3|^2 \le |a|^2 |w_2|^2 + \left( |b||w_2| + |c||w_3| \right)^2$$

$$\le |a|^2 |w_2|^2 + \max\{|b|^2, |c|^2\} \left( |w_2| + |w_3| \right)^2$$

$$\le |a|^2 |w_2|^2 + 2 \max\{|b|^2, |c|^2\} \left( |w_2|^2 + |w_3|^2 \right).$$

Hence, if

$$|a|^2 |w_2|^2 + 2 \max\{|b|^2, |c|^2\} \left( |w_2|^2 + |w_3|^2 \right) \le |w_2|^2 + |w_3|^2 \qquad (6.108)$$

then surely (6.107) is satisfied. Therefore, let us examine (6.108). From (6.108) one finds

$$\left( 1 - |a|^2 - 2 \max\{|b|^2, |c|^2\} \right) |w_2|^2 + \left( 1 - 2 \max\{|b|^2, |c|^2\} \right) |w_3|^2 \ge 0,$$

which is satisfied if one has

$$|a|^2 + 2\max\{|b|^2, |c|^2\} \le 1. \tag{6.109}$$

Now let us look at the condition (6.62). From (6.106), direct calculations show us that

$$\begin{cases} \mathbf{h}(\mathbf{w}) = (\overline{w}_2 w_3 - \overline{w}_3 w_2, 0, 0), \\ \gamma_{2,3} = (2ab|w_2|^2 + ac(\overline{w}_2 w_3 + w_2 \overline{w}_3), 0, 0), \\ \sum_{m=1}^{3} [\mathbf{x}_m, \overline{\mathbf{x}}_m] = 0. \end{cases} \tag{6.110}$$

Therefore, the left-hand side of (6.62) can be written as follows

$$\left\| \mathbf{h}(\mathbf{w}) - i\gamma_{2,3} + i\sum_{m=1}^{3} [\mathbf{x}_m, \overline{\mathbf{x}}_m] \right\| = \left| \overline{w}_2 w_3 (1 - iac) - \overline{w}_3 w_2 (1 + iac) - 2iab|w_2|^2 \right|.$$

Hence, the last equality with (6.106) reduces (6.62) to

$$\left| \overline{w}_2 w_3 (1 - iac) - \overline{w}_3 w_2 (1 + iac) - 2iab|w_2|^2 \right|$$
$$\le |w_2|^2 + |w_3|^2 - |a|^2 |w_2|^2 - |b\overline{w}_2 + c\overline{w}_3|^2.$$

Letting $w_3 = 0$ in the last inequality, one gets

$$2|ab||w_2|^2 \le |w_2|^2 (1 - |a|^2 - |b|^2),$$

which is equivalent to

$$|a| + |b| \le 1. \tag{6.111}$$

Consequently, if $|a| + |b| > 1$, then (6.111) is not satisfied, and this proves the desired assertion.

Now let us provide more concrete examples of the parameters. Take $c = 0$ and $a = b = 1/\sqrt{3}$. Then one can see that (6.100) and (6.109) are satisfied, but $|a| + |b| = 2/\sqrt{3} > 1$.

## 6.8  Example 2

In this section we are going to study one special class of quadratic operators.

Let $V$ be given by (6.87). In what follows, we assume that $b_{ij,k} = 0$ whenever $k \notin \{i,j\}$. Then $V$ has the following form

$$(V(\mathbf{x}))_k = x_k \left( b_{kk,k} x_k + \sum_{\substack{j=1 \\ j \neq k}}^{3} b_{kj,k} x_i \right), \quad \mathbf{x} = (x_1, x_2, x_3). \tag{6.112}$$

It is clear that the operator defined by (6.112) satisfies $V(\mathbf{S}) \subset \mathbf{S}$ if one has

$$\left| b_{kk,k} x_k + 2 \sum_{\substack{j=1 \\ j \neq k}}^{3} b_{kj,k} x_i \right| \leq 1, \quad k \in \{1,2,3\}. \tag{6.113}$$

Now from

$$\left| b_{kk,k} x_k + 2 \sum_{\substack{j=1 \\ j \neq k}}^{3} b_{kj,k} x_i \right| \leq |b_{kk,k} x_k| + 2 \sum_{\substack{j=1 \\ j \neq k}}^{3} |b_{kj,k} x_i|$$

$$\leq \left( b_{kk,k}^2 + \sum_{\substack{j=1 \\ j \neq k}}^{3} (2b_{kj,k})^2 \right) \sum_{i=1}^{3} x_i^2$$

$$\leq \left( b_{kk,k}^2 + \sum_{\substack{j=1 \\ j \neq k}}^{3} (2b_{kj,k})^2 \right) \tag{6.114}$$

we conclude that if

$$B_k := b_{kk,k}^2 + \sum_{\substack{j=1 \\ j \neq k}}^{3} (2b_{kj,k})^2 \leq 1 \qquad \forall k \in \{1,2,3\} \tag{6.115}$$

then $V$ is well defined.

We now consider several cases.

CASE 1.   Assume that $B_k < 1$ for $k = 1, 2, 3$. Now take any $\mathbf{x} \in \mathbf{S}$. Then using (6.114) and (6.115), from (6.112) we get

$$|(V(\mathbf{x}))_k| \leq \sqrt{B_k} |x_k|.$$

Hence

$$|(V^n(\mathbf{x}))_k| \leq (\sqrt{B_k})^n \tag{6.116}$$

for any $n \in \mathbb{N}$. According to our assumption one finds that $(V^n(\mathbf{x}))_k \to 0$ as $n \to \infty$ for every $\mathbf{x} \in \mathbf{S}$ (k=1,2,3).

CASE 2.    Assume that $B_{k_0} < 1$ and $B_j = 1$ for every $j \neq k_0$. Then similarly to the previous case (see (6.116)) we get

$$|(V^n(\mathbf{x}))_{k_0}| \leq (\sqrt{B_{k_0}})^n \tag{6.117}$$

for any $n \in \mathbb{N}$. So, $(V^n(\mathbf{x}))_{k_0} \to 0$ for every $\mathbf{x} \in \mathbf{S}$.

For fixed $j \neq k_0$ the equality $B_j = 1$ with (6.115) implies that there are three possibilities:

(a)  $|b_{jj,j}| = 1$;
(b)  $2|b_{ji,j}| = 1$ for $i \neq j$ and $i \neq k_0$;
(c)  $2|b_{jk_0,j}| = 1$.

Now consider these subcases separately.

Assume that (a) holds. Then from (6.112) we have $(V(\mathbf{x}))_j = b_{jj,j}x_j^2$, which implies that

$$(V^n(\mathbf{x}))_j = b_{jj,j}x^{2n}. \tag{6.118}$$

Hence, from (6.118) we conclude that

$$(V^n(\mathbf{x}))_j \to \begin{cases} 0, & \text{if } |x_j| < 1, \\ b_{jj,j} & \text{if } |x_j| = 1. \end{cases} \tag{6.119}$$

Assume that (c) holds. Then from (6.112) we have $(V(\mathbf{x}))_j = 2b_{jk_0,j}x_jx_{k_0}$, which with (6.117) implies that

$$|(V^{n+1}(\mathbf{x}))_j| \leq (\sqrt{B_{k_0}})^n. \tag{6.120}$$

Hence $(V^n(\mathbf{x}))_j \to 0$ for every $\mathbf{x} \in \mathbf{S}$.

Now finally assume that (b) holds, i.e. $|2b_{ji,j}| = 1$ for $i \neq j$. Then we have

$$(V(\mathbf{x}))_j = 2b_{ji,j}x_ix_j. \tag{6.121}$$

One can see that $b_{ik,i}$ also satisfies the above given conditions (a)–(c). Therefore, we again consider these three subcases.

So, let us consider when $|b_{ii,i}| = 1$. Then from (6.121) and case (a), it is obvious that if $|x_i| = 1$ then

$$(V^n(\mathbf{x}))_j = 0, \quad (V^n(\mathbf{x}))_i = b_{ii,i}.$$

If $|x_i| < 1$ then from case (a) we know $(V^n(\mathbf{x}))_i \to 0$. Hence, from (6.121) one gets $|(V^{n+1}(\mathbf{x}))_j| \leq |V^n(\mathbf{x}))_i|$, which implies $(V^n(\mathbf{x}))_j \to 0$.

Let us assume that $2|b_{ik_0,i}| = 1$, then from case (b) we know that $|(V^{n+1}(\mathbf{x}))_i| \leq (\sqrt{B_{k_0}})^n$, which with (6.121) implies that $|(V^{n+2}(\mathbf{x}))_j| \leq (\sqrt{B_{k_0}})^n$. Hence, $(V^n(\mathbf{x}))_j \to 0$ for every $\mathbf{x} \in \mathbf{S}$.

Finally, suppose that $2|b_{ij,i}| = 1$, then we have

$$(V(\mathbf{x}))_i = 2b_{ij,i}x_ix_j. \tag{6.122}$$

Now using

$$x_ix_j \leq \frac{1}{2}(x_i^2 + x_j^2) \leq \frac{1}{2},$$

from (6.121) and (6.122) one gets

$$|(V(\mathbf{x}))_j| \leq \frac{1}{2}, \quad |(V(\mathbf{x}))_i| \leq \frac{1}{2}.$$

Consequently, iterating (6.121) and (6.122), we obtain

$$|(V^n(\mathbf{x}))_j| \leq \frac{1}{2^n}, \quad |(V^n(\mathbf{x}))_i| \leq \frac{1}{2^n}.$$

So, $(V^n(\mathbf{x}))_k \to 0$ for every $\mathbf{x} \in \mathbf{S}$ and $k = i, j$.

CASE 3.   In this case we assume that $B_k = 1$ for $k = 1, 2, 3$. If $|b_{kk,k}| = 1$ for $k = 1, 2, 3$, then taking into account subcase (a) we conclude that

$$(V^n(\mathbf{x}))_k \to \begin{cases} 0, & \text{if } |x_k| < 1 \\ b_{kk,k} & \text{if } |x_k| = 1 \end{cases} \quad k \in \{1, 2, 3\}.$$

The rest of this case can easily be reduced to subcase (b) of case 2.

Summarizing all cases we can formulate the following.

**Theorem 6.8.1** *Let $V$ be a q.o. given by* (6.112). *Assume that* (6.115) *holds. The following statements hold:*

(i) *If there is a $k_0 \in \{1, 2, 3\}$ such that $|b_{k_0k_0,k_0}| = 1$, then $V$ has fixed points $(0, 0, 0)$ and $b_{k_0k_0,k_0}e_{k_0}$ such that $V^n(\mathbf{x}) \to 0$ for every $\mathbf{x} \in \mathbf{S}$ with $|x_{k_0}| < 1$, and $V^n(\mathbf{x}) \to b_{k_0k_0,k_0}e_{k_0}$ for $\mathbf{x} \in \mathbf{S}$ with $|x_{k_0}| = 1$.*

(ii) *If $|b_{kk,k}| < 1$ for all $k \in \{1, 2, 3\}$, then $(0, 0, 0)$ is a unique fixed point such that $V^n(\mathbf{x}) \to 0$ for every $\mathbf{x} \in \mathbf{S}$.*

# 6.9   Comments and References

As mentioned earlier, one of the central problems in the theory of quantum entanglement is to distinguish the difference between separable and entangled states of composite quantum systems. There are many papers devoted to finding the separability criterions for a given state (see [102] for review). The most general approach to characterizing quantum entanglement uses a notion of an entanglement witness [28, 101]. This uses the positivity of some mappings [27, 31, 96, 200]. Therefore, it would interesting to find some conditions for the positivity of the given mappings. Some characterizations of positive maps defined on $M_2(\mathbb{C})$ were considered in [144, 241]. In [219] the characterization of completely positive mappings from $M_2(\mathbb{C})$ into itself with an invariant state $\tau$ was established. A more general construction of positive mappings on $M_n(\mathbb{C})$ algebras has been investigated in [29–31]. In this direction there are many works (see for example [5, 17, 125, 200]).

From a mathematical point of view this leads to the characterization of positive and completely positive maps on $C^*$-algebras. There are many papers devoted to this problem (see for example [26, 143, 219, 241]). In the literature completely positive maps have proved to be of great importance in the structure theory of $C^*$-algebras (see [203]). However, general positive (order-preserving) linear maps are very intractable [143, 145, 241]. Therefore, it is interesting to study conditions stronger than positivity but weaker than complete positivity. Such a condition is called the *Kadison–Schwarz property*. Note that every unital completely positive map satisfies this inequality, and a famous result of Kadison states that any positive unital map satisfies the inequality for self-adjoint elements. In [208] relations between the $n$-positivity of a map $\phi$ and the Kadison–Schwarz property of a certain map are established. Certain relations between complete positivity, positivity and the Kadison–Schwarz property have been considered in [19, 21, 22]. Some spectral and ergodic properties of Kadison–Schwarz maps were investigated in [92, 93, 209].

The material of this chapter is taken from the papers [172, 180–183, 185]. The trace-preserving Kadison–Schwarz operators from $M_2(\mathbb{C})$ into itself are described in [179]. Such a description allowed us to compare these operators with completely positive ones (trace-preserving completely positive operators are described in [219]). One can see that to study the dynamics of quadratic operators, one needs to investigate the dynamics of quadratic mappings defined on the unit ball of $\mathbb{R}^3$. In general, it is impossible to study such operators. Some particular cases have been studied in this chapter. Other operators are investigated in [13, 148, 149, 244].

On the other hand, it is important to study extreme positive mappings. It is known [241] that for each extreme mapping $\phi$ of $M_2(\mathbb{C})$, there is a pure state $\rho$ of $M_2(\mathbb{C})$ such that $\rho \circ \phi$ is pure as well. Therefore, a mapping $\phi : A \to B$, where $A$ and $B$ are $C^*$-algebras, is called *pure* if it maps each pure state to a pure state. In [182] we have introduced a weaker condition than purity. This condition is called *q-purity*. This notion is based on quadratic operators acting on the state space of the algebra. In the mentioned paper we studied maps defined on $M_2(\mathbb{C})$, and for such maps the $q$-purity is equivalent to the invariance of the unit sphere in $\mathbb{R}^3$. In [178] a criterion of $q$-purity of quasi q.q.s.o.s is provided in terms of quadratic operators which map unit circles into the sphere (see also [223]).

# Chapter 7
# Infinite-Dimensional Quadratic Operators

In this chapter we study a class of q.q.s.o.s defined on the commutative algebra $\ell^\infty$. Essentially, we deal with conjugate quadratic operators. We define the notion of a Volterra quadratic operator and study its properties. Such operators have been studied by many authors (see for example [74, 252]) in the finite-dimensional setting.

## 7.1 Infinite-Dimensional Quadratic Stochastic Operators

In this chapter we will consider the case when the von Neumann algebra $M$ is an infinite-dimensional commutative discrete algebra, i.e.

$$M = \ell^\infty = \{\mathbf{x} = (x_n) : x_n \in \mathbb{R}, \ \|\mathbf{x}\|_\infty = \sup_{n \in \mathbb{N}} |x_i|\}.$$

Then the set of all normal functionals defined on $\ell^\infty$ coincides with

$$\ell^1 = \{\mathbf{x} = \{x_n\} : \|\mathbf{x}\|_1 = \sum_{k=1}^{\infty} |x_k| < \infty\}$$

(i.e. $\ell^1$ is a pre-dual space to $\ell^\infty$, namely $(\ell^1)^* = \ell^\infty$) and $S(\ell^\infty)$ with

$$S = \{\mathbf{x} = (x_n) \in \ell^1 : x_i \geq 0, \sum_{n=1}^{\infty} x_n = 1\}.$$

It is known [210] that $S = \overline{convh(ExtrS)}$, where $Extr(S)$ is the set of extremal points of $S$ and $convh(A)$ is the convex hull of a set $A$.

© Springer International Publishing Switzerland 2015
F. Mukhamedov, N. Ganikhodjaev, *Quantum Quadratic Operators and Processes*,
Lecture Notes in Mathematics 2133, DOI 10.1007/978-3-319-22837-2_7

Any extremal point $\varphi$ of $S$ has the following form

$$\varphi = (\underbrace{0, 0, \ldots, 1}_{n}, 0, \ldots)$$

for some $n \in \mathbb{N}$. Such an element will be denoted $e^{(n)}$.

In this section we study quadratic operators defined on $S$. Firstly, we need to describe conjugate quadratic operators. Note that, due to Theorem 5.2.3, each conjugate quadratic operator uniquely defines a q.q.s.o. The following theorem describes conjugate quadratic operators(c.q.o.s) when $M = \ell^\infty$.

**Theorem 7.1.1** *Every c.q.o. $\tilde{V}$ defines an infinite-dimensional matrix $(p_{ij,k})_{i,j,k\in\mathbb{N}}$ such that*

$$p_{ij,k} \geq 0, \;\; p_{ij,k} = p_{ji,k}, \;\; \sum_{k=1}^{\infty} p_{ij,k} = 1, \;\; i,j \in N. \tag{7.1}$$

*Conversely, every such matrix defines a c.q.o. $\tilde{V}$ as follows:*

$$(\tilde{V}(\mathbf{x}, \mathbf{y}))_k = \sum_{i,j=1}^{\infty} p_{ij,k} x_i y_j, \;\; k \in \mathbb{N}, \; \mathbf{x} = (x_i), \mathbf{y} = (y_i) \in S. \tag{7.2}$$

*Proof* Let $\tilde{V}$ be a c.q.o. For every $e^{(n)}, e^{(m)} \in Extr(S)$ we put

$$p_{mn,k} = (\tilde{V}(e^{(m)}, e^{(n)}))_k, \;\; m, n, k \in \mathbb{N}.$$

According to the positivity of $e^{(n)}, n \in \mathbb{N}$, and (ii) of Definition 5.2.1 we get $p_{mn,k} \geq 0$. It follows from (5.11) that $\tilde{V}(e^{(m)}, e^{(n)}) = \tilde{V}(e^{(n)}, e^{(m)})$, which implies that $p_{mn,k} = p_{nm,k}$. Since $\tilde{V}(e^{(m)}, e^{(n)}) \in S$ we find $\sum_{k=1}^{\infty} p_{mn,k} = 1$. Note that one has

$$(\tilde{V}(\mathbf{x}, \mathbf{y}))_k = \sum_{i,j=1}^{\infty} p_{ij,k} x_i y_j, \;\; k \in \mathbb{N},$$

for every $\mathbf{x} = (x_i), \mathbf{y} = (y_i) \in S$.

Conversely, let $(p_{ij,k})$ be a matrix satisfying (7.1). Define $\Delta : \ell^\infty \to \ell^\infty \otimes \ell^\infty$ as follows

$$(\Delta \mathbf{f})_{ij} = \sum_{k=1}^{\infty} p_{ij,k} f_k, \;\; i, j \in \mathbb{N},$$

for every $\mathbf{f} = (f_k) \in \ell^\infty$. The condition (7.1) implies that $P$ is a q.q.s.o. In particulary, one gets

$$\Delta(e^{(k)}) = \sum_{i,j \in \mathbb{N}} p_{ij,k} e^{(i)} \otimes e^{(j)}. \tag{7.3}$$

Let $\tilde{V}$ be the c.q.o. associated with $\Delta$. Take arbitrary $\mathbf{x}, \mathbf{y} \in S$. Then using (7.3) we find

$$(\tilde{V}(\mathbf{x}, \mathbf{y}))_k = x \otimes y(\Delta(e^{(k)})) = \sum_{k=1}^\infty p_{ij,k} x_i y_j.$$

Here $\mathbf{x} \otimes \mathbf{y} = (x_i y_j) \in S(\ell^\infty \otimes \ell^\infty)$. Thus, the theorem is proved.

We note that in this case the q.o. $V$ defined by (5.12) has the following form:

$$(V(\mathbf{x}))_k = \sum_{i,j=1}^\infty p_{ij,k} x_i x_j \quad k \in \mathbb{N}, \ \mathbf{x} = (x_i) \in S. \tag{7.4}$$

The constructed matrix $(p_{ij,k})_{i,j,k \in \mathbb{N}}$ is called the *determining matrix of the q.o. V*.

**Observation 7.1.1** *Let $T : \ell^\infty \to \ell^\infty$ be a positive identity preserving operator. Then it is easy to see that this operator can be represented as an infinite-dimensional stochastic matrix $(p_{ij})_{i,j \in \mathbb{N}}$, i.e. $p_{ij} \geq 0$, $\sum_{j=1}^\infty p_{ij} = 1$ for every $i, j \in \mathbb{N}$.*

*Then the determining matrix $(p_{ij,k})_{i,j,k \in \mathbb{N}}$ corresponding to the q.o. given by (5.13) is defined as*

$$p_{ij,k} = \frac{p_{ik} + p_{jk}}{2}, \quad i, j, k \in \mathbb{N}.$$

Recall that an element $\mathbf{x} \in S$ is called a *fixed point* of $V$ if $V(\mathbf{x}) = \mathbf{x}$. The set of all fixed points of $V$ belonging to $S$ is denoted by *Fix(V)*.

**Observation 7.1.2** *It is known that the set $S$ is not compact in the norm topology of $\ell^1$, or even in the $\sigma(\ell^1, \ell^\infty)$-topology. This is the difference between the finite and infinite-dimensional cases. In the finite-dimensional setting every q.o. $V : S^{n-1} \to S^{n-1}$ has at least one fixed point. In the infinite-dimensional case, not every q.o. has a fixed point. Indeed, let us define a linear operator $T : \ell^\infty \to \ell^\infty$ as follows*

$$T(x_1, x_2, \cdots, x_n, \cdots) = (x_2, \cdots, x_{n+1}, \cdots),$$

$(x_n) \in \ell^\infty$. *It is clear that $T$ is positive and $T\mathbb{1} = \mathbb{1}$. Now consider the q.q.s.o. defined by (5.13). Then by Observation 7.1.1, the q.o. $V$ acts as follows*

$$V(\varphi_1, \varphi_2, \cdots, \varphi_n, \cdots) = (0, \varphi_1, \varphi_2, \cdots, \varphi_n, \cdots)$$

where $(\varphi_n) \in S$. It is easy to see that this operator has no fixed points belonging to $S$, i.e. $Fix(V) = \emptyset$.

## 7.2  Volterra Operators

In this section we define Volterra operators and give some of their properties.

Recall that a convex set $C \subset S$ is called a *face* if $\lambda x + (1 - \lambda)y \in C$, where $x, y \in S$ è $\lambda \in (0, 1)$, implies that $x, y \in C$. For $\varphi, \psi \in S$ we define

$$\Gamma(\varphi, \psi) = \{\lambda\varphi + (1 - \lambda)\psi : \lambda \in [0, 1]\}.$$

**Definition 7.2.1** An operator $V$ defined by (5.12) (see Chap. 5) is called a *Volterra operator* if $\tilde{V}(\varphi, \psi) \in \Gamma(\varphi, \psi)$ for every $\varphi, \psi \in Extr(S)$.

By $\mathscr{QV}$ we denote the set of all quadratic operators defined on $S$, and the set of all Volterra operators is denoted by $\mathscr{V}$.

**Proposition 7.2.1** *Let $V \in \mathscr{QV}$ be a q.o. Then $V$ is Volterra if and only if the determining matrix $(p_{ij,k})$ of this operator satisfies*

$$p_{ij,k} = 0, \quad if \ k \notin \{i, j\}. \tag{7.5}$$

*Proof* Let $V$ be a Volterra operator. Then from Definition 7.2.1 we infer that

$$\tilde{V}(e^{(i)}, e^{(j)}) = p_{ij,i}e^{(i)} + p_{ij,j}e^{(j)}.$$

This yields that $p_{ij,i} + p_{ij,j} = 1$, so (7.5) is valid. The converse implication easily follows from Theorem 7.1.1. The proposition is proved.

Condition (7.5) biologically means that each individual can inherit only the species of the parents.

From Theorem 7.1.1 and Proposition 7.2.1, we immediately get the following.

**Proposition 7.2.2** *Let $V_1, V_2 \in \mathscr{V}$ be two Volterra operators such that for every $e^{(i)}, e^{(j)}, i, j \in \mathbb{N}$, the equality $\tilde{V}_1(e^{(i)}, e^{(j)}) = \tilde{V}_2(e^{(i)}, e^{(j)})$ holds, then $V_1 = V_2$.*

**Theorem 7.2.3** *Let $V \in \mathscr{QV}$ be a q.o. Then $V$ is a Volterra operator if and only if it can be represented as follows:*

$$(V(x))_k = x_k \left( 1 + \sum_{i=1}^{\infty} a_{ki} x_i \right), \quad k \in \mathbb{N}, \tag{7.6}$$

*where*

$$a_{ki} = -a_{ik}, \ |a_{ki}| \leq 1 \ for \ every \ k, i \in \mathbb{N}. \tag{7.7}$$

*Proof* From Definition 7.2.1 and Proposition 7.2.1, one gets $p_{kk,k} = 1, k \in \mathbb{N}$. Then from (7.4) we obtain

$$(V(x))_k = p_{kk,k} x_k^2 + \sum_{i=1, i \neq k} p_{ik,k} x_i x_k + \sum_{j=1, j \neq k} p_{kj,k} x_k x_j, \quad k \in \mathbb{N},$$

whence, keeping in mind $p_{ij,k} = p_{ji,k}$, we infer that

$$(V(x))_k = x_k(1 + 2 \sum_{i=1, i \neq k}^{\infty} p_{ik,k} x_i), \quad k \in \mathbb{N}.$$

Using $\sum_{i=1}^{\infty} x_i = 1$ one finds

$$(V(x))_k = x_k \left( 1 + \sum_{i=1, i \neq k}^{\infty} (2p_{ik,k} - 1) x_i \right), \quad k \in \mathbb{N}.$$

Setting $a_{ki} = 2p_{ik,k} - 1$ if $i \neq k$, and $a_{kk} = 0$, yields (7.6). The inequality $0 \leq p_{ik,k} \leq 1$ implies that $|a_{ki}| \leq 1$. Taking into account $p_{ik,k} + p_{ik,i} = 1$, we obtain

$$a_{ki} + a_{ik} = 2p_{ik,k} - 1 + 2p_{ki,i} - 1 = 2(p_{ik,k} + p_{ik,i} - 1) = 0.$$

Therefore $a_{ki} = -a_{ik}$.

The converse implication is obvious. This completes the proof.

**Corollary 7.2.4** *Let $V \in \mathscr{Q}\mathscr{V}$ be a q.o. Then $V$ is a Volterra operator if and only if $\tilde{V}$ can be represented as follows:*

$$(\tilde{V}(x, y))_k = \frac{1}{2} \left( x_k(1 + \sum_{i=1}^{\infty} a_{ki} y_i) + y_k(1 + \sum_{i=1}^{\infty} a_{ki} x_i) \right), \quad k \in \mathbb{N}. \qquad (7.8)$$

For a given subset $K$ of $\mathbb{N}$ we put

$$S^K = \{ \mathbf{x} \in S : x_i = 0, \ \forall i \in \mathbb{N} \setminus K \}.$$

**Corollary 7.2.5** *For every Volterra operator $V$ the following assertions hold:*

*(i) every face of $S$ invariant with respect to $V$;*
*(ii) $Extr(S) \subset Fix(V)$.*

The proof immediately follows from Theorem 7.2.3 since every face of $S$ is $S^K$, for some $K \subset \mathbb{N}$, and $\{e^{(i)}\} = S^{\{i\}}$ for every $e^{(i)} \in Extr(S)$.

Put

$$riS^K = \{\mathbf{x} \in S^K : x_i > 0, \quad \forall i \in K\}.$$

**Corollary 7.2.6** *Let V be a Volterra operator, then $V(riS^K) \subset riS^K$ holds, for every $K \subset \mathbb{N}$.*

*Proof* Let $x_k > 0, k \in K$, then according to the equality $a_{kk} = 0$ and (7.6) we have

$$
\begin{aligned}
(V(x))_k &= x_k(1 + a_{k1}x_1 + \ldots + a_{k,k-1}x_{k-1} + a_{k,k+1}x_{k+1} + \cdots) \\
&\geq x_k(1 - x_1 - \cdots - x_{k-1} - x_{k+1} - \cdots) \\
&= x_k^2 > 0.
\end{aligned}
$$

The corollary is proved.

*Remark 7.2.1* From Theorem 7.2.3, we see that the identity operator $Id : S \rightarrow S$, i.e.

$$(Id(x))_k = x_k, \quad k \in \mathbb{N}$$

is a Volterra operator. From Proposition 7.2.2 and Observations 7.1.1 and 7.1.2 one concludes that

$$\mathcal{QL}(\ell^\infty) \cap \mathcal{V} = Id.$$

**Theorem 7.2.7** *Let $V \in \mathcal{V}$ be a Volterra operator, then it is a bijection of S.*

*Proof* Let us first show that $V$ is injective. Assume that there are two elements $\mathbf{x}, \mathbf{y} \in S(x \neq y)$ such that

$$V(\mathbf{x}) = V(\mathbf{y}). \tag{7.9}$$

Without loss of generality we may assume that $x_i > 0, y_i > 0, \forall i \in \mathbb{N}$. If it is not true, then there is a face $S^K$, for some subset $K \subset \mathbb{N}$ of $S$, such that $\mathbf{x}, \mathbf{y} \in riS^K$, i.e. $x_i > 0, y_i > 0, \forall i \in K$. According to Corollaries 7.2.5 and 7.2.6, we have $V(S^K) \subset S^K$, therefore we may restrict $V$ to $S^K$. From (7.9) one gets that

$$x_k\left(1 + \sum_{i=1}^\infty a_{ki}x_i\right) = y_k\left(1 + \sum_{i=1}^\infty a_{ki}y_i\right),$$

or

$$(x_k - y_k)\left(1 + \sum_{i=1}^\infty a_{ki}y_i\right) = -x_k \sum_{i=1}^\infty a_{ki}(x_i - y_i). \tag{7.10}$$

Taking into account

$$1 + \sum_{i=1}^{\infty} a_{ki} y_i \geq 1 - y_1 - y_2 - \cdots - y_{k-1} - y_{k+1} - \cdots = y_k > 0,$$

and $x_k > 0$, from (7.10) we obtain that

$$sgn(x_k - y_k) = -sgn \sum_{i=1}^{\infty} a_{ki}(x_i - y_i). \qquad (7.11)$$

Hence,

$$(x_k - y_k) \sum_{i=1}^{\infty} a_{ki}(x_i - y_i) \leq 0, \quad k \in \mathbb{N}.$$

So, one gets

$$\sum_{k=1}^{\infty} (x_k - y_k) \sum_{i=1}^{\infty} a_{ki}(x_i - y_i) \leq 0.$$

Note that the last series absolutely converges, since

$$\left| \sum_{k=1}^{\infty} (x_k - y_k) \sum_{i=1}^{\infty} a_{ki}(x_i - y_i) \right| \leq \sum_{k=1}^{\infty} |x_k - y_k| \sum_{i=1}^{\infty} |a_{ki}| |x_i - y_i|$$

$$\leq \sum_{k=1}^{\infty} (x_k + y_k) \sum_{i=1}^{\infty} (x_i + y_i)$$

$$= 4 < \infty.$$

According to $a_{ki} = -a_{ik}$, we find

$$\sum_{k=1}^{\infty} (x_k - y_k) \sum_{i=1}^{\infty} a_{ki}(x_i - y_i) = 0.$$

Consequently,

$$(x_k - y_k) \sum_{i=1}^{\infty} a_{ki}(x_i - y_i) = 0, \quad k \in \mathbb{N}.$$

Equality (7.11) together with the last equality implies that $\mathbf{x} = \mathbf{y}$. Thus, $V : S \to S$ is injective.

Now let us show that $V$ is onto. Define

$$A_1 = \{[1,n] \subset \mathbb{N} : n \in \mathbb{N}\},$$

$$A_2 = \{a \subset [1,n] : |[1,n] \setminus a| \geq 2, n \in \mathbb{N}\},$$

$$A_3 = \{b \subset \mathbb{N} : a \subset b, a \in A_1 \cup A_2, |\mathbb{N} \setminus b| < \infty, \},$$

$$A = A_1 \cup A_2 \cup A_3.$$

Let us introduce an order on $A$ by inclusion, i.e. $a \leq b$ means that $a \subset b$ for $a, b \in A$. It is clear that $A$ is a completely ordered set. To prove that $V$ is surjective we will use a transfer induction method with respect to the set $A$. It is obvious that, for the first element $\{1\}$ of the set $A$, the operator $V$ on $S^{\{1\}}$ is surjective (see [74]). Assume that for an element $a \in A$ the operator $V$ is surjective on $S^b$ for every $b < a$. Let us show that it is surjective on $S^a$. Suppose that $V(S^a) \neq S^a$. For the boundary $\partial S^a$ of $S^a$ we have $\partial S^a = \bigcup_{c \in A : c < a} S^c$. According to the assumption of induction one gets

$$V(\partial S^a) = \partial S^a. \tag{7.12}$$

On the other hand, there exist $\mathbf{x}, \mathbf{y} \in riS^a$ such that $x \in V(S^a)$ and $y \notin V(S^a)$. The segment $[\mathbf{x}, \mathbf{y}]$ contains at least one boundary point $\mathbf{z}$ of the set $V(S^a)$. Since $V : S^a \to V(S^a)$ is a continuous bijection, the boundary point is mapped to a boundary point. Therefore, for $\mathbf{z} \in riS^a$ one has $V^{-1}(z) \in \partial S^a$, which contradicts (7.12). Thus the theorem is proved.

## 7.3  The Set of Volterra Operators

In this section we are going prove that the set $\mathcal{V}$ is compact, and describe its extremal points.

Now we endow $\mathcal{QV}$ with a topology which is defined by the following system of semi-norms:

$$p_{\varphi, \psi, k}(V) = |(V(\varphi, \psi))_k|, \quad V \in \mathcal{QV},$$

where $\varphi, \psi \in S$ and $k \in \mathbb{N}$. This topology is called the *weak topology* and is denoted by $\tau_w$.

A net $\{V_\mathbb{N}\}$ of quadratic operators converges to $V$ with respect to $\tau_w$ if for every $\varphi, \psi \in S$ and $k \in \mathbb{N}$ one has

$$(V_\nu(\varphi, \psi))_k \to (V(\varphi, \psi))_k.$$

Since $\mathcal{V} \subset \mathcal{QV}$, on $\mathcal{V}$ we consider the topology induced by $\mathcal{QV}$.

In Chap. 5, Sect. 5.2 we proved that the set of all quantum quadratic stochastic operators defined on von Neumann algebra forms a weakly compact convex set. In the present situation we cannot apply the mentioned result, since in this chapter q.q.s.o.s corresponding to Volterra operators are normal (i.e. ultraweakly continuous). In general, the set of all normal q.q.s.o.s is not weakly compact.[1] Therefore, here we use another method to prove the weak compactness of $\mathscr{V}$.

By $\mathbb{A}$ we denote the set of all matrices $(a_{ki})$ satisfying (7.7). It is clear that $\mathbb{A}$ is convex. The set $\mathbb{A}$ can be considered as a subset of the space

$$\ell^{\infty}(\mathbb{N} \times \mathbb{N}) = \{\mathbf{x} = (x_{n,m}) : x_{n,m} \in \mathbb{R}, \ n, m \in \mathbb{N}, \ \|\mathbf{x}\|_{\infty} = \sup_{n,m \in \mathbb{N}} |x_{n,m}|\}.$$

It is well known [25] that the space

$$\ell^{1}(\mathbb{N} \times \mathbb{N}) = \{\mathbf{x} = (x_{n,m}) : x_{n,m} \in \mathbb{R}, \ n, m \in \mathbb{N}, \ \|\mathbf{x}\|_{1} = \sum_{n,m \in \mathbb{N}} |x_{n,m}| < \infty\}$$

is pre-dual to $\ell^{\infty}(\mathbb{N} \times \mathbb{N})$, i.e. $\ell^{1}(\mathbb{N} \times \mathbb{N})^{*} = \ell^{\infty}(\mathbb{N} \times \mathbb{N})$. Therefore on $\ell^{\infty}(\mathbb{N} \times \mathbb{N})$ we can consider the $\sigma(\ell^{\infty}(\mathbb{N} \times \mathbb{N}), \ell^{1}(\mathbb{N} \times \mathbb{N}))$-topology. In the sequel, we will denote it by $\tau$. According to the Banach–Alaoglu theorem the set $\mathbb{A}$ is $\sigma(\ell^{\infty}(\mathbb{N} \times \mathbb{N}), \ell^{1}(\mathbb{N} \times \mathbb{N}))$-weakly compact in $\ell^{\infty}(\mathbb{N} \times \mathbb{N})$.

From Theorem 7.2.3 we conclude that every $(a_{ki})$ matrix with the property (7.7) defines a Volterra operator $V$ of the form (7.6) (see also (7.8)). So, one can define a mapping $T : \mathbb{A} \to \mathscr{V}$. It is clear that Theorem 7.2.3 and Proposition 7.2.2 imply that this map is a bijection and convex.

**Theorem 7.3.1** *The map* $T : (\mathbb{A}, \tau) \to (\mathscr{V}, \tau_w)$ *is continuous.*

*Proof* Let a net $(a_{ki}^{(\upsilon)}) \subset \mathbb{A}$ converge to $(a_{ki})$ in the weak topology. This means that for an arbitrary $\varepsilon > 0$ and every $k, i \in \mathbb{N}$ there is a $\nu_0(ki)$ such that $|a_{ki}^{(\upsilon)} - a_{ki}| < \varepsilon$ for every $n \geq \nu_0(ki)$. Let $V^{(\upsilon)} = T((a_{ki}^{\upsilon}))$ and $V = T((a_{ki}))$.

Take any $\mathbf{x}, \mathbf{y} \in S$. Then there is a number $N_0 \in \mathbb{N}$ such that

$$\sum_{i=N_0+1}^{\infty} x_i < \varepsilon, \qquad \sum_{i=N_0+1}^{\infty} y_i < \varepsilon. \tag{7.13}$$

Now consider two separate cases.

---

[1]Each state $\omega \in S(M)$ defines a linear positive operator as $T(x) = \omega(x)\mathbb{1}$. So, according to Observation 7.1.1 the set of all normal states can be included in $\mathscr{QV}$. Therefore, we can consider the induced weak topology (defined as above) on $S(M)$. It is clear that this topology coincides with the $*$-topology on $S(M)$, but in this topology $S(M)$ is not compact. Hence, $\mathscr{QV}$ is not weakly compact.

Case (i). In this case, we assume that $1 \leq k \leq N_0$. Then according to Corollary 7.2.4 and using (7.7) and (7.13) we infer that

$$\left| (\tilde{V}^{(v)}(\mathbf{x}, \mathbf{y}))_k - (\tilde{V}(\mathbf{x}, \mathbf{y}))_k \right| \leq \frac{1}{2} \left( \sum_{i \in \mathbb{N}} (y_k x_i + x_k y_i) |a_{ki}^{(v)} - a_{ki}| \right)$$

$$\leq \frac{1}{2} \left( \sum_{i=1}^{N_0} (x_i + y_i) |a_{ki}^{(v)} - a_{ki}| \right) + \sum_{i=N_0+1}^{\infty} (x_i + y_i)$$

$$< 3\varepsilon$$

for every $v \geq \max\{v_0(ki) : k, i \leq N_0\}$. Here we have used that

$$\sum_{i=1}^{N_0} (x_i + y_i) \leq \sum_{i=1}^{\infty} (x_i + y_i) = 2.$$

Case (ii). Now assume that $k \geq N_0 + 1$. Using the above argument one gets

$$\left| (\tilde{V}^{(v)}(\mathbf{x}, \mathbf{y}))_k - (\tilde{V}(\mathbf{x}, \mathbf{y}))_k \right|$$

$$\leq \frac{1}{2} \left( \sum_{i \in \mathbb{N}} (y_k x_i + x_k y_i) |a_{ki}^{(v)} - a_{ki}| \right)$$

$$\leq y_k + x_k \leq \sum_{i=N_0+1}^{\infty} (x_i + y_i) < 2\varepsilon$$

for every $v \geq \max\{v_0(ki) : k, i \leq N_0\}$. Hence, the map $T$ is continuous. The theorem is proved.

**Corollary 7.3.2** *The set $\mathscr{V}$ is convex and weakly compact.*

The proof immediately comes from the fact that $\mathbb{A}$ is compact and $T$ is continuous.

We say that a q.o. $V \in \mathscr{QV}$ is *pure* if for every $\varphi, \psi \in S$ one has

$$\tilde{V}(\varphi, \psi) \in Extr \Gamma(\varphi, \psi) = \{\varphi, \psi\}.$$

It is clear that pure q.o.s are Volterra.

**Proposition 7.3.3** *The set $\mathscr{V}$ is convex. Moreover, $V$ is an extreme point of $\mathscr{V}$ if and only if it is pure.*

*Proof* The convexity of $\mathscr{V}$ is obvious. Let $V$ be a pure q.o. Let us assume that there exist $\lambda \in (0, 1)$ and operators $V_1, V_2 \in \mathscr{V}$ such that $V = \lambda V_1 + (1 - \lambda)V_2$.

Let $\varphi\,\psi \in Extr(S)$, then we have

$$\tilde{V}(\varphi, \psi) = \lambda \tilde{V}_1(\varphi, \psi) + (1 - \lambda)\tilde{V}_2(\varphi, \psi). \tag{7.14}$$

Without loss of generality, we may suppose that $\tilde{V}(\varphi, \psi) = \varphi$, since $V$ is pure. Therefore, the extremity of $\varphi$ with (7.14) implies that $V_i(\varphi, \psi) = \varphi$, $i = 1, 2$. Hence, $V_1(\varphi, \psi) = V_2(\varphi, \psi)$ for every $\varphi, \psi \in Extr(S)$. According to Proposition 7.2.2 one gets $V = V_1 = V_2$. Hence, $V \in Extr(\mathscr{V})$.

We now let $V \in Extr(\mathscr{V})$ and show that $V$ is pure. Assume that $V$ is not pure, i.e. there are $\varphi_0, \psi_0 \in Extr(S)$ and a number $\lambda \in (0, 1)$ such that $\tilde{V}(\varphi_0, \psi_0) = \lambda\varphi_0 + (1 - \lambda)\psi_0$. Define q.o.s $V_1$ and $V_2$ as follows:

$$\tilde{V}_1(\varphi_0, \psi_0) = \varphi_0, \quad \tilde{V}_2(\varphi_0, \psi_0) = \psi_0,$$

$$\tilde{V}_i(\varphi, \psi) = \tilde{V}(\varphi, \psi), \quad \forall \varphi, \psi \in Extr(S), \; \varphi, \psi \notin \{\varphi_0, \psi_0\}.$$

Then again using Proposition 7.2.2 we get $V = \lambda V_1 + (1 - \lambda)V_2$, which contradicts the extremity of $V$. This completes the proof.

We note that Proposition 7.3.3 can also be proved by means of Theorem 7.2.3 and Corollary 7.2.4.

From Corollary 7.3.2 and Proposition 7.3.3 we have the following

**Corollary 7.3.4** *A Volterra operator $V \in \mathscr{V}$ is extremal if and only if for the associated skew-symmetric matrix $(a_{ki})$ one has $|a_{ki}| = 1$, for every $k, i \in \mathbb{N}$.*

The proof immediately follows since the extremal points of $\mathbb{A}$ satisfy the last condition and the map $T$ is convex and a bijection.

## 7.4 The Limit Behavior of Volterra Operators

In this section we give some limit theorems concerning the trajectories of Volterra operators.

Let $V : S \rightarrow S$ be a Volterra operator. According to Theorem 7.2.3 it has the form (7.6).

Define

$$\mathbf{Q}_V = \{\mathbf{y} \in S : \sum_{i=1}^{\infty} a_{ki}y_i \leq 0, \; k \in \mathbb{N}\}. \tag{7.15}$$

It is clear that $\mathbf{Q}$ is convex subset of $S$.

**Proposition 7.4.1** *For every Volterra operator $V$ one has $\mathbf{Q}_V \subset Fix(V)$.*

*Proof* Let $\mathbf{y} \in \mathbf{Q}_V$. Then

$$(V(\mathbf{y}))_k = y_k(1 + \sum_{i=1}^{\infty} a_{ki}y_i) \leq y_k, \quad k \in \mathbb{N}. \tag{7.16}$$

According to equality $\sum_{i=1}^{\infty} y_i = \sum_{i=1}^{\infty} (V(\mathbf{y}))_i = 1$, from (7.16) we find $(V(\mathbf{y}))_k = y_k$ for every $k \in \mathbb{N}$, i.e. $V\mathbf{y} = \mathbf{y}$.

**Theorem 7.4.2** *Let $V$ be a Volterra operator such that $\mathbf{Q}_V \neq \emptyset$. Suppose $\mathbf{x}^0 \in riS$ (i.e. $x_i^0 > 0, \forall i \in \mathbb{N}$) such that $V\mathbf{x}^0 \neq \mathbf{x}^0$ and the limit $\lim\limits_{n\to\infty} V^n\mathbf{x}^0$ exists. Then $\lim\limits_{n\to\infty} V^n\mathbf{x}^0 \in \mathbf{Q}_V$.*

*Proof* Let $\mathbf{x}^0 \in riS$ and $\lim\limits_{n\to\infty} \mathbf{x}^{(n)} = \tilde{\mathbf{x}}$, where $\mathbf{x}^{(n)} = V^n\mathbf{x}^0$, $n \in \mathbb{N}$. Let $\tilde{\mathbf{x}} = (q_1, q_2, \ldots, q_n, \ldots)$. It is clear that $V\tilde{\mathbf{x}} = \tilde{\mathbf{x}}$. Hence

$$q_k = q_k\left(1 + \sum_{i=1}^{\infty} a_{ki}q_i\right), \quad k \in \mathbb{N}. \tag{7.17}$$

Set $I_+ = \{i \in \mathbb{N}|q_i > 0\}$ and $I_0 = \{i \in \mathbb{N}|q_i = 0\}$. If $k \in I_+$, then from (7.17) we get

$$\sum_{i=1}^{\infty} a_{ki}q_i = 0, \quad k \in I_+.$$

Assume that there is a $k_0 \in I_0$ such that

$$\sum_{i=1}^{\infty} a_{k_0i}q_i > 0.$$

Since $x_k^{(m)} \to q_k$, there is an $m_0 \in \mathbb{N}$ such that

$$\sum_{i=1}^{\infty} a_{k_0i}x_i^{(m)} > 0, \quad \text{for every } m \geq m_0. \tag{7.18}$$

According to Corollary 7.2.6 we have $\mathbf{x}^{(m)} \in riS$, $\forall m \in \mathbb{N}$, i.e. $x_k^{(m)} > 0$, $\forall m, k \in \mathbb{N}$. The inequality (7.18) with

$$x_{k_0}^{(m+1)} = x_{k_0}^{(m)}\left(1 + \sum_{i=1}^{\infty} a_{k_0i}x_i^{(m)}\right) > x_{k_0}^{(m)} \quad \forall m \geq m_0$$

implies that $x_{k_0}^{(m+1)} > x_{k_0}^{(m)}$, which contradicts $x_{k_0}^{(m)} \to q_{k_0} = 0$. Therefore if $k \in I_0$, then $\sum_{i=1}^{\infty} a_{ki}q_i \leq 0$. Thus $\tilde{\mathbf{x}} \in \mathbf{Q}_V$. The theorem is proved.

Given a Volterra operator $V$ and $K \subset \mathbb{N}$ we set $V_K = V|_{S^K}$. Let $\mathbf{Q}_{V_K}$ be the set $\mathbf{Q}_V$ corresponding to $V_K$. Then from Theorem 7.4.2 and Corollary 7.2.6 we immediately get

**Corollary 7.4.3** *Let $\mathbf{Q}_{V_K} \neq \emptyset$ and $\mathbf{x}^0 \in riS^K$ (i.e. $x_i^0 > 0, \forall i \in K$) such that $V\mathbf{x}^0 \neq \mathbf{x}^0$ and the limit $\lim_{n\to\infty} V^n\mathbf{x}^0$ exists. Then $\lim_{n\to\infty} V^n\mathbf{x}^0 \in \mathbf{Q}_{V_K}$.*

**Corollary 7.4.4** *Assume that a Volterra operator $V$ has an isolated fixed point $\mathbf{x}^0 \in Fix(V)$ (i.e. there is a weak neighbor $U(\mathbf{x}^0) \subset S$ of $\mathbf{x}^0$ such that $U(\mathbf{x}^0) \cap Fix(V) = \{\mathbf{x}^0\}$) such that $\mathbf{x}^0 \in riS$. Then for any $\mathbf{x} \in riS, \mathbf{x} \notin Fix(V)$ the limit $\lim_{n\to\infty} V^n\mathbf{x}$ does not exist.*

*Proof* Suppose that $\lim_{n\to\infty} V^n\mathbf{x} = \bar{\mathbf{x}}$ exists. Then according to Theorem 7.4.2 we have $\bar{\mathbf{x}} \in \mathbf{Q}_V$. On the other hand, from $\mathbf{x}^0 \in Fix(V)$ and $\mathbf{x}^0 \in riS$ one finds $\mathbf{x}^0 \in \mathbf{Q}_V$. Therefore, the convexity of $\mathbf{Q}_V$ yields that $\lambda\bar{\mathbf{x}} + (1-\lambda)\mathbf{x}^0 \in \mathbf{Q}$ for every $\lambda \in [0, 1]$. However, this contradicts the fact that $\mathbf{x}^0$ is isolated. This completes the proof.

*Remark 7.4.1* It is known [74] that the set $\mathbf{Q}_V$ is not empty for any Volterra operator defined on a finite-dimensional space. But unfortunately, in our situation $\mathbf{Q}_V$ might be empty.

Let us give some more examples of q.o.s for which $\mathbf{Q}_V$ is empty and non empty.

*Example 7.4.1* Let us consider a Volterra operator defined as follows:

$$\begin{cases} (Vx)_{2k-1} = x_{2k-1}(1 - a^{(k)}x_{2k}), \\ (Vx)_{2k} = x_{2k}(1 + a^{(k)}x_{2k-1}), \end{cases} \quad k \in \mathbb{N}$$

where $0 < a^{(k)} \leq 1$.

Let us describe $\mathbf{Q}_V$ for the defined $V$. To this end we should find solutions of the system:

$$\begin{cases} -a^{(k)}x_{2k} \leq 0, \\ a^{(k)}x_{2k-1} \leq 0, \end{cases} \quad k \in \mathbb{N}$$

One easily gets that $\mathbf{Q}_V = \{x \in S : x_{2k-1} = 0, k \in \mathbb{N}\}$. So, $\mathbf{Q}_V \neq \emptyset$.

Let $\mathbf{x} \in riS$, then the trajectory of $\mathbf{x}$ is defined by the following recurrent relations

$$\begin{cases} x_{2k-1}^{(m+1)} = x_{2k-1}^{(m)}(1 - a^{(k)}x_{2k}^{(m)}), \\ x_{2k}^{(m+1)} = x_{2k}^{(m)}(1 + a^{(k)}x_{2k-1}^{(m)}), \end{cases} \quad k \in \mathbb{N}, \ m \in \mathbb{N}.$$

According to $a^{(k)} > 0$ we find $1 + a^{(k)}x_{2k-1}^{(m)} > 0$, hence we have $x_{2k}^{(m+1)} \geq x_{2k}^{(m)}$, therefore $\{x_{2k}^{(m)}\}$ is a non-decreasing sequence. From $0 \leq 1 - a^{(k)}x_{2k}^{(m)} \leq 1$ it follows that $\{x_{2k-1}^{(m)}\}$ is a non-increasing sequence such that $0 \leq x_{2k}^{(m)}, x_{2k-1}^{(m)} \leq 1$. So the limits

$$\lim_{m\to\infty} x_{2k-1}^{(m)} = \alpha_{2k-1}, \quad \lim_{m\to\infty} x_{2k}^{(m)} = \beta_{2k}$$

exist.

According to Theorem 7.4.2 we infer that $\alpha_{2k-1} = 0$ for every $k \in \mathbb{N}$.

Now let $\mathbf{x} \notin riS$ and $I_x = \{k \in \mathbb{N} : x_k = 0\}$. Then using Corollary 7.2.5 we find $V(S^{\mathbb{N}\backslash I_x}) = S^{\mathbb{N}\backslash I_x}$. The restriction of $V$ to $S^{\mathbb{N}\backslash I_x}$ is denoted by $V_{\mathbb{N}\backslash I_x}$. From definition of $S^{\mathbb{N}\backslash I_x}$ we find that $x \in riS^{\mathbb{N}\backslash I_x}$, whence according to Corollary 7.4.3 we obtain

$$\lim_{m\to\infty} x_{2k-1}^{(m)} = 0, \quad \lim_{m\to\infty} x_{2k}^{(m)} = \begin{cases} \beta_{2k}, & 2k \in \mathbb{N} \setminus I_x, \\ 0, & 2k \in I_x. \end{cases}$$

*Example 7.4.2* Let us define a Volterra operator as follows:

$$(V(x))_k = x_k\left(1 + \sum_{i=1}^{\infty} a_{ki}x_i\right), \quad k \in \mathbb{N}, \tag{7.19}$$

where $a_{ki} = (-1)^i$, $a_{ik} = -a_{ki}$ at $i \geq k + 1$.

Then it is not hard to check that the set $\mathbf{Q}_V$ consists of solutions of the following system

$$\begin{cases} \sum_{k=2}^{\infty} (-1)^{k+1}x_k \leq 0, \\ x_1 + \sum_{k=3}^{\infty} (-1)^k x_k \leq 0, \\ -x_1 + x_2 + \sum_{k=4}^{\infty} (-1)^{k+1}x_k \leq 0, \\ \dots\dots\dots\dots\dots\dots\dots\dots\dots \\ \sum_{k=2}^{n-1} (-1)^{n+k}x_k + \sum_{k=n+1}^{\infty} (-1)^{n+k+1}x_k \leq 0, \\ \dots\dots\dots\dots\dots\dots\dots\dots\dots \end{cases} \tag{7.20}$$

Whence one gets $x_n \leq x_{n+1}$ for every $k \in \mathbb{N}$. Since $x_1 \geq 0$ and $x_n \to 0$ at $n \to \infty$, we obtain $x_n = 0$, $\forall n \in \mathbb{N}$, which is impossible, because $\sum_{k=1}^{\infty} x_k = 1$. Consequently, $\mathbf{Q}_V = \emptyset$.

Now let us look for the set $Fix(V)$. Let $\mathbf{x}^0 \in riS$, i.e. $x_k^0 > 0$, $\forall k \in \mathbb{N}$, be a fixed point of $V$. It follows from (7.19) and (7.20) that

$$x_1^0 = x_2^0 = \ldots x_k^0 = \cdots, \quad k \in \mathbb{N},$$

but this equality is impossible since $x_1^0 \neq 0$ and $x_n^0 \to 0$. Hence, interior fixed points for $V$ do not exist. So, there is a subset $I \subset \mathbb{N}$ such that $I = \{k \in \mathbb{N}|x_k^0 = 0\}$. The set $\mathbb{N} \setminus I$ is finite. Indeed, assume that $|\mathbb{N} \setminus I| = \infty$, then consider a face $S^{\mathbb{N}\setminus I}$. Then according to Corollary 7.2.5, $V_{\mathbb{N}\setminus I}$ is a Volterra operator. It is clear that the point $\mathbf{x}^{0,\mathbb{N}\setminus I} = \{x_k^0|\ k \in \mathbb{N} \setminus I\}$ is a fixed point of $V_{\mathbb{N}\setminus I}$. Using the same argument as above, from (7.19) we find that the set $J = \{k \in \mathbb{N}|x_k^{0,\mathbb{N}\setminus I} = 0\}$ is non-empty, which contradicts the choice of $I$. Consequently, we infer that all fixed points of $V$ lie on faces $S^I$ such that $|\mathbb{N} \setminus I| < \infty$. Hence, we conclude that the set $\mathbf{Q}_V$ turns out to be empty while the set $Fix(V)$ is not. Therefore, Theorem 7.4.2 implies that if $\mathbf{x} \in riS$ then the limit $\lim_{n\to\infty} V^n\mathbf{x}$ does not exist. Now let $\mathbf{x} \notin riS$, then for the set $I_x$ there are two possibilities. First assume that $|\mathbb{N} \setminus I_x| = \infty$, then $\mathbf{x} \in riS^{\mathbb{N}\setminus I_x}$. From condition $|\mathbb{N} \setminus I_x| = \infty$, reasoning analogously as above, one can show that the set $\mathbf{Q}_{V_I}$ is empty. According to Corollary 7.4.3, we infer that the limit $\lim_{n\to\infty} V^n\mathbf{x}$ does not exist. Now suppose that $|\mathbb{N}\setminus I_x| < \infty$. Then the operator $V_I$ reduces to a finite-dimensional operator, therefore the set $\mathbf{Q}_{V_I}$ is not empty (see [74]). So, the limit $\lim_{n\to\infty} V^n\mathbf{x}$ exists, since $|a_{ik}| = 1$ (see [74]).

Now we will give a sufficient condition for $V$ which ensures that the set $\mathbf{Q}_V$ is not empty.

Let $V : S \to S$ be a Volterra operator which has the form (7.6). Let $A = (a_{ki})$ be the corresponding skew-symmetric matrix. Furthermore, we will assume that $A$ acts on $\ell^1$. A matrix $A$ is called *finite-dimensional* if $A(\ell^1)$ is finite-dimensional. We say that $A$ is *finitely generated* if there is a sequence of finite-dimensional matrices $\{A_n\}$ such that $\sup_n \|A_n\| < \infty$ and

$$A = A_1 \oplus A_2 \oplus \cdots \oplus A_n \oplus \cdots$$

**Proposition 7.4.5** *Let* $A = (a_{ki})$ *be a skew-symmetric, finitely generated matrix corresponding to a Volterra operator (see (7.6)). Then the system*

$$\sum_{i=1}^{\infty} a_{ki}y_i \geq 0, \quad k \in \mathbb{N} \tag{7.21}$$

*has at least one element belonging to S.*

*Proof* First assume that $A$ is finite-dimensional, i.e. there is an $n \in \mathbb{N}$ such that $A(\ell^1) = \mathbb{R}^n$. According to the skew-symmetricity of $A$, we find that $a_{ij} = 0$ for

$i, j \geq n + 1$. Therefore we may assume that $A$ acts on $\mathbb{R}^n$. Then (7.21) is rewritten as follows

$$\sum_{j=1}^{n} a_{kj} y_j \geq 0, \quad k = 1, \ldots, n. \tag{7.22}$$

According to [74] this system has a solution $y = \{y_k\}_{k=1}^{n} \in S^{n-1}$ such that (7.22) holds. Now define an element $\tilde{y} = \{\tilde{y}_k\}_{k=1}^{\infty} \in S$ by

$$\tilde{y}_k = \begin{cases} y_k, & \text{if } 1 \leq k \leq n, \\ 0, & \text{if } k \geq n + 1. \end{cases}$$

It is evident that $A\tilde{y} \geq 0$.

Now let us assume that $A$ is finitely generated, i.e. $A = A_1 \oplus A_2 \oplus \cdots \oplus A_n \oplus \cdots$. Since the operators $A_n$ are finite-dimensional, suppose that for every $n \in \mathbb{N}$ there is an $m_n \in \mathbb{N}$ such that $A_n$ acts on $\mathbb{R}^{m_n}$, i.e. $A_n : \mathbb{R}^{m_n} \to \mathbb{R}^{m_n}$. Consider the system

$$A_n \mathbf{y}^{(n)} \geq 0, \quad n \in \mathbb{N}.$$

According to the argument above, for every $n \in \mathbb{N}$, there is an element $\mathbf{z}^{(n)} \in S^{m_n - 1}$ such that $A_n \mathbf{z}^{(n)} \geq 0$. Define $\mathbf{z} = (z_k)_{k=1}^{\infty}$ by

$$\mathbf{z} = \frac{1}{2} \mathbf{z}^{(1)} \oplus \frac{1}{2^2} \mathbf{z}^{(2)} \oplus \cdots \oplus \frac{1}{2^n} \mathbf{z}^{(n)} \oplus \cdots$$

From

$$\sum_{k=1}^{\infty} z_k = \frac{1}{2} \sum_{k=1}^{m_1} z_k^{(1)} + \frac{1}{2^2} \sum_{k=1}^{m_2} z_k^{(2)} + \ldots + \frac{1}{2^n} \sum_{k=1}^{m_n} z_k^{(n)} + \ldots$$

$$= \frac{1}{2} + \frac{1}{2^2} + \ldots + \frac{1}{2^n} + \ldots$$

$$= 1$$

we see that $\mathbf{z} \in S$. From

$$A\mathbf{z} = \frac{1}{2} A_1 \mathbf{z}^{(1)} \oplus \cdots \oplus \frac{1}{2^n} A_n \mathbf{z}^{(n)} \oplus \cdots \geq 0$$

we conclude that the element $\mathbf{z}$ is a solution of (7.21). The proposition is proved.

**Corollary 7.4.6** *Let the condition of the previous proposition hold. Then the set $\mathbf{Q}_V$ is not empty.*

The proof immediately comes from Proposition 7.4.5 by changing the matrix $A$ to $-A$, since $-A$ is also skew-symmetric.

## 7.5 Extension of Finite-Dimensional Volterra Operators

In this section we are going to construct infinite-dimensional Volterra operators by means of finite-dimensional ones.

Let $K_n = [1, n] \cap \mathbb{N}$ for every $n \in \mathbb{N}$. Consider a sequence $V_{n]} : S^{K_n} \to S^{K_n}$ of finite-dimensional Volterra operators, i.e.

$$(V_{n]}(x))_k = x_k \left( 1 + \sum_{i=1}^{n} a_{ki}^{n]} x_i \right) \quad k = 1, \cdots, n, \quad n \in \mathbb{N}. \tag{7.23}$$

Here $(a_{ki}^{n]})$ is a skew-symmetric matrix.

We say that the sequence $\{V_{n]}\}$ of Volterra operators is *compatible* if

$$V_{n+1]} \upharpoonright S^{K_n} = V_{n]} \tag{7.24}$$

for every $n \in \mathbb{N}$. The compatibility condition with (7.23) implies that

$$a_{ki}^{n+1]} = a_{ki}^{n]}, \quad \forall k, i \in \{1, \cdots, n\}. \tag{7.25}$$

Define

$$S^{[n} = \left\{ x = (x_n, x_{n+1}, \cdots) : x_k \geq 0, \forall k \geq n, \sum_{k=n}^{\infty} x_k = 1 \right\}, \quad n \in \mathbb{N}.$$

Let $\{W_{[n} : S^{[n} \to S^{[n} : n \in \mathbb{N}\}$ be a sequence of Volterra operators

$$(W_{[n}(x))_k = x_k (1 + \sum_{i=n}^{\infty} a_{ki}^{[n} x_i) \quad k \geq n, \quad n \in \mathbb{N}. \tag{7.26}$$

Define a sequence $\{\mathbf{W}_n : S \to S, \ n \in \mathbb{N}\}$ of infinite-dimensional operators as follows

$$(\mathbf{W}_n(x))_k = \begin{cases} (V_{n]}(x))_k, & \text{if } n \leq k, \\ (W_{[n+1}(x))_k, & \text{if } k \geq n+1, \end{cases} \quad n \in \mathbb{N}. \tag{7.27}$$

According to Theorem 7.2.3 the defined operators are Volterra.

**Theorem 7.5.1** *The sequence $\{\mathbf{W}_n\}$ of Volterra operators weakly converges to a Volterra operator $\mathbf{W}$. Moreover, if each $V_{n]}$ is pure, then so is $\mathbf{W}$.*

*Proof* Let $x \in S$. If there is a finite subset $K$ of $\mathbb{N}$ such that $x \in S^K$, then according to the compatibility condition (7.24) we get $\mathbf{W}(x) = \mathbf{W}_n(x)$ for all $n \geq \max\{m : m \in K\}$.

Now assume that $x_i > 0$ for all $i \in \mathbb{N}$. Let us prove that $\{\mathbf{W}_n(x)\}$ is a Cauchy sequence with respect to the weak topology. Let $\varepsilon > 0$ be an arbitrary number. Since $x \in S$ there is a number $n_0 \in \mathbb{N}$ such that

$$\sum_{j=n+1}^{\infty} x_j < \varepsilon, \quad \forall n \geq n_0. \tag{7.28}$$

We consider several cases:

Case (i).   Suppose that $1 \leq k \leq n$. Using (7.27), (7.25), (7.23), (7.7) and (7.28) we have

$$|(\mathbf{W}_n(x))_k - (\mathbf{W}_{n+p}(x))_k| = |(V_{n]}(x))_k - (V_{n+p]}(x))_k|$$

$$= \left| x_k \left( 1 + \sum_{i=1}^{n} a_{ki}^{n]} x_i \right) - x_k \left( 1 + \sum_{j=1}^{n+p} a_{kj}^{n+p]} x_j \right) \right|$$

$$\leq x_k \left( \sum_{j=n+1}^{n+p} x_j \right)$$

$$\leq \sum_{j=n+1}^{\infty} x_j < \varepsilon \tag{7.29}$$

for all $n \geq n_0$.

Case (ii).   Assume $n + 1 \leq k \leq n + p$. It then follows from (7.26) and (7.27) that

$$|(\mathbf{W}_n(x))_k - (\mathbf{W}_{n+p}(x))_k| = |(W_{[n+1}(x))_k - (V_{n+p]}(x))_k|$$

$$= \left| x_k \left( 1 + \sum_{j=n+1}^{\infty} a_{kj}^{[n} x_j \right) - x_k \left( 1 + \sum_{j=1}^{n} a_{kj}^{n+p]} x_j \right) \right|$$

$$\leq x_k \sum_{j=1}^{\infty} |\gamma_{ki}| x_j$$

$$\leq 2x_k < 2\varepsilon \tag{7.30}$$

for all $n \geq n_0$. Here

$$\gamma_{kj} = \begin{cases} a_{kj}^{n+p]} & \text{if } j \leq n, \\ a_{kj}^{n+p]} + a_{kj}^{[n} & \text{if } n + 1 \leq j \leq n + p, \\ a_{kj}^{[n+1} & \text{if } j \geq n + p + 1. \end{cases}$$

Case (iii). Now assume that $k \geq n + p + 1$. Then from (7.27) we have

$$
\begin{aligned}
&|(W_{[n}(x))_k - (W_{[n+p}(x))_k| \\
&= \left| x_k \left( 1 + \sum_{j=n+1}^{\infty} a_{ki}^{[n+1]} x_j \right) - x_k \left( 1 + \sum_{j=n+p+1}^{\infty} a_{kj}^{[n+p+1]} x_j \right) \right| \\
&\leq 2x_k \sum_{j=n+1}^{\infty} x_j \\
&< 2\varepsilon^2.
\end{aligned}
\tag{7.31}
$$

Hence the sequence $(W_n(x))$ is Cauchy, therefore $W_n(x) \to W(x)$. In the same way, we can show that $\tilde{W}_n(x, y) \to \tilde{W}(x, y)$. Because of $W_n(e^{(i)}, e^{(j)}) \in \Gamma(e^{(i)}, e^{(j)})$ and the compatibility condition we find that $W$ is Volterra. According to (7.24), for every $e^{(i)}$ and $e^{(j)}$ there is an $n_0 \in \mathbb{N}$ such that $\tilde{W}(e^{(i)}, e^{(j)}) = \tilde{V}_{n]}(e^{(i)}, e^{(j)})$ for all $n \geq n_0$. If each $V_{n]}$ is pure for all $n \in \mathbb{N}$ then $W$ is also pure. The theorem is proved.

Let $\{V_n : S \to S, \ n \in \mathbb{N}\}$ be a sequence of operators associated with (see (7.27))

$$(W_{[n}(x))_k = x_k, \quad k \geq n, \ n \in \mathbb{N}.$$

According to Theorem 7.5.1 the defined sequence $\{V_n\}$ converges to a Volterra operator $V$.

There naturally arises a question: are the operators $V$ and $W$ equal? The next theorem gives an affirmative answer to this question.

**Theorem 7.5.2** *The operators $V$ and $W$ are equal.*

*Proof* Let $\varepsilon > 0$ be an arbitrary number and $x \in S$ be fixed. To prove the assertion, it is enough to show, for every $k \in \mathbb{N}$, that the following relation holds

$$|(W_n(x))_k - (V_n(x))_k| < \varepsilon.$$

There is a number $n_0 \in \mathbb{N}$ such that (7.28) holds. Consider two cases.

case (i). Let $1 \leq k \leq n$. Then (7.27) implies that

$$|(W_n(x))_k - (V_n(x))_k| = 0,$$

for every $n \geq n_0$.

case (ii). Let $k \geq n + 1$, then it follows from (7.28) that

$$|(W_n(x))_k - (V_n(x))_k| = \left| x_k \left( 1 + \sum_{j=n+1}^{\infty} a_{kj}^{[n]} x_j \right) - x_k \right| \leq x_k \sum_{j=n+1}^{\infty} x_j < \varepsilon.$$

Hence, we have proved the desired relation. This completes the proof.

Thus according to the last theorem, we will consider only the sequence $\{\mathbf{V}_n\}$. Now we are interested in the convergence of powers of the sequence $\{\mathbf{V}_n\}$.

Let $V$ be an arbitrary Volterra operator. By $V^m$ we will denote the $m$th iteration of $V$, i.e. $V^m(x) = \underbrace{V(V \cdots (V(x)) \cdots )}_{m}$. Before formulating the result we need an auxiliary fact.

**Lemma 7.5.3** *Let $V$ be an arbitrary Volterra operator. Then $(V^m(x))_k \leq 2^m x_k$, for every $k, m \in \mathbb{N}$ and $x \in S$.*

*Proof* According to Theorem 7.2.3 we have

$$(V^m(x))_k = (V^{m-1}(x))_k \left( 1 + \sum_{i=1}^{\infty} a_{ki}(V^{m-1}(x))_i \right)$$

$$\leq 2(V^{m-1}(x))_k \leq \cdots \leq 2^m x_k,$$

which is the required relation.

**Theorem 7.5.4** *For every $m \in \mathbb{N}$ the sequence $\{\mathbf{V}_n^m\}$ converges.*

*Proof* To show the convergence it is enough to prove that $\{\mathbf{V}_n^m(x)\}$ is a Cauchy sequence for every $x \in S$. Without loss of generality, we may assume that $x_k > 0$ for all $k \in \mathbb{N}$.

Let $\varepsilon > 0$ be an arbitrary number. Since $x \in S$ there is a number $n_0 \in \mathbb{N}$ such that (7.28) holds. We consider several cases.

Case (i). Let $1 \leq k \leq n$ and $p \in \mathbb{N}$ be an arbitrary number. For the sake of brevity we will put

$$a_k^{(s)} = (V_{n]}^s(x))_k, \quad b_k^{(s)} = (V_{n+p]}^s(x))_k, \tag{7.32}$$

where $s, k \in \mathbb{N}$. Then from (7.25), (7.27) and (7.32) we have

$$|(\mathbf{V}_n^m(x))_k - (\mathbf{V}_{n+p}^m(x))_k|$$

$$= |a_k^{(m)} - b_k^{(m)}|$$

$$= \left| a_k^{(m-1)} \left( 1 + \sum_{i=1}^{\infty} a_{ki}^{n]} a_i^{(m-1)} \right) - b_k^{(m-1)} \left( 1 + \sum_{i=1}^{\infty} a_{ki}^{n+p]} b_i^{(m-1)} \right) \right|$$

$$\leq |a_k^{(m-1)} - b_k^{(m-1)}| + \left| \sum_{i=1}^{n} a_{ki}^{n]} \left( a_k^{(m-1)} a_i^{(m-1)} - b_k^{(m-1)} b_i^{(m-1)} \right) \right|$$

$$+b_k^{(m-1)} \sum_{j=n+1}^{n+p} |a_{kj}^{n+p]}| b_j^{(m-1)}$$

$$\leq |a_k^{(m-1)} - b_k^{(m-1)}| + |a_k^{(m-1)} - b_k^{(m-1)}| \sum_{i=1}^{n} |a_{ki}^{n]}| a_i^{(m-1)}$$

$$+b_k^{(m-1)} \sum_{i=1}^{n} |a_{ki}^{n]}||a_i^{(m-1)} - b_i^{(m-1)}| + b_k^{(m-1)} \sum_{j=n+1}^{n+p} b_j^{(m-1)}$$

$$\leq 2|a_k^{(m-1)} - b_k^{(m-1)}| + b_k^{(m-1)} \sum_{i=1}^{n} |a_i^{(m-1)} - b_i^{(m-1)}|$$

$$+b_k^{(m-1)} \sum_{j=n+1}^{n+p} b_j^{(m-1)}. \tag{7.33}$$

Now we need the following.

**Lemma 7.5.5** *For every $m \in \mathbb{N}$ the following inequality holds*

$$|a_k^{(m)} - b_k^{(m)}| \leq \alpha_m x_k \sum_{j=n+1}^{n+p} x_j, \tag{7.34}$$

*where*

$$\alpha_1 = 1, \ \alpha_m = \alpha_{m-1}(2 + 2^{m-1}) + 2^{2(m-1)}, \quad m \geq 2.$$

*Proof* Let us first consider the case $m = 1$. We have

$$|a_k^{(1)} - b_k^{(1)}| = |(V_{n]}(x))_k - (V_{n+p]}(x))_k$$

$$= \left| x_k \left( \sum_{j=n+1}^{n+p} a_{kj}^{n+p]} x_j \right) \right|$$

$$\leq x_k \sum_{j=n+1}^{n+p} x_j.$$

This shows that $\alpha_1 = 1$. Now assume that (7.34) is valid for $m - 1$. Show that it is true for $m$. Indeed, it follows from (7.33) and Lemma 7.5.3 that

$$|a_k^{(m)} - b_k^{(m)}| \leq 2\alpha_{m-1}x_k \sum_{j=n+1}^{n+p} x_j + \alpha_{m-1}2^{m-1}x_k \sum_{i=1}^{n} x_i \sum_{j=n+1}^{n+p} x_j$$

$$+ 2^{2(m-1)}x_k \sum_{j=n+1}^{n+p} x_j$$

$$\leq (\alpha_{m-1}(2 + 2^{m-1}) + 2^{2(m-1)})x_k \sum_{j=n+1}^{n+p} x_j,$$

which proves the lemma.

Now we continue the proof of Theorem 7.5.4. According to Lemma 7.5.5 we find that $|(\mathbf{V}_n^m(x))_k - (\mathbf{V}_{n+p}^m(x))_k| < \varepsilon$, for every $n \geq n_0$.

Case (ii). Let $n + 1 \leq k \leq n + p$. We have

$$|(\mathbf{V}_n^m(x))_k - (\mathbf{V}_{n+p}^m(x))_k| = \left| x_k - b_k^{(m-1)}\left(1 + \sum_{i=1}^{\infty} a_{ki}^{n+p]}b_i^{(m-1)}\right)\right|$$

$$\leq |x_k - b_k^{(m-1)}| + b_k^{(m-1)} \sum_{i=1}^{n+p} b_i^{(m-1)}$$

$$\leq |x_k - b_k^{(m-1)}| + 2^{m-1}x_k. \tag{7.35}$$

Now consider

$$|x_k - b_k^{(m-1)}| \leq |x_k - b_k^{(m-2)}| + b_k^{(m-2)} \sum_{i=1}^{n+p} b_i^{(m-2)}$$

$$\leq \cdots \leq |x_k - b_k^{(1)}| + \sum_{j=1}^{m-2} b_k^{(j)} \sum_{i=1}^{n+p} b_i^{(j)}$$

$$\leq x_k \sum_{i=1}^{n+p} x_i + x_k \sum_{j=1}^{m-2} 2^j$$

$$\leq x_k \sum_{j=0}^{m-2} 2^j.$$

Hence, from (7.35) we infer that

$$|(\mathbf{V}_n^m(x))_k - (\mathbf{V}_{n+p}^m(x))_k| \le \left(\sum_{j=0}^{m-1} 2^j\right) \varepsilon$$

for every $n \ge n_0$.

Now let $k \ge n + p$. Then $(\mathbf{V}_n^m(x))_k = (\mathbf{V}_{n+p}^m(x))_k$.

Thus we have proved that $\{\mathbf{V}_n^m(x)\}$ is a Cauchy sequence. The limit of this sequence we denote by $\mathbf{W}_m(x)$. The theorem is proved.

From this theorem, there naturally arises a question: does the equality $\mathbf{W}_m = \mathbf{V}^m$ hold?

Before answering this question we should prove an auxiliary fact.

**Lemma 7.5.6** *Let $V$ be an arbitrary Volterra operator. Then the following inequality holds*

$$\|V(x) - V(y)\|_1 \le 3\|x - y\|_1$$

*for every $x, y \in S$.*

*Proof* We have

$$\|V(x) - V(y)\|_1 = \sum_{k=1}^{\infty} |(V(x))_k - (V(y))_k|$$

$$\le \sum_{k=1}^{\infty} \left(\left(1 + \sum_{i=1}^{\infty} |a_{ki}| x_i\right) |x_k - y_k| + x_k \sum_{i=1}^{\infty} |a_{ki}| |x_i - y_i|\right)$$

$$\le 2 \sum_{k=1}^{\infty} |x_k - y_k| + \sum_{i=1}^{\infty} |x_i - y_i|$$

$$= 3\|x - y\|_1.$$

This completes the proof.

**Theorem 7.5.7** *For every $m \in \mathbb{N}$ the equality $\mathbf{W}_m = \mathbf{V}^m$ is valid.*

*Proof* Let $x \in S$ be an arbitrary element. Then given $\varepsilon > 0$ there is a number $n \in \mathbb{N}$ and $y \in S^{K_n}$ such that $\|x - y\|_1 < \varepsilon$. According to the compatibility condition (7.24) we have $\mathbf{V}(y) \in S^{K_n}$ and hence $\mathbf{V}^m(y) = V_{n]}^m(y)$. Therefore $\mathbf{W}_m(y) = \mathbf{V}^m(y)$. Using this we have

$$|(\mathbf{W}_m(x))_k - (\mathbf{V}^m(x))_k| \le |(\mathbf{W}_m(x))_k - (\mathbf{W}_m(y))_k| + |(\mathbf{V}^m(x))_k - (\mathbf{V}^m(x))_k| \qquad (7.36)$$

for every $k \in \mathbb{N}$.

According to Theorem 7.5.4 we know that there is an $n_0 \in \mathbb{N}$ such that

$$|(\mathbf{W}_m(x))_k - (\mathbf{V}_n^m(x))_k| < \varepsilon \tag{7.37}$$

for every $n \geq n_0$.

Using Lemma 7.5.6 one gets

$$|(\mathbf{V}_n^m(x))_k - (\mathbf{V}_n^m(y))_k| \leq \|\mathbf{V}_n^m(x) - \mathbf{V}_n^m(y)\|_1 \leq 3^m \|x - y\|_1 < 3^m \varepsilon. \tag{7.38}$$

It follows from (7.37) and (7.38) that

$$|(\mathbf{W}_m(x))_k - (\mathbf{W}_m(y))_k| \leq |(\mathbf{W}_m(x))_k - (\mathbf{V}_n^m(x))_k| + |(\mathbf{V}_n^m(x))_k - (\mathbf{V}_n^m(y))_k|$$
$$+ |(\mathbf{V}_n^m(y))_k - (\mathbf{W}_m(y))_k| \leq (1 + 3^m)\varepsilon. \tag{7.39}$$

Here we have used the equality $\mathbf{W}_m(y) = \mathbf{V}_n^m(y)$.

Now, again using Lemma 7.5.6, we find

$$|(\mathbf{V}^m(x))_k - (\mathbf{V}^m(y))_k| \leq 3^m \|x - y\|_1 < 3^m \varepsilon. \tag{7.40}$$

Consequently, the inequalities (7.39) and (7.40) with (7.36) imply that

$$|(\mathbf{W}_m(x))_k - (\mathbf{V}^m(x))_k| < (1 + 2 \cdot 3^m)\varepsilon.$$

As $\varepsilon$ was arbitrary, this completes the proof.

This theorem gives us information on how to investigate the limiting behavior of infinite-dimensional Volterra operators by means of finite-dimensional ones.

## 7.6  Comments and References

In Chap. 5 we defined q.q.s.o.s on arbitrary von Neumann algebras (vNas). In the present Chap. 7 we have introduced a class of q.q.s.o.s defined on the commutative vNa $\ell^\infty$. We have investigated infinite-dimensional Volterra q.q.s.o.s. The results of this chapter are taken from [163, 184]. Recently in [198] we have provided another alternative definition of Volterra operators (which is based on the absolutely continuity of the measures). We think this definition will allow us to investigate noncommutative analogues of Volterra operators.

Infinite-dimensional analogues of the results relating to Sects. 2.3–2.4 of Chap. 2 can be found in [15] (see also [16]). Some extensions of these operators have been studied in [190]. Recently, some infinite-dimensional q.s.o.s have been discussed in [53]. Finite-dimensional Volterra quadratic operators are investigated in [74, 76, 80, 130, 191, 220, 222, 251]. In [52, 54, 55, 69, 221, 254] it is proved that the trajectory of two-dimensional Volterra quadratic operators may behave chaotically, i.e. the averages of the trajectory is not convergent. This gives a negative answer to a hypothesis of Ulam [250]. Some generalizations of Volterra operators have been investigated in [186, 192, 193, 211, 213, 214]. Note that any permutation of Volterra operator corresponds to an automorphism of the simplex. Such operators have been investigated in [77–79, 81, 83]. Some other kinds of Lotka–Volterra type operators are investigated in [37, 99, 100, 150, 152, 248].

# Chapter 8
# Quantum Quadratic Stochastic Processes and Their Ergodic Properties

Quadratic stochastic processes describe the physical systems defined above, but they do not cover the cases at a quantum level. So, it is natural to define a concept of a quantum quadratic process. In this chapter we will define a quantum (noncommutative) analogue of quadratic stochastic processes. In our case, such a process will be defined on a von Neumann algebra, and we will study its ergodic properties such as the ergodic principle, regularity, etc. From the physical point of view, such a principle means that for sufficiently large values of time a system described by the process does not depend on the initial state of the system.

## 8.1 Quantum Quadratic Stochastic Processes

Let $\mathcal{M}$ be a von Neumann algebra. The set of all continuous (resp. ultraweakly continuous) functionals on $\mathcal{M}$ is denoted by $\mathcal{M}^*$ (resp. $\mathcal{M}_*$). As before, by $S$ and $S^2$ we denote the set of all normal states on $\mathcal{M}$ and $\mathcal{M} \otimes \mathcal{M}$, respectively. Recall a mapping $U : \mathcal{M} \otimes \mathcal{M} \to \mathcal{M} \otimes \mathcal{M}$ is a linear operator such that $U(x \otimes y) = y \otimes x$ for all $x, y \in \mathcal{M}$.

Recall that given a state $\varphi \in S$ the conditional expectation operator $E_\varphi : \mathcal{M} \otimes \mathcal{M} \to \mathcal{M}$ is defined by

$$E_\varphi(a \otimes b) = \varphi(a)b \tag{8.1}$$

and extend it by linearity and continuity to $\mathcal{M} \otimes \mathcal{M}$. Clearly, such an operator is completely positive and $E_\varphi \mathbb{1}_{\mathcal{M} \otimes \mathcal{M}} = \mathbb{1}_{\mathcal{M}}$.

Let $\{P^{s,t} : \mathcal{M} \to \mathcal{M} \otimes \mathcal{M}, s, t \in \mathbb{R}_+, t - s \geq 1\}$ be a family of linear operators.

© Springer International Publishing Switzerland 2015
F. Mukhamedov, N. Ganikhodjaev, *Quantum Quadratic Operators and Processes*,
Lecture Notes in Mathematics 2133, DOI 10.1007/978-3-319-22837-2_8

**Definition 8.1.1** We say that a pair $(\{P^{s,t}\}, \omega_0)$, where $\omega_0 \in S$ is an initial state, forms a *quantum quadratic stochastic process (q.q.s.p.)* if each operator $P^{s,t}$ is ultraweakly continuous and the following conditions hold:

(i) each operator $P^{s,t}$ is a completely positive q.q.s.o.;
(ii) An analogue of the Kolmogorov–Chapman equation is satisfied: for initial state $\omega_0 \in S$ and arbitrary numbers $s, \tau, t \in \mathbb{R}_+$ with $\tau - s \geq 1$, $t - \tau \geq 1$ one has either

(ii)$_A$   $P^{s,t}x = P^{s,\tau}(E_{\omega_\tau}(P^{\tau,t}x))$,   $x \in \mathcal{M}$

or

(ii)$_B$   $P^{s,t}x = E_{\omega_s}P^{s,\tau} \otimes E_{\omega_s}P^{s,\tau}(P^{\tau,t}x)$,   $x \in \mathcal{M}$,

where $\omega_\tau(x) = \omega_0 \otimes \omega_0(P^{0,\tau}x)$, $x \in \mathcal{M}$.

If a q.q.s.p. satisfies the fundamental equations (ii)$_A$ (resp. (ii)$_B$) then we say that the q.q.s.p. is of *type (A)* (resp. *type (B)*).

The unital complete positivity (ucp) of each $P^{s,t}$ implies that

$$\|P^{s,t}x\| \leq \|x\| \qquad \text{for any } x \in \mathcal{M}. \tag{8.2}$$

*Remark 8.1.1* By means of q.q.s.p.s one can specify a law of interaction of states $\varphi, \psi \in S$ by

$$V^{s,t}(\varphi, \psi)(x) = \varphi \otimes \psi(P^{s,t}x), \qquad x \in \mathcal{M}.$$

This equality gives a rule according to which the state $V^{s,t}(\varphi, \psi)$ appears at time $t$ as a result of the interaction of states $\varphi$ and $\psi$ at time $s$. From the physical point of view, the interaction of states can be explained as follows: Consider two physical systems separated by a barrier and assume that one of these systems is in the state $\varphi$ and the other one is in the state $\psi$. Upon the removal of the barrier, the new physical system is in the state $\varphi \otimes \psi$ and, as a result of the action of the operator $P^{s,t}$, a new state is formed. This state is exactly the result of the interaction of the states $\varphi$ and $\psi$.

*Remark 8.1.2* If $\mathcal{M}$ is a commutative algebra, i.e. $\mathcal{M} = L^\infty(X, \mathfrak{F})$, then a quantum quadratic stochastic process coincides with a quadratic stochastic process. Indeed, we set

$$P(s, u, v, t, A) = P^{s,t}(\chi_A)(u, v), \quad u, v \in X,$$

where $\chi_A$ is the indicator of the set $A \in \mathfrak{F}$. Then, by Definition 8.1.1, the family of functions $\{P(s, u, v, t, A)\}$ forms a quadratic stochastic process (see Chap. 3). Conversely, if we have a quadratic stochastic process $(\{P(s, u, v, t, A)\}, \mu_0)$ on a measurable space $(E, \mathfrak{F})$ then one can define a quantum quadratic stochastic process on $L^\infty(X, \mathfrak{F})$ as follows:

$$(P_p^{s,t}f)(u, v) = \int_E f(w)P(s, u, v, t, dw), \ f \in L^\infty(X, \mathfrak{F}).$$

As the initial state, we take the following state

$$\omega_0(f) = \int_E f(u)\mu_0(du),$$

where $\mu_0$ is the initial measure for the quadratic stochastic process. Note that in this case, the expectation $E_\mu$ of a function $f(u, v)$ is defined by

$$(E_\mu(f))(v) = \int_E f(u, v)\mu(du).$$

Here $\mu$ is a measure on $(E, \mathfrak{F})$. Taking this into consideration, one can easily check that the conditions of Definition 8.1.1 are satisfied. Thus, the notion of a quantum quadratic stochastic process generalizes the notion of a quadratic stochastic process.

As in the commutative case (see Chap. 3), q.q.s.p.s can be divided into the following classes:

(i) *homogeneous*, i.e. such that each $P^{s,t}$ depends only on the difference $t - s$ for every $s$ and $t$ such that $t - s \geq 1$.
(ii) *homogeneous for unit interval of time*, i.e. satisfying $P^{s,s+1} = P^{0,1}$ for all $s \geq 1$, but not homogeneous.
(iii) *non-homogeneous*, which do not belong to the classes (i) or (ii).

*Example 8.1.1* Let $\mathcal{M}$ be a von Neumann algebra and let $\{T_t\}_{t \geq 1} : \mathcal{M} \to \mathcal{M}$ be a semigroup of ultraweakly continuous ucp maps. Then define

$$P^{s,t}x = \frac{1}{2}(T^{s,t}x \otimes \mathbb{1} + \mathbb{1} \otimes T^{s,t}x), \tag{8.3}$$

where

$$T^{s,t}x = \frac{1}{2^{t-s-1}}(T^{t-s}x + (2^{t-s-1} - 1)\omega_0(T^t x)\mathbb{1}), \ x \in \mathcal{M}, \ t - s \geq 1.$$

One can check that $(P^{s,t}, \omega_0)$ is a non-homogeneous q.q.s.p. of both types A and B.

If one assumes that a state $\omega_0 \in S$ is invariant with respect to $T_t$, that is $\omega_0(T_t x) = \omega_0(x)$ for all $x \in \mathcal{M}, t \geq 1$, then $(P^{s,t}, \omega_0)$ is a homogeneous q.q.s.p.

*Example 8.1.2* Let $\{T^{s,t} : s, t \in \mathbb{R}_+, t - s \geq 1\}$ be a family of ultraweakly continuous ucp maps $\mathcal{M}$ and let $\omega_0 \in S$ be a state. Define $P^{s,t}$ by (8.3).

If the family $\{T^{s,t}\}$ satisfies the condition

$$T^{s,t}x = \frac{1}{2}(T^{s,\tau}T^{\tau,t}x + \omega_0(T^{0,\tau}T^{\tau,t}x)\mathbb{1}), \ x \in \mathcal{M},$$

for any $\tau - s \geq 1$, $t - \tau \geq 1$. Then $(\{P^{s,t}\}, \omega_0)$ is a q.q.s.p. of type (A).

If the family $\{T^{s,t}\}$ satisfies the condition

$$T^{s,t}x = \frac{1}{2}(T^{s,\tau}T^{\tau,t}x + \omega_0(T^{0,s}T^{s,\tau}T^{\tau,t}x)\mathbb{1}), \quad x \in \mathcal{M},$$

and $T^{s,s+1} = T^{0,1}$, for any $\tau - s \geq 1$, $t - \tau \geq 1$. Then $(\{P^{s,t}\}, \omega_0)$ is a q.q.s.p. of type (B) with the initial state $\omega_0$.

By $P_*^{s,t}$ we denote the linear operator mapping from $(\mathcal{M} \otimes \mathcal{M})_*$ into $\mathcal{M}_*$ given by

$$P_*^{s,t}(\varphi)(x) = \varphi(P^{s,t}x), \quad \varphi \in (\mathcal{M} \otimes \mathcal{M})_*, \ x \in \mathcal{M}. \tag{8.4}$$

From Definition 8.1.1 we have the following.

**Observation 8.1.1** *The defined family* $(\{P_*^{s,t}\}, \omega_0)$ *satisfies the following properties*

(i) $P_*^{s,t}(S^2) \subset S$;

(ii) $P_*^{s,t}(\varphi \otimes \psi) = P_*^{s,t}(\psi \otimes \varphi)$ *for all* $\varphi, \psi \in S$;

(iii) *Let* $s, \tau, t \in \mathbb{R}_+$ *such that* $\tau - s \geq 1$ *and* $t - \tau \geq 1$, *then*

(iii)$_A$ $P_*^{s,t}(\varphi) = P_*^{\tau,t}(P_*^{s,\tau}(\varphi) \otimes \omega_\tau)$, $\varphi \in S^2$;

(iii)$_B$ $P_*^{s,t}(\varphi \otimes \psi) = P_*^{\tau,t}(P_*^{s,\tau}(\omega_s \otimes \varphi) \otimes P_*^{s,\tau}(\omega_s \otimes \psi))$, $\varphi, \psi \in \mathcal{M}_*$,

*where* $\omega_s = P_*^{0,s}(\omega_0 \otimes \omega_0)$. *Here we have used the equality* $\varphi(E_\psi(x)) = \varphi \otimes \psi(x)$, $\varphi, \psi \in S$. *Moreover, from* (8.2) *we get*

$$\|P_*^{s,t}f\|_1 \leq \|f\|_1 \qquad \text{for any } f \in (\mathcal{M} \otimes \mathcal{M})_*. \tag{8.5}$$

In what follows, we need the following auxiliary fact.

**Lemma 8.1.1** *Let* $(\{P^{s,t}\}, \omega_0)$ *be a homogeneous q.q.s.p. on a von Neumann algebra* $\mathcal{M}$. *Then* $\omega_2 = \omega_t$ *holds for every* $t \geq 2$.

*Proof* First we assume that the q.q.s.p. $(\{P^{s,t}\}, \omega_0)$ has type (A). From (iii)$_A$ of Observation 8.1.1 and the homogeneity of the q.q.s.p., one gets

$$\begin{aligned}
\omega_t &= P_*^{0,t}(\omega_0 \otimes \omega_0) = P_*^{s-t,s}(\omega_0 \otimes \omega_0) \\
&= P_*^{s-1,s}(P_*^{s-t,s-1}(\omega_0 \otimes \omega_0) \otimes \omega_{s-1}) \\
&= P_*^{0,1}(P_*^{0,t-1}(\omega_0 \otimes \omega_0) \otimes \omega_{s-1}) \\
&= P_*^{0,1}(\omega_{t-1} \otimes \omega_{s-1}),
\end{aligned} \tag{8.6}$$

for any $t \geq 1$ and $s \geq t$.

In particular, from (8.6) we find

$$\omega_t = P^{0,1}_*(\omega_{t-1} \otimes \omega_{t-1}) \tag{8.7}$$

$$\omega_2 = P^{0,1}_*(\omega_1 \otimes \omega_{s-1}) \tag{8.8}$$

for any $t \geq 1$ and $s \geq 1$.

Again using (iii)$_A$ of Observation 8.1.1 we obtain

$$\omega_t = P^{t-2,t}_*(P^{0,t-2}_*(\omega_0 \otimes \omega_0) \otimes \omega_{t-2})$$
$$= P^{0,2}_*(\omega_{t-2} \otimes \omega_{t-2}) \quad \text{for all } t \geq 2.$$

The last equality with (8.7) and the homogeneity of the q.q.s.p. yield that

$$\omega_t = P^{0,2}_*(\omega_{t-2} \otimes \omega_{t-2}) = P^{s-2,s}_*(\omega_{t-2} \otimes \omega_{t-2})$$
$$= P^{s-1,s}_*(P^{s-2,s-1}_*(\omega_{t-2} \otimes \omega_{t-2}) \otimes \omega_{s-1})$$
$$= P^{0,1}_*(P^{0,1}_*(\omega_{t-2} \otimes \omega_{t-2}) \otimes \omega_{s-1})$$
$$= P^{0,1}_*(\omega_{t-1} \otimes \omega_{s-1}),$$

for any $t \geq 2$ and $s \geq 2$. So, the last equality with $s = 2$ implies that

$$\omega_t = P^{0,1}_*(\omega_{t-1} \otimes \omega_1),$$

which together with (8.8) and (ii) of Observation 8.1.1 yields $\omega_t = \omega_2$ for all $t \geq 2$.

Now let us assume that the q.q.s.p. $(\{P^{s,t}\}, \omega_0)$ has type (B). From (iii)$_B$ of Observation 8.1.1 and the homogeneity of the q.q.s.p. one gets

$$\omega_t = P^{0,t}_*(\omega_0 \otimes \omega_0) = P^{s-t,s}_*(\omega_0 \otimes \omega_0)$$
$$= P^{s-1,s}_*(P^{s-t,s-1}_*(\omega_{s-t} \otimes \omega_0) \otimes P^{s-t,s-1}_*(\omega_{s-t} \otimes \omega_0))$$
$$= P^{0,1}_*(P^{0,t-1}_*(\omega_{s-t} \otimes \omega_0) \otimes P^{0,t-1}_*(\omega_{s-t} \otimes \omega_0)) \tag{8.9}$$

for any $t \geq 1$ and $s \geq t$.

On the other hand, again using (iii)$_B$ of Observation 8.1.1 one finds

$$\omega_t = P^{t-2,t}_*(P^{0,t-2}_*(\omega_0 \otimes \omega_0) \otimes P^{0,t-2}_*(\omega_0 \otimes \omega_0))$$
$$= P^{0,2}_*(\omega_{t-2} \otimes \omega_{t-2}). \tag{8.10}$$

In (8.9) by taking $t = 2$, we obtain

$$\omega_2 = P^{0,1}_*(P^{0,1}_*(\omega_{s-2} \otimes \omega_0) \otimes P^{0,1}_*(\omega_{s-2} \otimes \omega_0))$$
$$= P^{1,2}_*(P^{0,1}_*(\omega_0 \otimes \omega_{s-2}) \otimes P^{0,1}_*(\omega_0 \otimes \omega_{s-2}))$$
$$= P^{0,2}_*(\omega_{s-2} \otimes \omega_{s-2}). \tag{8.11}$$

Hence, the last equality with (8.10) yields that $\omega_2 = \omega_s$ for all $s \geq 2$. Consequently, the lemma is proved.

## 8.2   Analytic Methods for q.q.s.p.s

In this section we are going to derive partial differential equations for q.q.s.p.s, as in Chap. 4.

In this section, for the sake of transparency, we consider only q.q.s.p.s ($\{P^{s,t}\}, \omega_0$) of type (A) on a von Neumann algebra $\mathcal{M}$. In what follows, we suppose that the q.q.s.p. is taken from the class (ii). Furthermore, we will assume that $P^{s,t}$ is continuous with respect to both parameters $s$ and $t$, i.e. for every $x \in \mathcal{M}$ the following relations hold

$$\lim_{t \to t_0} \|P^{s,t}x - P^{s,t_0}x\| = 0, \quad \lim_{s \to s_0} \|P^{s,t}x - P^{s_0,t}x\| = 0.$$

In addition, we also assume that $P^{s,t}$ is differentiable with respect to $s$ and $t$ when $t - s \geq 2$.

For $t > s + 2$, according to (ii)$_A$ we have

$$P^{s,t+\delta}x - P^{s,t}x = P^{s,t-1}E_{\omega_{t-1}}(P^{t-1,t+\delta}x) - P^{s,t-1}E_{\omega_{t-1}}(P^{t-1,t}x)$$
$$= P^{s,t-1}E_{\omega_{t-1}}[(P^{t-1,t+\delta} - P^{t-1,t+\delta})x], \quad x \in \mathcal{M}. \quad (8.12)$$

Put

$$\tilde{A}(t)x = \lim_{\delta \to 0+} \frac{(P^{t-1,t+\delta} - P)x}{\delta}, \quad (8.13)$$

where $P := P^{0,1}$. We denote by $D(\tilde{A})$ the set of all $x \in \mathcal{M}$ for which (8.13) exists. Dividing both sides of (8.12) by $\delta$ and passing to the limit we get the first partial differential equation

$$\frac{\partial P^{s,t}x}{\partial t} = P^{s,t-1}E_{\omega_{t-1}}(\tilde{A}(t)x), \quad x \in D(\tilde{A}). \quad (8.14)$$

According to the definition of $\omega_{t-1}$, (8.14) can be rewritten as

$$\frac{\partial P^{s,t}x}{\partial t} = (P^{s,t-1} \otimes P_*^{0,t-1}(\omega_0 \otimes \omega_0))(\tilde{A}(t)x), \quad x \in D(\tilde{A}). \quad (8.15)$$

We analogously get

$$P^{s,t}x - P^{s+\delta,t}x = [(P^{s,s+1+\delta} - P)E_{\omega_{s+1+\delta}}](P^{s+1\delta,t}x) \quad (8.16)$$

for $t > s + 2$. Now dividing both sides of (8.16) by $\delta$ and passing to the limit, we get the second partial differential equation:

$$\frac{\partial P^{s,t}x}{\partial s} = -\tilde{A}(s+1)E_{\omega_{s+1}}(P^{s+1,t}x).$$

Taking into account the definition of $\omega_{s+1}$ one finds

$$\frac{\partial P^{s,t}x}{\partial s} = -(\tilde{A}(s+1) \otimes P_*^{0,s+1}(\omega_0 \otimes \omega_0))(P^{s+1,t}x). \tag{8.17}$$

**Theorem 8.2.1** *Let* $(\{P^{s,t}\}, \omega_0)$ *be a q.q.s.p. of type (A) from the class (ii) on a von Neumann algebra* $\mathcal{M}$. *Then it satisfies the Eqs.* (8.15) *and* (8.17).

*Remark 8.2.1* Using the same argument as above one can derive partial differential equations for type (B) processes.

*Remark 8.2.2* We note that if a von Neumann algebra is commutative then the obtained Eqs. (8.17) and (8.17) coincide with known equations derived in Chap. 4.

In the sequel, we are going to consider only homogeneous q.q.s.p.s. That means $P^{s,t}$ is replaced by $P^{t-s}$. Then the Eqs. (8.15) and (8.17) are reduced to the following system:

$$\begin{cases} \frac{\partial P^{t-s}x}{\partial t} = (P^{t-s-1} \otimes P_*^{t-1}(\omega_0 \otimes \omega_0))(\tilde{A}x), \\ \frac{\partial P^{t-s}x}{\partial s} = -(\tilde{A} \otimes P_*^{s+1}(\omega_0 \otimes \omega_0))(P^{t-s-1}x), \end{cases} \tag{8.18}$$

where $\tilde{A} = \tilde{A}(1) = \tilde{A}(t)$, $t \geq 1$.

If we put $s = 0$ in (8.18) we then get

$$\frac{dP^t x}{dt} = (P^{t-1} \otimes P_*^{t-1}(\omega_0 \otimes \omega_0))(\tilde{A}x). \tag{8.19}$$

According to the homogeneity of the q.q.s.p. it is enough to study the Eq. (8.19).

Thanks to Lemma 8.1.1, the Eq. (8.19) reduces to

$$\begin{cases} \frac{dP^t x}{dt} = (P^{t-1} \otimes P_*^{t-1}(\omega_0 \otimes \omega_0))(\tilde{A}x), & 2 < t \leq 3 \\ \frac{dP^t x}{dt} = (P^{t-1} \otimes \omega_2)(\tilde{A}x), & t \geq 3. \end{cases} \tag{8.20}$$

An important problem in the theory of analytic methods is how to find the process $P^t$ from the Eq. (8.20) with given initial condition

$$P^t = Q^t, \quad 1 \leq t \leq 2,$$

where $Q^t$ is the given process, such that $\frac{dQ^t}{dt}\big|_{t=1} = \tilde{A}$.

Let us consider a simple problem:

$$\begin{cases} \frac{dP^t x}{dt} = (P^{t-1} \circ P_*^{t-1}(\omega_0 \otimes \omega_0))(\tilde{A}x), & 2 \le t \\ P^t = Q^t & 1 \le t \le 2, \end{cases} \qquad (8.21)$$

where

$$Q^{(t)} = \frac{1}{2}(\beta^{\{t\}}x \otimes \mathbb{1} + \mathbb{1} \otimes \beta^{\{t\}}x), \; 1 \le t \le 2$$

$$\beta^{\{t\}} = \frac{1}{2^{t-1}}(T^t x + (2^{t-1} - 1)\omega_0(x)\mathbb{1}), \; x \in \mathcal{M},$$

and $T^t$ is a $C^1$-semigroup of ucp maps with an invariant state $\omega_0$. An infinitesimal operator $\tilde{A}$ is defined by

$$\tilde{A}x = \frac{dQ^t x}{dt} \big|_{t=1}$$

$$= \frac{1}{2}(ATx \otimes \mathbb{1} + \mathbb{1} \otimes ATx - (Tx \otimes \mathbb{1} + \mathbb{1} \otimes Tx)\ln 2 + 2\omega_0(x)(\ln 2)\mathbb{1}),$$

where $A$ is the generator of the semigroup $T^t$.

**Theorem 8.2.2** *For the problem* (8.21) *there exists a unique solution which has the following form:*

$$P^{(t)} = \frac{1}{2}(\beta^{\{t\}}x \otimes \mathbb{1} + \mathbb{1} \otimes \beta^{\{t\}}x), \; 1 \le t \qquad (8.22)$$

$$\beta^{\{t\}} = \frac{1}{2^{t-1}}(T^t x + (2^{t-1} - 1)\omega_0(x)\mathbb{1}), \; x \in \mathcal{M}.$$

*Proof* We are going to solve the problem (8.21) by means of an iteration method. Namely, substituting $Q^t$ into the right-hand side of the Eq. (8.21) and integrating the obtained expression w.r.t. $\tau$ from 2 to $t$, $\tau \in [2, 3]$, i.e. one gets

$$P^{(t)}x - P^{(2)}x = \int_2^t (Q^{\tau-1} \circ \omega_0)(\tilde{A}x)d\tau,$$

where we have used the equality $\omega_0 \otimes \omega_0(Q^t x) = \omega_0(x)$, $t \in [1, 2]$, $x \in \mathcal{M}$, which easily follows from the invariance of $\omega_0$ w.r.t. $T^t$.

One can see the following equalities:

1. $\frac{dT^t x}{dt} = \frac{1}{2^{t-1}}(AT^t x - (\ln 2)T^t x + \ln 2\omega_0(x)\mathbb{1})$;
2. $\omega_0(Ax) = 0$;
3. $\beta^{\{t-1\}}(AT^t x) = \frac{T^t Ax}{2^{t-2}}$;

4. $\beta^{\{t\}}(Tx) = \frac{1}{2^{t-2}}(T^t x + (2^{t-1} - 1)\omega_0(x)\mathbb{1});$
5. $P^{(t)}(ATx) = \frac{1}{2^{t-1}}(T^t Ax \otimes \mathbb{1} + \mathbb{1} \otimes AT^t x);$
6. $P^{(t-1)}(ATx) = \frac{1}{2^{t-1}}(T^t x \otimes \mathbb{1} + \mathbb{1} \otimes T^t x + 2(2^{t-2} - 1)\omega_0(x)\mathbb{1} \otimes \mathbb{1}).$

Hence, according to 2–6 we have

$$(Q^{\tau-1}E_{\omega_0})(\tilde{A}x) = \frac{1}{2^\tau}\left( AT^\tau x \otimes \mathbb{1} + \mathbb{1} \otimes AT^\tau x \right.$$
$$\left. - \ln 2(T^\tau x \otimes \mathbb{1} + \mathbb{1} \otimes T^\tau x) + 2\ln 2\omega_0(x)\mathbb{1} \otimes \mathbb{1} \right),$$

where we have used

$$AT^t = T^t A, \ t \geq 0.$$

Take into account the equalities 1 and 3, we easily obtain

$$P^{(t)}x = \frac{1}{2}(\beta^{\{t\}}x \otimes \mathbb{1} + \mathbb{1} \otimes \beta^{\{t\}}x), \ 2 \leq t \leq 3.$$

Similarly iterating this procedure we obtain that (8.22) is a unique solution.

## 8.3   The Ergodic Principle

This section is devoted to the ergodic principle for q.q.s.p.s. From the physical point of view, such a principle means that for sufficiently large values of time a system described by the process does not depend on the initial state of the system.

**Definition 8.3.1**  A q.q.s.p. $(\{P^{s,t}\}, \omega_0)$ is said to satisfy *the ergodic principle*, or $L_1$-*weak ergodicity*, if for every $\varphi, \psi \in S^2$ and $s \in \mathbb{R}_+$ one has

$$\lim_{t\to\infty} \|P_*^{s,t}\varphi - P_*^{s,t}\psi\|_1 = 0,$$

where $\|\cdot\|_1$ is the norm on $\mathcal{M}^*$.

If the von Neumann algebra $\mathcal{M} = L^\infty(E, \mathfrak{F})$ is commutative, then the definition of the ergodic principle can be stated as follows

$$\lim_{t\to\infty} |P(s, x, y, t, A) - P(s, u, v, t, A)| = 0$$

for any $x, y, u, v \in E$ and $A \in \mathfrak{F}$.

*Remark 8.3.1*  The concept of the ergodic principle was first introduced by A.N. Kolmogorov for Markov processes (see, for example, [121]). For quadratic stochastic processes this concept was introduced in [177, 236]. In [174, 176, 177] this ergodic principle was described as the $L_1$-*weak ergodicity* of non-homogeneous Markov processes.

Similarly as in Chap. 3, we can define the weak ergodicity of a q.q.s.p.

**Definition 8.3.2** A q.q.s.p. $(\{P^{s,t}\}, \omega_0)$ is said to satisfy *weak ergodicity* if for every $s \in \mathbb{R}_+$ one has

$$\lim_{t \to \infty} \sup_{\varphi, \psi \in S^2} \|P_*^{s,t}\varphi - P_*^{s,t}\psi\|_1 = 0.$$

We note that weak ergodicity is a stronger condition than the ergodic principle (see [176]).

Recall that for a given von Neumann algebra $\mathcal{N}$ with unit $\mathbb{1}_{\mathcal{N}}$, by $\mathcal{N}_*^h$ we denote the set of all Hermitian functionals. Let us define

$$\mathcal{F}_0(\mathcal{N}_*^h) = \{f \in \mathcal{N}_*^h : f(\mathbb{1}_{\mathcal{N}}) = 0\}. \tag{8.23}$$

Furthermore, for the sake of brevity let us put

$$\mathcal{F}_1 = \mathcal{F}_0(\mathcal{M}_*^h), \quad \mathcal{F}_2 = \mathcal{F}_0(\mathcal{M}_*^h \otimes_{\alpha_0^*} \mathcal{M}_*^h). \tag{8.24}$$

**Lemma 8.3.1** *The following equality holds*

$$\overline{(\mathcal{M}_*^h \otimes \mathcal{F}_1) \oplus (\mathcal{F}_1 \otimes \mathcal{M}_*^h)}^{\|\cdot\|_1} = \mathcal{F}_2.$$

*Proof* It is known [122] that the subspaces $\mathcal{F}_1$ and $\mathcal{F}_2$ have co-dimensionality 1. Therefore, one has $\mathcal{M}_*^h = \mathbb{R}g_1 \oplus \mathcal{F}_1$ with $g_1(\mathbb{1}) \neq 0$. Hence, we can write

$$\mathcal{M}_*^h \otimes_{\alpha_0^*} \mathcal{M}_*^h = \mathbb{R}(g_1 \otimes g_1) \oplus \mathcal{F}_2 \tag{8.25}$$

since $g_1 \otimes g_1(\mathbb{1} \otimes \mathbb{1}) \neq 0$.

Let $f \in \mathcal{M}_*^h \otimes_{\alpha_0^*} \mathcal{M}_*^h$. Then

$$f = \sum_{i=1}^{\infty} \lambda_i \varphi_i \otimes \psi_i,$$

where $\{\varphi_i\}, \{\psi_i\} \subset \mathcal{M}_*^h$ and $\{\lambda_i\} \subset \mathbb{R}$. Let us put

$$f_N = \sum_{i=1}^{N} \lambda_i \varphi_i \otimes \psi_i, \quad N \in \mathbb{N}.$$

Then one sees that $f_N \to f$ in the $\| \cdot \|_1$-norm. From the equality $\mathcal{M}_*^h = \mathbb{R}g_1 \oplus \mathcal{F}_1$ we have $\varphi_i = \mu_i g_1 + \xi_i$, $\psi_i = \gamma_i g_1 + \eta_i$, where $\{\xi_i\}, \{\eta_i\} \subset \mathcal{F}_1$, hence

$$f_N = \left( \sum_{i=1}^{N} \lambda_i \mu_i \gamma_i \right) g_1 \otimes g_1 + \sum_{i=1}^{N} \lambda_i (\mu_i g_1 \otimes \eta_i + \gamma_i \xi_i \otimes g_1 + \xi_i \otimes \eta_i).$$

This means that $\{f_N\} \subset \mathbb{R}(g_1 \otimes g_1) \oplus \left( (\mathcal{M}_*^h \otimes \mathcal{F}_1) \oplus (\mathcal{F}_1 \otimes \mathcal{M}_*^h) \right)$. Therefore, $f \in \mathbb{R}(g_1 \otimes g_1) \oplus \overline{(\mathcal{M}_*^h \otimes \mathcal{F}_1) \oplus (\mathcal{F}_1 \otimes \mathcal{M}_*^h)}^{\| \cdot \|_1}$, since $(\mathcal{M}_*^h \otimes \mathcal{F}_1) \oplus (\mathcal{F}_1 \otimes \mathcal{M}_*^h) \subset \mathcal{F}_2$. This with (8.25) proves the assertion.

**Lemma 8.3.2** Let $f \in \mathcal{F}_1$. Then there are states $\varphi, \psi \in S$ such that

$$f = \frac{\|f\|_1}{2} (\varphi - \psi). \tag{8.26}$$

*Proof* Since $f$ is Hermitian, there exist positive functionals $f_+, f_- \in (\mathcal{M} \otimes \mathcal{M})_{*,+}$ (see [246, Theorem 4.2, p.140]) such that $f = f_+ - f_-$. The equality $f(\mathbb{1}) = 0$ and positivity of $f_+, f_-$ imply that $\|f_+\|_1 = \|f_-\|_1$. The last one with $\|f\|_1 = \|f_+\|_1 + \|f_-\|_1$ yields that $\|f_+\|_1 = \|f\|_1/2$. Putting

$$\varphi = \frac{f_+}{\|f_+\|_1}, \quad \psi = \frac{f_-}{\|f_-\|_1}$$

we get the required assertion.

*Remark 8.3.2* Note that an analogous assertion also holds for $\mathcal{F}_2$.

**Theorem 8.3.3** Let $(\{P^{s,t}\}, \omega_0)$ be a q.q.s.d.p. on a von Neumann algebra $\mathcal{M}$. The following statements are equivalent:

(i) for every $\sigma, \varphi, \psi \in S$ and $s \in \mathbb{R}_+$

$$\lim_{t \to \infty} \|P_*^{s,t}(\sigma \otimes \varphi) - P_*^{s,t}(\sigma \otimes \psi)\|_1 = 0;$$

(ii) for every $f \in (\mathcal{M}_*^h \otimes \mathcal{F}_1) \oplus (\mathcal{F}_1 \otimes \mathcal{M}_*^h)$ and $s \in \mathbb{R}_+$

$$\lim_{t \to \infty} \|P_*^{s,t}(f)\|_1 = 0;$$

(iii) for every $f \in \mathcal{F}_2$ and $s \in \mathbb{R}_+$

$$\lim_{t \to \infty} \|P_*^{s,t}(f)\|_1 = 0;$$

(iv) $(\{P^{s,t}\}, \omega_0)$ satisfies the ergodic principle.

*Proof* (i)⇒(ii). Let $f \in (\mathcal{M}_*^h \otimes \mathcal{F}_1) \oplus (\mathcal{F}_1 \otimes \mathcal{M}_*^h)$, then one has

$$f = \sum_{n=1}^{N} \lambda_n g_n \otimes \xi_n + \sum_{m=1}^{M} \mu_n \eta_n \otimes h_m, \qquad (8.27)$$

where $\{\xi_n\}_{n=1}^{N}, \{\eta_m\}_{m=1}^{M} \subset \mathcal{F}_1$ and $\{g_n\}_{n=1}^{N}, \{h_m\}_{m=1}^{M} \subset \mathcal{M}_*^h$. Then according to Lemma 8.3.2, for every $n$ and $m$ we get

$$\xi_n = \alpha_n(\xi_{1,n} - \xi_{2,n}), \quad \eta_m = \beta_m(\eta_{1,m} - \eta_{2,m}).$$

Here $\{\xi_{k,n}\}, \{\eta_{k,m}\} \subset S$ and $\alpha_n, \beta_m$ are non-negative numbers. Hence, substituting the last relations into (8.27) we obtain

$$f = \sum_{n=1}^{N} \lambda_n \alpha_n g_n \otimes (\xi_{1,n} - \xi_{2,n}) + \sum_{m=1}^{M} \mu_m \beta_n(\eta_{1,m} - \eta_{2,m}) \otimes h_m. \qquad (8.28)$$

From (i) one finds that

$$\|P_*^{s,t} g_n \otimes (\xi_{1,n} - \xi_{2,n})\|_1 \to 0, \quad \|P_*^{s,t} h_m \otimes (\eta_{1,n} - \eta_{2,m})\|_1 \to 0,$$

for every $n \in \overline{\{1, N\}}, m \in \overline{\{1, M\}}$. Hence, (8.28) implies the required assertion.

(ii)⇒(iii). Take any $f \in \mathcal{F}_2$, then thanks to Lemma 8.3.1 there exists a $g \in (\mathcal{M}_*^h \otimes \mathcal{F}_1) \oplus (\mathcal{F}_1 \otimes \mathcal{M}_*^h)$ such that $\|f - g\|_1 < \epsilon/2$. Condition (iii) implies the existence of a number $t_0 = t_0(s, g)$ such that

$$\|P_*^{s,t} g\|_1 < \epsilon/2 \text{ for every } t \geq t_0.$$

The last inequalities with (8.5) imply

$$\|P_*^{k,n} f\|_1 \leq \|P_*^{k,n}(f - g)\|_1 + \|P_*^{k,n} g\|_1$$
$$< \|f - g\|_1 + \epsilon/2$$
$$< \epsilon \qquad \text{for every } t \geq t_0.$$

The implications (iii)⇒(iv)⇒(i) are obvious.

We say that a q.q.s.p. $(\{P^{s,t}\}, \omega_0)$ satisfies *condition ($A_1$)* on a subset $\mathcal{N} \subset S^2$ if there exists a number $\lambda \in [0, 1)$ such that for every pair $\varphi, \psi \in S$ and $s \in \mathbb{R}_+$ we have

$$\|P_*^{s,t}\varphi - P_*^{s,t}\psi\|_1 \leq \lambda \|\varphi - \psi\|_1 \qquad (8.29)$$

for at least one $t = t(s, \varphi, \psi)$.

We say that a q.q.s.p. ($\{P^{s,t}\}, \omega_0$) satisfies *condition ($A_1$) uniformly* on a subset $\mathcal{N} \subset S^2$ if there exists a number $\lambda \in [0, 1)$ and, for every $s \in \mathbb{R}_+$, there exists a $t = t(s)$ such that

$$\|P_*^{s,t}\varphi - P_*^{s,t}\psi\|_1 \leq \lambda \|\varphi - \psi\|_1 \qquad \text{for all } \varphi, \psi \in \mathcal{N}. \tag{8.30}$$

**Theorem 8.3.4** *Let* ($\{P^{s,t}\}, \omega_0$) *be a q.q.s.p. on a von Neumann algebra $\mathcal{M}$. Then the following statements are equivalent:*

*(i)* ($\{P^{s,t}\}, \omega_0$) *satisfies the ergodic principle;*
*(ii)* ($\{P^{s,t}\}, \omega_0$) *satisfies condition ($A_1$) on $S^2$;*
*(iii)* ($\{P^{s,t}\}, \omega_0$) *satisfies condition ($A_1$) on a dense subset $\mathcal{R}$ of $S^2$.*

*Proof* The implications (i)$\Rightarrow$(ii)$\Rightarrow$(iii) are obvious.

(iii)$\Rightarrow$(i). We are going to consider two distinct cases with respect to the type of the q.q.s.p.

Case (A). In this case, assume that ($\{P^{s,t}\}, \omega_0$) has type (A). Assume that states $\varphi, \psi \in S^2$ and a number $s \in \mathbb{R}_+$ are fixed. Let $t_0 \in \mathbb{R}_+$ be an arbitrary number such that $t_0 \geq s + 1$. Let

$$\varphi_0 = P_*^{s,t_0}\varphi \otimes \omega_{t_0}, \quad \psi_0 = P_*^{s,t_0}\psi \otimes \omega_{t_0}.$$

Then one can see that $\varphi_0, \psi_0 \in S^2$, and (8.5) implies that

$$\|\varphi_0 - \psi_0\|_1 \leq \|\varphi - \psi\|_1. \tag{8.31}$$

Due to the density of $\mathcal{R}$ we can find states $\mu_0, \nu_0 \in \mathcal{R}$ such that

$$\|\varphi_0 - \mu_0\|_1 < \epsilon 2^{-4}, \quad \|\psi_0 - \nu_0\|_1 < \epsilon 2^{-4},$$

where $\epsilon > 0$ is an arbitrary fixed number. For these states, according to condition ($A_1$), one can find a number $t_1 \in \mathbb{R}_+$ such that

$$\|P_*^{t_0, t_0+t_1}\mu_0 - P_*^{t_0, t_0+t_1}\nu_0\|_1 \leq \lambda \|\mu_0 - \nu_0\|_1. \tag{8.32}$$

Now using (iii)$_A$, (8.5), (8.32) and (8.31) one gets

$$\|P_*^{s,t_0+t_1}\varphi - P_*^{s,t_0+t_1}\psi\|_1$$
$$= \|P_*^{t_0,t_0+t_1}(P_*^{s,t_0}\varphi \otimes \omega_{t_0}) - P_*^{t_0,t_0+t_1}(P_*^{s,t_0}\psi \otimes \omega_{t_0})\|_1$$
$$\leq \|P_*^{t_0,t_0+t_1}(\varphi_0 - \mu_0)\|_1 + \|P_*^{t_0,t_0+t_1}(\psi_0 - \nu_0)\|_1 + \|P_*^{t_0,t_0+t_1}(\mu_0 - \nu_0)\|_1$$
$$\leq \|\varphi_0 - \mu_0\|_1 + \|\psi_0 - \nu_0\|_1 + \lambda\|\mu_0 - \nu_0\|_1$$
$$\leq \epsilon 2^{-3} + \lambda(\|\mu_0 - \varphi_0\|_1 + \|\nu_0 - \psi_0\|_1 + \|\varphi_0 - \psi_0\|_1)$$
$$\leq \epsilon 2^{-2} + \lambda\|\varphi - \psi\|_1.$$

Hence, we have

$$\|P_*^{s,t_0+t_1}\varphi - P_*^{s,t_0+t_1}\psi\|_1 \le \epsilon 2^{-2} + \lambda\|\varphi - \psi\|_1.$$

Now assume that numbers $\{t_i\}_{i=0}^m$ have been found such that

$$\|P_*^{s,T_m}\varphi - P_*^{s,T_m}\psi\|_1 \le \epsilon 2^{-2}(1 + 2^{-1} + \cdots + 2^{-(m-1)})$$
$$+ \lambda^m\|\varphi - \psi\|_1, \tag{8.33}$$

where $T_m = \sum_{i=0}^m t_i$. We claim that this inequality holds for $m+1$. Let

$$\varphi_m = P_*^{s,T_m}\varphi \otimes \omega_{T_m}, \quad \psi_m = P_*^{s,T_m}\psi \otimes \omega_{T_m}.$$

Then $\varphi_m, \psi_m \in S^2$ and one can find states $\mu_m, \nu_m \in \mathscr{R}$ such that

$$\|\varphi_m - \mu_m\|_1 < \epsilon 2^{-(m+4)}, \quad \|\psi_m - \nu_m\|_1 < \epsilon 2^{-(m+4)}. \tag{8.34}$$

By condition (A$_1$) one can find a number $t_{m+1} \in \mathbb{R}_+$ such that

$$\|P_*^{T_m,T_{m+1}}\mu_m - P_*^{T_m,T_{m+1}}\nu_m\|_1 \le \lambda\|\mu_m - \nu_m\|_1. \tag{8.35}$$

Hence, using our assumption (see (8.33)) with (iii)$_A$, (8.5), (8.34) and (8.35) one gets

$$\|P_*^{s,T_{m+1}}(\varphi - \psi)\|_1 \le \|P_*^{T_m,T_{m+1}}(P_*^{s,T_m}(\varphi - \psi) \otimes \omega_{T_m})\|_1$$
$$\le \|\varphi_m - \mu_m\|_1 + \|\psi_m - \nu_m\|_1 + \lambda\|\mu_m - \nu_m\|_1$$
$$\le \epsilon 2^{-(m+3)} + \lambda(\|\mu_m - \varphi_m\|_1 + \|\nu_m - \psi_m\|_1 + \|\varphi_m - \psi_m\|_1)$$
$$\le \epsilon 2^{-(m+2)} + \lambda(\epsilon 2^{-2}(1 + 2^{-1} + \cdots + 2^{-(m-1)}) + \lambda^m\|\varphi - \psi\|_1)$$
$$\le \epsilon 2^{-2}(1 + 2^{-1} + \cdots + 2^{-m}) + \lambda^{m+1}\|\varphi - \psi\|_1.$$

Thus we have proved that (8.33) holds for $m+1$. So, by induction it holds for all $m \in \mathbb{N}$.

Now choose $m$ such that $\lambda^m\|\varphi - \psi\|_1 < \epsilon/2$. Then for $t \ge T_m + 1$ we have

$$t = T_m + r, \quad 1 \le r < t_{m+1}. \tag{8.36}$$

So, by (8.5) and (8.33) one gets

$$\|P_*^{s,t}\varphi - P_*^{s,t}\psi\|_1 = \|P_*^{T_m,t}(P_*^{s,T_m}(\varphi - \psi) \otimes \omega_{T_m})\|_1$$
$$\le \|P_*^{s,T_m}(\varphi - \psi)\|_1$$
$$\le \epsilon 2^{-2}(1 + 2^{-1} + \cdots + 2^{-(m-1)}) + \lambda^m\|\varphi - \psi\|_1 < \epsilon.$$

Consequently, this yields that the q.q.s.p. satisfies the ergodic principle.

Case (B). Now suppose that the q.q.s.p. $(\{P^{s,t}\}, \omega_0)$ has type (B). Assume that states $\varphi, \psi, \sigma \in S$ and $s \in \mathbb{R}_+$ are fixed. Let

$$\begin{cases} h^\varphi = \sigma \otimes \varphi, \ h^\psi = \sigma \otimes \psi, \ w = \omega_s \otimes \omega_s, \\ f^\varphi = \omega_s \otimes \varphi \ f^\psi = \omega_s \otimes \psi, \ g = \sigma \otimes \omega_s \end{cases} \tag{8.37}$$

and let

$$\varphi_0 = P_*^{s,t_0} w \otimes P_*^{s,t_0} f^\varphi, \ \psi_0 = P_*^{s,t_0} w \otimes P_*^{s,t_0} f^\psi \tag{8.38}$$

for $t_0 \geq s + 1$.

Then $\varphi_0, \psi_0 \in S^2$. Take an arbitrary $\epsilon > 0$. Thanks to the density of $\mathcal{R}$ one finds states $\mu_0, \nu_0 \in \mathcal{R}$ such that

$$\|\varphi_0 - \mu_0\|_1 < \epsilon 2^{-4}, \ \|\psi_0 - \nu_0\|_1 < \epsilon 2^{-4}.$$

By condition $(A_1)$ for $\mu_0$ and $\nu_0$ one can find a number $t_1 \in \mathbb{R}_+$ such that

$$\|P_*^{t_0,t_0+t_1}(\mu_0 - \nu_0)\|_1 \leq \lambda \|\mu_0 - \nu_0\|_1. \tag{8.39}$$

Hence using $(iii)_B$ with (8.37), (8.38) and (8.39) one gets

$$\|P_*^{s,t_0+t_1} f^\varphi - P_*^{s,t_0+t_1} f^\psi\|_1$$
$$\leq \|P_*^{t_0,t_0+t_1}(P_*^{s,t_0} w \otimes P_*^{s,t_0} f^\varphi - P_*^{s,t_0} w \otimes P_*^{s,t_0} f^\psi)\|_1$$
$$\leq \|\varphi_0 - \mu_0\|_1 + \|\psi_0 - \nu_0\|_1 + \|P_*^{t_0,t_0+t_1}(\mu_0 - \nu_0)\|_1$$
$$\leq \epsilon 2^{-2} + \lambda \|\varphi - \psi\|_1.$$

Repeating the arguments used in case (A), we can show the existence of numbers $\{t_i\}_{i=0}^m$ such that

$$\|P_*^{s,T_m} f^\varphi - P_*^{s,T_m} f^\psi\|_1 \leq \epsilon 2^{-2}(1 + 2^{-1} + \cdots + 2^{-(m-1)})$$

$$+ \lambda^m \|\varphi - \psi\|_1. \tag{8.40}$$

Let us choose $m$ such that $\lambda^m \|\varphi - \psi\|_1 < \epsilon/2$. For $t \geq T_m + 1$ we have the decomposition (8.36), hence using $(iii)_B$, (8.5) and (8.40) we find

$$\|P_*^{s,t} h^\varphi - P_*^{s,t} h^\psi\|_1 = \|P^{T_m,t}(P_*^{s,T_m}(f^\varphi - f^\psi) \otimes P_*^{s,T_m} g)\|_1$$
$$\leq \|P_*^{s,T_m}(f^\varphi - f^\psi)\|_1$$
$$\leq \epsilon 2^{-2}(1 + 2^{-1} + \cdots + 2^{-(m-1)}) + \lambda^m \|\varphi - \psi\|_1 < \epsilon.$$

Therefore,

$$\lim_{t \to \infty} \|P_*^{s,t} h^\varphi - P_*^{s,t} h^\psi\|_1 = 0.$$

Equation (8.37) and Theorem 8.3.3 imply the required assertion. This completes the proof.

*Remark 8.3.3* A similar kind of theorem was proved for homogeneous Markov process with discrete time defined on an ordered space in [239].

Now using the same argument in the proof of Theorem 8.3.4, we can prove the following result.

**Theorem 8.3.5** *Let* $(\{P^{s,t}\}, \omega_0)$ *be a q.q.s.p. on a von Neumann algebra* $\mathscr{M}$. *Then the following statements are equivalent:*

*(i)* $(\{P^{s,t}\}, \omega_0)$ *satisfies weak ergodicity;*

*(ii)* $(\{P^{s,t}\}, \omega_0)$ *satisfies condition* $(A_1)$ *uniformly on* $S^2$.

*Remark 8.3.4* We note that the weak ergodicity of a q.q.s.o. can be investigated by means of the Dobrushin ergodicity coefficient [175, 176] of noncommutative Markov operators.


## 8.4  Regularity of q.q.s.p.s

In this section we study the regularity of q.q.s.p.s.

**Definition 8.4.1** A q.q.s.p. $(\{P^{s,t}\}, \omega_0)$ on $\mathscr{M}$ is said to satisfy

(i) the *regularity* condition if there is a state $\mu_1 \in S$ such that for every $\varphi \in S^2$ and $s \in \mathbb{R}_+$ one has

$$\lim_{t \to \infty} \|P_*^{s,t} \varphi - \mu_1\|_1 = 0;$$

(ii) *the exponential regularity* condition if there is a state $\mu_1 \in S$ such that

$$\|P_*^{s,t} \varphi - \mu_1\|_1 \le d \exp(-bt),$$

for all $\varphi \in S^2$ and $t \ge t_0$ with some $t_0 \in \mathbb{R}_+$, where $d, b > 0$.

*Remark 8.4.1* In the case when the von Neumann algebra is commutative, the regularity condition can be stated as follows: there is a probability (that is, normalized) measure $\mu_1$ on $(E, \mathfrak{F})$ such that

$$\lim_{t \to \infty} |P(s, x, y, t, A) - \mu_1(A)| = 0$$

for any $x, y \in E$, $A \in \mathfrak{F}$ and $s \in \mathbb{R}_+$.

We say that a q.q.s.p. ($\{P^{s,t}\}, \omega_0$) satisfies the condition (A$_2$) (resp. uniform) on $\mathcal{N} \subset S^2$ if there is a number $\lambda \in (0, 1]$ and a state $\mu \in S$ such that for any $\varphi \in \mathcal{N}$ and $s \in \mathbb{R}_+$ one can find a family $\{\gamma_t^{(s)} : t \geq s + 1\} \subset \mathcal{M}_{*,+}$ and a number $t_0 \in \mathbb{R}_+$ that satisfy the following conditions:

(i) $\|\gamma_t^{(s)}\|_1 \to 0$ (resp. $\sup_{\varphi \in \mathcal{N}} \|\gamma_t^{(s)}\|_1 \to 0$) as $t \to \infty$,

(ii) $P_*^{s,t}\varphi + \gamma_t^s \geq \lambda\mu$, for all $t \geq t_0$.

**Lemma 8.4.1** *Assume that a q.q.s.p. ($\{P^{s,t}\}, \omega_0$) on $\mathcal{M}$ satisfies the condition (A$_2$) on $\mathcal{N}$. Then it satisfies (A$_2$) on the convex hull $\mathcal{N}^{ch}$ of $\mathcal{N}$.*

*Proof* Let $\{\varphi_i\}_{i=1}^m \subset \mathcal{N}$, $m \in \mathbb{N}$ and let $\{\lambda_i\}_{i=1}^m$ be numbers such that $\lambda_i \geq 0$, $\sum_{i=1}^m \lambda_i = 1$. By condition (A$_2$), for each state $\varphi_i$ ($i = \overline{1, m}$) one can find $\{\gamma_{t,i}^{(s)} : t \geq s + 1\}$ and a number $t_{0,i} \in \mathbb{R}_+$ such that

$$P_*^{s,t}\varphi_i + \gamma_{t,i}^{(s)} \geq \lambda\mu, \quad \forall t \geq t_{0,i}, \tag{8.41}$$

where $\gamma_{t,i}^{(s)} \to 0$ as $t \to \infty$. For a state $\sum_{i=1}^m \lambda_i\varphi_i \in \mathcal{N}^{ch}$ we put

$$\gamma_t^{(s)} = \sum_{i=1}^m \lambda_i\gamma_{t,i}^{(s)}.$$

It is clear that $\gamma_t^{(s)} \to 0$ as $t \to \infty$ and

$$P_*^{s,t}\left(\sum_{i=1}^m \lambda_i\varphi_i\right) + \gamma_t^{(s)} = \sum_{i=1}^m \lambda_i(P_*^{s,t}\varphi_i + \gamma_{t,i}^{(s)})$$

$$\geq \lambda\mu, \qquad \forall t \geq \max_{1 \leq i \leq m}\{t_{0,i}\},$$

which completes the proof.

**Theorem 8.4.2** *Assume that a q.q.s.p. ($\{P^{s,t}\}, \omega_0$) on $\mathcal{M}$ satisfies the condition (A$_2$) on $\mathcal{N}$, whose convex hull $\mathcal{N}^{ch}$ is $\sigma((\mathcal{M} \otimes \mathcal{M})_*, \mathcal{M} \otimes \mathcal{M})$-weakly dense in $S^2$. Then ($\{P^{s,t}\}, \omega_0$) satisfies the ergodic principle.*

*Proof* According to Lemma 8.4.1 the q.q.s.p. ($\{P^{s,t}\}, \omega_0$) satisfies the condition (A$_2$) on $\mathcal{N}^{ch}$. Furthermore, without loss of generality we may assume that $\lambda < 1$. If $\lambda = 1$ then the ergodic principle holds automatically.

Let $\varphi, \psi \in \mathcal{N}^{ch}$ be arbitrary states and $s \in \mathbb{R}_+$. By Lemma 8.3.2 for $\varphi$ and $\psi$ one can find states $\zeta, \xi \in S^2$ such that the decomposition

$$\varphi - \psi = \frac{\|\varphi - \psi\|_1}{2}(\zeta - \xi) \tag{8.42}$$

holds. Take an arbitrary $\epsilon > 0$. The density of $\mathcal{N}^{ch}$ implies the existence of states $\zeta_1, \xi_1 \in \mathcal{N}^{ch}$ such that

$$\|\zeta - \zeta_1\|_1 < \epsilon, \quad \|\xi - \xi_1\|_1 < \epsilon.$$

Thanks to condition $(A_2)$ for $\zeta_1$ and $\xi_1$ one finds a number $t_0 = t_0(s, \zeta_1, \xi_1) \in \mathbb{R}_+$ such that

$$P_*^{s,t}\zeta_1 + \gamma_t^{(s)} \geq \lambda\mu, \ P_*^{s,t}\xi_1 + \gamma_t^{(s)} \geq \lambda\mu, \ \|\gamma_t^{(s)}\|_1 \leq \lambda/2, \ \forall t \geq t_0.$$

Hence, we have

$$\|P_*^{s,t}\zeta_1 + +\gamma_t^{(s)} - \lambda\mu\|_1 = P_*^{s,t}\zeta_1(\mathbb{1}) + +\gamma_t^{(s)}(\mathbb{1}) - \lambda\mu(\mathbb{1}) = 1 - c_1 \leq 1 - \lambda/2,$$
$$\|P_*^{s,t}\xi_1 + +\gamma_t^{(s)} - \lambda\mu\|_1 = P_*^{k,n}\xi_1(\mathbb{1}) + +\gamma_t^{(s)}(\mathbb{1}) - \lambda\mu(\mathbb{1}) = 1 - c_1 \leq 1 - \lambda/2,$$

for all $t \geq t_0$. Let

$$\mu_1^{s,t} = (1 - c_1)^{-1}(P_*^{s,t}\zeta_1 + \gamma_t^{(s)} - \lambda\mu),$$
$$\nu_1^{s,t} = (1 - c_1)^{-1}(P_*^{s,t}\xi_1 + \gamma_t^{(s)} - \lambda\mu).$$

It is obvious that $\mu_1^{s,t}, \nu_1^{s,t} \in S$ and

$$P_*^{s,t}\zeta_1 - P_*^{k,n}\xi_1 = (1 - c_1)(\mu_1^{s,t} - \nu_1^{s,t}). \tag{8.43}$$

Using (8.5), (8.42) and (8.43) one gets

$$\|P_*^{s,t}\varphi - P_*^{s,t}\psi\|_1$$
$$\leq \frac{\|\varphi - \psi\|_1}{2}\left(\|P_*^{s,t}(\zeta - \zeta_1)\|_1 + \|P_*^{s,t}(\xi - \xi_1)\|_1 + \|P_*^{s,t}\zeta_1 - P_*^{s,t}\xi_1\|_1\right)$$
$$\leq \frac{\|\varphi - \psi\|_1}{2}(\|\zeta - \zeta_1\|_1 + \|\xi - \xi_1\|_1 + (1 - c_1)\|\mu_1^{s,t} - \nu_1^{s,t}\|_1)$$
$$\leq \frac{\|\varphi - \psi\|_1}{2}(2\epsilon + 2(1 - c_1))$$
$$= (\epsilon + 1 - c_1)\|\varphi - \psi\|_1.$$

The arbitrariness of $\epsilon$ yields that

$$\|P_*^{s,t}\varphi - P_*^{s,t}\psi\|_1 \leq q\|\varphi - \psi\|_1, \ \forall t \geq t_0,$$

where $q$ is an arbitrary number that satisfies the condition $1 - \lambda/2 < q < 1$. So, by Theorem 8.3.4 the q.q.s.p. $(\{P^{s,t}\}, \omega_0)$ satisfies the ergodic principle.

Now Theorem 8.3.4 together with Lemma 8.1.1 immediately imply the following result.

**Theorem 8.4.3** *Assume that* $(\{P^{s,t}\}, \omega_0)$ *is a homogeneous q.q.s.p. on the von Neumann algebra* $\mathcal{M}$. *The following statements are equivalent:*

(i)  $(\{P^{s,t}\}, \omega_0)$ *is a regular process;*
(ii)  $(\{P^{s,t}\}, \omega_0)$ *satisfies condition* $(A_1)$ *on* $S^2$;
(iii)  $(\{P^{s,t}\}, \omega_0)$ *satisfies condition* $(A_1)$ *on a dense subset* $\mathcal{R}$ *of* $S^2$.

**Theorem 8.4.4** *Assume that* $(\{P^{s,t}\}, \omega_0)$ *is a homogeneous q.q.s.p. on the von Neumann algebra* $\mathcal{M}$. *The following statements are equivalent:*

(i)  $(\{P^{s,t}\}, \omega_0)$ *is exponentially regular;*
(ii)  $(\{P^{s,t}\}, \omega_0)$ *satisfies condition* $(A_1)$ *uniformly on* $S^2$;
(iii)  $(\{P^{s,t}\}, \omega_0)$ *satisfies the condition* $(A_1)$ *uniformly on a dense subset* $\mathcal{R}$ *of* $S^2$.

*Proof* The implications (i)$\Rightarrow$ (ii)$\Rightarrow$ (iii) are obvious. We claim that (iii) $\Rightarrow$ (i). First consider the case when the q.q.s.p. has type (A). Repeating the argument used in the proof of Theorem 8.3.4 and taking into account the homogeneity of the process, we obtain the following inequality for arbitrary $\epsilon > 0$ and $m \in \mathbb{N}$:

$$\|P_*^{0,mt_0}\varphi - P_*^{0,mt_0}\psi\|_1 \le \epsilon 2^{-2}(1 + 2^{-1} + \cdots + 2^{-(m-2)}) + \lambda^{m-1}\|\varphi - \psi\|_1,$$

for all $\varphi, \psi \in S^2$. Hence,

$$\|P_*^{0,t}\varphi - P_*^{0,t}\psi\|_1 \le \epsilon 2^{-2}(1 + 2^{-1} + \cdots + 2^{-(m-2)}) + \lambda^{m-1}\|\varphi - \psi\|_1,$$

for any $t \ge t_0$, where $m = [t/t_0]$. The arbitrariness of $\epsilon$ together with Lemma 8.1.1 implies that the process is exponentially regular.

The case when the q.q.s.p. has type (B) can be considered likewise.

**Proposition 8.4.5** *Assume that* $(\{P^{s,t}\}, \omega_0)$ *is a regular (resp. exponentially regular) q.q.s.p. on the von Neumann algebra* $\mathcal{M}$. *Then it satisfies condition* $(A_2)$ *(resp. uniformly) on* $S^2$.

*Proof* Let $\mu_1 \in S$ be a limit state for $P_*^{s,t}\varphi$, where $\varphi \in S^2$. Let

$$\gamma_t^{(s)} = (P_*^{s,t}\varphi - \mu_1)_-,$$

where

$$P_*^{s,t}\varphi - \mu_1 = (P_*^{s,t}\varphi - \mu_1)_+ - (P_*^{s,t}\varphi - \mu_1)_-$$

is the Jordan decomposition. We claim that $\|\gamma_t^{(s)}\|_1 \to 0$ (resp. $\sup_{\varphi \in S^2} \|\gamma_t^{(s)}\|_1 \to 0$) as $t \to \infty$. Indeed,

$$\|\gamma_t^{(s)}\|_1 = \|(P_*^{s,t}\varphi - \mu_1)_-\|_1 \le \|(P_*^{s,t}\varphi - \mu_1)\|_1 \to 0 \text{ as } t \to \infty,$$

(respectively,

$$\sup_{\varphi \in S^2} \|\gamma_t^{(s)}\|_1 \leq d \exp(-bt) \to 0, \quad \text{as } t \to \infty).$$

since $\{P^{s,t}\}$ is regular (resp. exponentially regular). Due to

$$P_*^{s,t}\varphi + \gamma_t^{(s)} = \mu_1 + (P_*^{s,t}\varphi - \mu_1)_+ \geq \mu_1, \quad \text{for any } t \geq t_0,$$

where $t_0 = s + 1$, we find that the condition (A$_2$) (resp. uniformly) is satisfied on $S^2$.

As a consequence of Theorems 8.4.2, 8.4.4 and Proposition 8.4.5 we obtain the following regularity criterion.

**Theorem 8.4.6** *Assume that* $(\{P^{s,t}\}, \omega_0)$ *is a homogeneous q.q.s.p. on the von Neumann algebra* $\mathscr{M}$. *The following statements are equivalent:*

*(i)* $(\{P^{s,t}\}, \omega_0)$ *is regular (resp. exponentially regular);*
*(ii)* $(\{P^{s,t}\}, \omega_0)$ *satisfies condition (A$_2$) (resp. uniformly) on* $S^2$;
*(iii)* $(\{P^{s,t}\}, \omega_0)$ *satisfies condition (A$_2$) (resp. uniformly) on a dense subset* $\mathscr{N}$ *of* $S^2$.

## 8.5   Expansion of Quantum Quadratic Stochastic Processes

In this section we give an expansion of quantum quadratic stochastic processes into fibrewise Markov processes.

Consider a family of maps $\{H^{s,t} : \mathscr{M}_* \otimes_{\alpha_0^*} \mathscr{M}_* \to \mathscr{M}_*, \ s, t \in \mathbb{R}_+, \ t - s \geq 1\}$, where as before $\alpha_0^*$ denotes the dual norm to the smallest $C^*$-cross-norm $\alpha_0$ on $\mathscr{M} \otimes \mathscr{M}$, such that for each $s$ and $t$, $H^{s,t}(\cdot \otimes \cdot)$ is a bilinear function on $\mathscr{M}_* \times \mathscr{M}_*$ and satisfies the following conditions:

(i)   $H^{s,t}(S^2) \subset S$ and its dual $(H^{s,t})^*$ is a completely positive map;
(ii)  $H^{s,t}(\varphi \otimes \psi) = H^{s,t}(\psi \otimes \varphi)$ for all $\varphi, \psi \in S$;
(iii) for an initial state $\omega_0 \in S$ and all numbers $s, \tau, t \in \mathbb{R}_+$ with $\tau - s \geq 1$ and $t - \tau \geq 1$ one of the following equations holds:

(iii)$_a$  $H^{s,t}(\varphi) = H^{\tau,t}(H^{s,\tau}(\varphi) \otimes \omega_\tau), \quad \varphi \in S^2$;
(iii)$_b$  $H^{s,t}(\varphi \otimes \psi) = H^{\tau,t}(H^{s,\tau}(\omega_s \otimes \varphi) \otimes H^{s,\tau}(\omega_s \otimes \psi)), \quad \varphi, \psi \in \mathscr{M}_*$,

where $\omega_s(x) = H^{0,s}(\omega_0 \otimes \omega_0)(x)$.

Such a family is denoted by $(H^{s,t}, \omega_0)$ and called a *quadratic process (q.p.)* of type (A) (resp. type (B)) when equation (iii)$_a$ (resp. (iii)$_b$) holds.

From Observation 8.1.1 we know that every q.q.s.p. $(\{P^{s,t}\}, \omega_0)$ determines an associated q.p. $(P_*^{s,t}, \omega_0)$ given by (8.4). It is clear that the q.p. $(P_*^{s,t}, \omega_0)$ is of type (A) or (B) if the q.q.s.p. $(\{P^{s,t}\}, \omega_0)$ has the corresponding type.

We are interested in the reverse question: can every q.p. define a q.q.s.p.? The following result responds to this question.

**Theorem 8.5.1** *Every quadratic process* $(H^{s,t}, \omega_0)$ *determines a q.q.s.p.* $(\{P^{s,t}\}, \omega_0)$ *of the same type. Moreover, one has*

$$H^{s,t}(\varphi)(x) = \varphi(P^{s,t}x), \quad \varphi \in \mathcal{M}_* \otimes_{\alpha_0^*} \mathcal{M}_*, \ x \in \mathcal{M}. \tag{8.44}$$

*Proof* Define a family of maps by

$$P^{s,t} := (H^{s,t})^*, \quad s,t \in \mathbb{R}_+, \ t-s \geq 1. \tag{8.45}$$

Since $(\mathcal{M}_* \otimes_{\alpha_0^*} \mathcal{M}_*)^* = \mathcal{M} \otimes \mathcal{M}$, $(\mathcal{M}_*)^* = \mathcal{M}$ (see [226, Def. 1.22.10]) it is clear that $P^{s,t} : \mathcal{M} \to \mathcal{M} \otimes \mathcal{M}$. Thanks to condition (i), $P^{s,t}$ is completely positive. Using the dual notation, we may rewrite (8.45) as

$$\langle P^{s,t}x, \varphi \rangle = \langle x, H^{s,t}\varphi \rangle, \qquad x \in \mathcal{M}, \varphi \in \mathcal{M}_* \otimes_{\alpha_0^*} \mathcal{M}_*. \tag{8.46}$$

This yields Eq. (8.44).

Let us verify that $(\{P^{s,t}\}, \omega_0)$ is a q.q.s.p. The ultraweak continuity of $P^{s,t}$ follows from condition (i) in the definition of a quadratic process. Suppose that $\langle \mathbb{1}, \varphi \rangle \in \mathcal{M}, \varphi \in \mathcal{M}_*$. Then

$$\langle P^{s,t}\mathbb{1}, \psi \rangle = \langle \mathbb{1}, H^{s,t}(\psi) \rangle = H^{s,t}(\psi)(\mathbb{1}) = 1,$$

where $\psi \in S^2$. It follows that $P^{s,t}\mathbb{1} = \mathbb{1} \otimes \mathbb{1}$.

Condition (ii) implies that

$$\langle P^{s,t}x, \varphi \otimes \psi \rangle = \langle P^{s,t}x, \psi \otimes \varphi \rangle,$$

whence $UP^{s,t}x = P^{s,t}$. We distinguish two cases when the q. p. $(H^{s,t}, \omega_0)$ has either type (A) or type (B).

Suppose that $(H^{s,t}, \omega_0)$ is of type (A). For every $s, \tau, t \in \mathbb{R}_+$ with $\tau - s \geq 1$ and $t - \tau \geq 1$, from the equation (iii)$_a$ we get

$$\begin{aligned}
\langle P^{s,t}x, \psi \rangle &= \langle x, H^{s,t}(\psi) \rangle = H^{s,t}(\psi)(x) \\
&= H^{\tau,t}(H^{s,\tau}(\psi) \otimes \omega_\tau)(x) \\
&= \langle P^{\tau,t}x, H^{s,\tau}(\psi) \otimes \omega_\tau \rangle \\
&= \langle E_{\omega_\tau}(P^{\tau,t}x), H^{s,\tau}(\psi) \rangle \\
&= \langle P^{s,\tau}(E_{\omega_\tau}(P^{\tau,t}x)), \psi \rangle
\end{aligned}$$

for all $x \in \mathcal{M}, \psi \in S^2$. It follows that

$$P^{s,t}x = P^{s,\tau}(E_{\omega_\tau}(P^{\tau,t}x)), \quad x \in \mathcal{M}.$$

Thus, $(\{P^{s,t}\}, \omega_0)$ has type (A).

Now assume that $(H^{s,t}, \omega_0)$ has type (B). Then using equation (iii)$_b$ with $s, \tau, t \in \mathbb{R}_+$ as above, one gets

$$
\begin{aligned}
\langle P^{s,t}x, \varphi \otimes \psi \rangle &= \langle x, H^{s,t}(\varphi \otimes \psi) \rangle = H^{s,t}(\varphi \otimes \psi)(x) \\
&= H^{\tau,t}(H^{s,\tau}(\omega_s \otimes \varphi) \otimes H^{s,\tau}(\omega_s \otimes \psi))(x) \\
&= \langle P^{\tau,t}x, H^{s,\tau}(\omega_s \otimes \varphi) \otimes H^{s,\tau}(\omega_s \otimes \psi) \rangle \\
&= \langle P^{s,\tau} \otimes P^{s,\tau}(P^{\tau,t}x), (\omega_s \otimes \varphi) \otimes (\omega_s \otimes \psi) \rangle \\
&= \langle E_{\omega_s} \otimes E_{\omega_s}(P^{s,\tau} \otimes P^{s,\tau})(P^{\tau,t}x), \varphi \otimes \psi) \rangle,
\end{aligned}
$$

for all $x \in \mathcal{M}, \varphi, \psi \in \mathcal{M}_*$. Hence,

$$
(P^{s,t}x, g) = \langle E_{\omega_s} \otimes E_{\omega_s}(P^{s,\tau} \otimes P^{s,\tau})(P^{\tau,t}x), g \rangle
$$

for all $g \in \mathcal{M}_* \odot \mathcal{M}_*$. Since $\mathcal{M}_* \odot \mathcal{M}_*$ is strongly dense in $\mathcal{M}_* \otimes_{\alpha_0^*} \mathcal{M}_*$ we obtain

$$
(P^{s,t}x, f) = \langle E_{\omega_s} \otimes E_{\omega_s}(P^{s,\tau} \otimes P^{s,\tau})(P^{\tau,t}x), f \rangle \quad \forall f \in \mathcal{M}_* \otimes_{\alpha_0^*} \mathcal{M}_*.
$$

This yields equation (ii)$_B$ in Definition 8.1.1. Thus, $(\{P^{s,t}\}, \omega_0)$ is a q.q.s.p. of type (B). This completes the proof.

Let $\mathscr{T} = \{T^{s,t}(\varphi) : \mathcal{M} \to \mathcal{M} : s, t \in \mathbb{R}_+, t - s \geq 1, \varphi \in \mathcal{M}_*\}$ be a family of ultraweakly continuous linear maps on a von Neumann algebra $\mathcal{M}$.

**Definition 8.5.1** A pair $(\mathscr{T}, \omega_0)$, where $\omega_0 \in S$ is an initial state, is called a *fibrewise Markov process (f.m.p.)* if $T^{s,t}(\cdot)$ is a linear function on $\mathcal{M}_*$ for all $s, t$ and the following conditions hold:

(i)  $T^{s,t}(\varphi)\mathbb{1}_{\mathcal{M}} = \mathbb{1}_{\mathcal{M}}$ for $\varphi \in S$;
(ii)  $T^{s,t}(\varphi)$ is completely positive for $\varphi \in \mathcal{M}_{*,+}$;
(iii)  for the initial state $\omega_0 \in S$ and arbitrary $s, \tau, t \in R_+$ with $\tau - s \geq 1$ and $t - \tau \geq 1$ one of the following equations holds:
(iii)$_A$  $T^{s,t}(\varphi) = T^{s,\tau}(\varphi)T^{\tau,t}(\omega_\tau), \quad \varphi \in S$;
(iii)$_B$  $T^{s,t}(\varphi) = T^{s,\tau}(\omega_s)T^{\tau,t}(T_*^{s,\tau}(\omega_s)\varphi), \quad \varphi \in S$,

where $\omega_\tau(x) = \omega_0(T^{0,\tau}(\omega_0)x), x \in \mathcal{M}$ and $(T_*^{s,\tau}(\omega_s)\varphi)(x) = \varphi(T^{s,t}(\omega_s)x), x \in \mathcal{M}$.

We say that the f.m.p. $(\mathscr{T}, \omega_0)$ is of type (A) (resp. type (B)) if the fundamental equation (iii)$_A$ (resp. (iii)$_B$) holds.

**Theorem 8.5.2** *Every q.q.s.p.* $(\{P^{s,t}\}, \omega_0)$ *uniquely determines an f.m.p* $(\mathscr{T}_p, \omega_0)$ *of the corresponding type given by*

$$
T^{s,t}(\varphi)x = E_\varphi(P^{s,t}x), \quad \varphi \in \mathcal{M}_*, x \in \mathcal{M}. \tag{8.47}
$$

*The following equations hold for all $f, g \in \mathcal{M}_*$:*

$$T_*^{s,t}(f)g = T_*^{s,t}(g)f, \quad \|T_*^{s,t}(f)\|_1 \leq \|f\|_1. \tag{8.48}$$

*Moreover, we have*

$$P_*^{s,t}(f \otimes g) = T_*^{s,t}(f)g, \quad \forall f, g \in \mathcal{M}_*. \tag{8.49}$$

*Proof* It is clear from the definition of $T^{s,t}(\varphi)$ that conditions (i) and (ii) hold. Let us verify the fundamental equations (iii) in Definition 8.5.1. Suppose that $s, \tau, t \in \mathbb{R}_+$ satisfy $\tau - s \geq 1$, $t - \tau \geq 1$, and take $\varphi \in \mathcal{M}_*$. Using the fundamental equation (ii)$_A$ of q.q.s.p. one gets

$$
\begin{aligned}
T^{s,t}(\varphi)x &= E_\varphi(P^{s,t}x) = E_\varphi(P^{s,\tau}E_{\omega_\tau}(P^{\tau,t}x))) \\
&= T^{s,\tau}(\varphi)T^{\tau,t}(\omega_\tau)(x), \quad x \in \mathcal{M}.
\end{aligned}
$$

Similarly using (ii)$_B$ we obtain

$$
\begin{aligned}
T^{s,t}(\varphi)x &= E_\varphi(E_{\omega_s} \otimes E_{\omega_s}(P^{s,\tau} \otimes P^{s,\tau})(P^{\tau,t}x)) \\
&= E_\varphi((E_{\omega_s}P^{s,\tau} \otimes E_{\omega_s}P^{s,\tau})(P^{\tau,t}x)) \\
&= T^{s,\tau}(\omega_s) \otimes T_*^{s,\tau}(\omega_s)(\varphi)(P^{\tau,t}x) \\
&= T^{s,\tau}(\omega_s)E_{T_*^{s,\tau}(\omega_s)(\varphi)}(P^{\tau,t}x) \\
&= T^{s,\tau}(\omega_s)T^{\tau,t}(T_*^{s,\tau}(\omega_s))(x), \quad x \in \mathcal{M},
\end{aligned}
$$

where $\omega_s(x) = \omega_0 \otimes \omega_0(P^{0,s}x) = \omega_0(T^{0,s}(\omega_0)x)$.
Due to Proposition 5.2.2 we immediately obtain the Eqs. (8.48) and (8.49).

The fibrewise Markov process $(\mathcal{T}_p, \omega_0)$ defined by (8.47) is called the *expansion of the q.q.s.p.* $(\{P^{s,t}\}, \omega_0)$ into a fibrewise Markov process.

## 8.6   The Connection Between the Fibrewise Markov Process and the Ergodic Principle

Let $(\{P^{s,t}\}, \omega_0)$ be a q.q.s.p. on a von Neumann algebra $\mathcal{M}$, and $(\mathcal{T}, \omega_0)$ be its expansion into an f.m.p.

We say that the f.m.p. $(\mathcal{T}, \omega_0)$ satisfies the *ergodic principle* if for every $\sigma, \varphi, \psi \in S$ and $s \in \mathbb{R}_+$ one has

$$\lim_{t \to \infty} \|T_*^{s,t}(\sigma)\varphi - T_*^{s,t}(\sigma)\psi\|_1 = 0.$$

**Theorem 8.6.1** *Let* $(\{P^{s,t}\}, \omega_0)$ *be a q.q.s.p. on a von Neumann algebra $\mathcal{M}$ and let* $(\mathcal{T}, \omega_0)$ *be its expansion into an f.m.p. Then the following statements are equivalent:*

(i) $(\{P^{s,t}\}, \omega_0)$ *satisfies the ergodic principle;*
(ii) *The f.m.p.* $(\mathcal{T}, \omega_0)$ *satisfies the ergodic principle;*
(iii) *For* $\sigma, \varphi, \psi \in \mathcal{R}$, *where $\mathcal{R}$ is a dense subset of S, and for all $s \in \mathbb{R}_+$, one has*

$$\lim_{t \to \infty} \|T^{s,t}_*(\sigma)\varphi - T^{s,t}_*(\sigma)\psi\|_1 = 0.$$

*Proof* Due to Theorem 8.5.2 the implications (i)$\Leftrightarrow$(ii) are direct corollaries of Theorem 8.3.3. The implication (ii)$\Rightarrow$(iii) is obvious.

Now we consider the implication (iii)$\Rightarrow$(ii). Suppose that $\sigma, \varphi, \psi \in S$. Since $\mathcal{R}$ is dense, for any $\epsilon > 0$ there are states $\eta, \xi, \zeta \in \mathcal{R}$ such that

$$\|\sigma - \eta\|_1 < \epsilon, \quad \|\varphi - \xi\|_1 < \epsilon, \quad \|\psi - \zeta\|_1 < \epsilon.$$

By inequality (8.48) we have

$$\|T^{s,t}_*(f)g\| \leq 2\|f\|_1\|g\|_1, \quad f, g \in \mathcal{M}_*. \tag{8.50}$$

Hence,

$$\|T^{s,t}_*(\eta)\xi - T^{s,t}_*(\sigma)\varphi\|_1 < 4\epsilon, \quad \|T^{s,t}_*(\eta)\zeta - T^{s,t}_*(\sigma)\psi\|_1 < 4\epsilon. \tag{8.51}$$

Now according to condition (iii), for the states $\eta, \xi, \zeta \in \mathcal{R}$ and $s \in \mathbb{R}_+$ one can find $t_0 \in \mathbb{R}_+$ such that

$$\|T^{s,t}_*(\eta)\xi - T^{s,t}_*(\eta)\zeta\|_1 < \epsilon, \quad \text{for all } t \geq t_0. \tag{8.52}$$

So, inequalities (8.50)–(8.52) imply that

$$\|T^{s,t}_*(\sigma)\varphi - T^{s,t}_*(\sigma)\psi\|_1 < 9\epsilon,$$

which means that condition (ii) holds.

An f.m.p. $(\mathcal{T}, \omega_0)$ is said to *satisfy condition (E)* if there is a number $\lambda \in [0, 1)$ such that, given any $\varphi, \psi, \sigma \in S$ and $s \in \mathbb{R}_+$, we have

$$\|T^{s,t}_*(\sigma)\varphi - T^{s,t}_*(\sigma)\psi\|_1 \leq \lambda\|\varphi - \psi\|_1$$

for at least one $t = t(\varphi, \psi, \sigma, s) \in \mathbb{R}_+$.

**Theorem 8.6.2** *Let* $(\{P^{s,t}\}, \omega_0)$ *be a q.q.s.p. on a von Neumann algebra $\mathcal{M}$ and let* $(\mathcal{T}, \omega_0)$ *be its expansion into an f.m.p. Then the following statements are equivalent:*

(i)  $(\{P^{s,t}\}, \omega_0)$ *satisfies the ergodic principle;*
(ii)  *The f.m.p.*  $(\mathcal{T}, \omega_0)$ *satisfies condition* (E).

*Proof* (i)$\Rightarrow$(ii). Suppose that a q.q.s.p. $(\{P^{s,t}\}, \omega_0)$ satisfies the ergodic principle. Then by Theorem 8.3.4 there is a number $\lambda \in [0, 1)$ such that, given any $\xi, \eta \in S^2$ and $s \in \mathbb{R}_+$, we have

$$\|P_*^{s,t}(\xi) - P_*^{s,t}(\xi)\|_1 \leq \lambda \|\xi - \eta\|_1 \tag{8.53}$$

for at least one $t = t(\varphi, \psi, \sigma, s) \in \mathbb{R}_+$. For the states $\xi$ and $\eta$, we take $\sigma \otimes \varphi$ and $\sigma \otimes \psi$, respectively, where $\sigma, \varphi, \psi \in S$. Using Theorem 8.5.2, we see that (8.53) takes the form

$$\|T_*^{s,t}(\sigma)\varphi - T_*^{s,t}(\sigma)\psi\|_1 \leq \lambda \|\sigma \otimes (\varphi - \psi)\|_1 = \lambda \|\varphi - \psi\|_1.$$

Hence, the condition (E) holds.

(ii)$\Rightarrow$(i). Let us fix any states $\sigma, \varphi, \psi \in S$ and a number $s \in \mathbb{R}_+$.
Now consider two cases with respect to types of the f.m.p.
Case  (A) . Suppose that an f.m.p. $(\mathcal{T}, \omega_0)$ is of type (A). By condition (E), there is a number $t_1 \in \mathbb{R}_+$ such that

$$\|T_*^{s,t_1}(\sigma)\varphi - T_*^{s,t_1}(\sigma)\psi\|_1 \leq \lambda \|\varphi - \psi\|_1. \tag{8.54}$$

Applying condition (E) to the states $\omega_{t_1}, T_*^{s,t_1}(\sigma)\psi, T_*^{s,t_1}(\sigma)\psi$ and the number $t_1 \in \mathbb{R}_+$, one can find $t_2 \in \mathbb{R}_+$ such that

$$\|T_*^{t_1,t_1+t_2}(\omega_{t_1})(T_*^{s,t_1}(\sigma)\varphi) - T_*^{t_1,t_1+t_2}(\omega_{t_1})(T_*^{s,t_1}(\sigma)\psi)\|_1$$
$$\leq \lambda \|T_*^{s,t_1}(\sigma)\varphi - T_*^{s,t_1}(\sigma)\psi\|_1. \tag{8.55}$$

Condition (iii)$_A$ of Definition 8.5.1 yields that, for any $x \in \mathcal{M}$,

$$\begin{aligned}
T_*^{t_1,t_1+t_2}(\omega_{t_1})(T_*^{s,t_1}(\sigma)\varphi)(x) &= T_*^{s,t_1}(\sigma)\varphi(T^{t_1,t_1+t_2}(\omega_{t_1})x) \\
&= \varphi(T^{s,t_1}(\sigma)T^{t_1,t_1+t_2}(\omega_{t_1})x) \\
&= \varphi(T^{s,t_1+t_2}(\sigma)x) \\
&= T_*^{s,t_1+t_2}(\sigma)\varphi(x). \tag{8.56}
\end{aligned}$$

Using this together with (8.54) and (8.55), we get

$$\|T_*^{s,t_1+t_2}(\sigma)\varphi - T_*^{s,t_1+t_2}(\sigma)\psi\|_1 \leq \lambda \|T_*^{s,t_1}(\sigma)\varphi - T_*^{s,t_1}(\sigma)\psi\|_1$$
$$\leq \lambda^2 \|\varphi - \psi\|_1.$$

Assume that one can find numbers $\{t_i\}_{i=1}^m$ such that

$$\|T_*^{s,K_m}(\sigma)\varphi - T_*^{s,K_m}(\sigma)\psi\|_1 \leq \lambda^m \|\varphi - \psi\|_1, \tag{8.57}$$

where $K_m = \sum_{i=1}^m t_i$.

We claim that (8.57) also holds for $m + 1$. Indeed, by the hypothesis of the theorem, given the states $T_*^{s,K_m}(\sigma)\varphi$, $T_*^{s,K_m}(\sigma)\psi$, $\omega_{K_m}$ and the number $K_m \in \mathbb{R}_+$ we can find a number $t_{m+1} \in \mathbb{R}_+$ such that

$$\|T_*^{K_m,K_m+t_{m+1}}(\omega_{K_m})(T_*^{s,K_m}(\sigma)\varphi) - T_*^{K_m,K_m+t_{m+1}}(\omega_{K_m})(T_*^{s,K_m}(\sigma)\psi)\|_1$$
$$\leq \lambda \|T_*^{s,K_m}(\sigma)\varphi - T_*^{s,K_m}(\sigma)\psi\|_1. \tag{8.58}$$

By condition $(iii)_A$ one has

$$T_*^{K_m,K_m+t_{m+1}}(\omega_{K_m})(T_*^{s,K_m}(\sigma)\varphi) = T_*^{s,K_{m+1}}(\sigma)\varphi,$$
$$T_*^{K_m,K_m+t_{m+1}}(\omega_{K_m})(T_*^{s,K_m}(\sigma)\psi) = T_*^{s,K_{m+1}}(\sigma)\psi.$$

Using the last equalities together with (8.57) and (8.58), we find

$$\|T_*^{s,K_{m+1}}(\sigma)\varphi - T_*^{s,K_{m+1}}(\sigma)\psi\|_1 \leq \lambda^{m+1}\|\varphi - \psi\|_1,$$

where $K_{m+1} = K_m + t_{m+1}$. So, by induction, inequality (8.57) holds for all $m \in \mathbb{N}$.

Take any $\epsilon > 0$. Choose $m \in \mathbb{N}$ such that $\lambda^m \|\varphi - \psi\|_1 < \epsilon$. For $t \geq K_m + 1$ we have

$$t = K_m + r, \quad 1 \leq r < t_{m+1}. \tag{8.59}$$

Therefore,

$$\|T_*^{s,t}(\sigma)\varphi - T_*^{s,t}(\sigma)\psi\|_1 = \|T_*^{K_m,t}(\omega_{K_m})(T_*^{s,K_m}(\sigma)\varphi - T_*^{s,K_m}(\sigma)\psi)\|_1$$
$$\leq \|T_*^{s,K_m}(\sigma)\varphi - T_*^{s,K_m}(\sigma)\psi\|_1$$
$$\leq \lambda^m\|\varphi - \psi\|_1 < \epsilon.$$

Using Theorem 8.6.1, we conclude that the q.q.s.p. $\{P^{s,t}\}$ satisfies the ergodic principle.

Case (B). Now assume that the f.m.p. $(\mathscr{T}, \omega_0)$ is of type (B). By condition (E) there is a $t_1 \in \mathbb{R}_+$ such that

$$\|T_*^{s,t_1}(\omega_s)\varphi - T_*^{s,t_1}(\omega_s)\psi\|_1 \leq \lambda\|\varphi - \psi\|_1. \tag{8.60}$$

Applying condition (E) to the states $T_*^{s,t_1}(w_s)\omega_s$, $T_*^{s,t_1}(\omega_s)\varphi$, $T_*^{s,t_1}(\omega_s)\psi$ and the number $t_1$ one can find $t_2 \in \mathbb{R}_+$ such that

$$\|T_*^{t_1,t_1+t_2}(T_*^{s,t_1}(\omega_s)\omega_s)T_*^{s,t_1}(\omega_s)\varphi - T_*^{t_1,t_1+t_2}(T_*^{s,t_1}(\omega_s)\omega_s)T_*^{s,t_1}(\omega_s)\psi\|$$
$$\leq \lambda\|T_*^{s,t_1}(\omega_s)\varphi - T_*^{s,t_1}(\omega_s)\psi\|. \tag{8.61}$$

For any states $\varphi, \omega \in S$ and numbers $s, \tau, t \in \mathbb{R}_+$ with $\tau - s \geq 1$ and $t - \tau \geq 1$, from the condition (iii)$_B$ of Definition 8.5.1 we infer that

$$T_*^{s,t}(\varphi)\omega = T_*^{\tau,t}(T_*^{s,\tau}(\omega_s)\varphi)T_*^{s,t}(\omega_s)\omega. \tag{8.62}$$

Using (8.62) one gets

$$T_*^{t_1,t_1+t_2}(T_*^{s,t_1}(\omega_s)\omega_s)T_*^{s,t_1}(\omega_s)\varphi = T_*^{s,t_1+t_2}(\omega_s)\varphi,$$
$$T_*^{t_1,t_1+t_2}(T_*^{s,t_1}(\omega_s)\omega_s)T_*^{s,t_1}(\omega_s)\psi = T_*^{s,t_1+t_2}(\omega_s)\psi.$$

Substituting this into (8.61) and using (8.60) we obtain

$$\|T_*^{s,t_1+t_2}(\omega_s)\varphi - T_*^{s,t_1+t_2}(\omega_s)\psi\|_1 \leq \lambda^2\|\varphi - \psi\|_1.$$

As in case (A), one can show that there are $\{t_i\}_{i=1}^m \subset \mathbb{R}_+$, such that

$$\|T_*^{s,K_m}(\omega_s)\varphi - T_*^{s,K_m}(\omega_s)\psi\|_1 \leq \lambda^m\|\varphi - \psi\|_1. \tag{8.63}$$

Take any $\epsilon > 0$ and choose $m \in \mathbb{N}$ such that $\lambda^m\|\varphi - \psi\|_1 < \epsilon$. A number $t \geq K_m + 1$ can be represented as in (8.59), hence for any state $\sigma \in S$ we can use (8.62) and (8.63) to get

$$\|T_*^{s,t}(\sigma)\varphi - T_*^{s,t}(\sigma)\psi\|_1$$
$$= \|T_*^{K_m,t}(T_*^{s,K_m}(\omega_s)*\sigma)T_*^{s,K_m}(\omega_s)*\varphi - T_*^{K_m,t}(T_*^{s,K_m}(\omega_s)*\sigma)T_*^{s,K_m}(\omega_s)*\psi\|_1$$
$$\leq \|T_*^{s,K_m}(\omega_s)*\varphi - T_*^{s,K_m}(\omega_s)*\psi\|_1$$
$$\leq \lambda^m\|\varphi - \psi\|_1 < \epsilon.$$

Using Theorem 8.6.1, we conclude that the ergodic principle holds. This completes the proof.

**Proposition 8.6.3** *Let $(\{P^{s,t}\}, \omega_0)$ be a q.q.s.p. on a von Neumann algebra $\mathcal{M}$ and let $(\mathcal{T}, \omega_0)$ be the corresponding f.m.p. Then for any numbers $s, \tau, t \in \mathbb{R}_+$ with $\tau - s \geq 1, t - \tau \geq 1$ and any $x \in \mathcal{M}$ one has*

$$T^{s,t}(\omega_s)x = T^{s,\tau}(\omega_s)T^{\tau,t}(\omega_\tau)x. \tag{8.64}$$

*Proof* If the q.q.s.p. $P^{s,t}$ is of type (A), (8.64) follows directly from (iii)$_A$ of Definition 8.5.1.

Now suppose that the q.q.s.p. is of type (B). Then, for all $x \in \mathcal{M}$ we have

$$T_*^{s,\tau}(\omega_s)\omega_s(x) = \omega_s(T^{s,\tau}(\omega_s)x)$$
$$= T_*^{0,s}(\omega_0)\omega_0(T^{s,\tau}(\omega_s)x)$$

$$= \omega_0(T^{0,s}(\omega_0)T^{s,\tau}(\omega_s)x)$$

$$= \omega_0(T^{0,s}(\omega_0)T^{s,\tau}(T_*^{0,s}(\omega_0)\omega_0)x)$$

$$= \omega_0(T^{0,\tau}(\omega_0)x)$$

$$= \omega_\tau(x). \tag{8.65}$$

Here we have used equation (iii)$_B$ of Definition 8.5.1.

Hence, again from equation (iii)$_B$ one gets

$$T^{s,t}(\omega_s) = T^{s,\tau}(\omega_s)T^{\tau,t}(T^{s,\tau}(\omega_s)\omega_s)$$

$$= T^{s,\tau}(\omega_s)T^{\tau,t}(\omega_\tau).$$

The proof is complete.

**Corollary 8.6.4** *Let* $(\{P^{s,t}\}, \omega_0)$ *be a q.q.s.p. on a von Neumann algebra* $\mathcal{M}$ *and let* $(\mathcal{T}, \omega_0)$ *be its expansion into an f.m.p. Then the following statements are equivalent:*

*(i)* $(\{P^{s,t}\}, \omega_0)$ *satisfies the ergodic principle;*
*(ii)* *The f.m.p.* $(\mathcal{T}, \omega_0)$ *satisfies condition (E);*
*(iii)* *There is a number* $\lambda \in [0, 1)$ *such that, given any states* $\varphi, \psi \in S^2$ *and a number* $s \in \mathbb{R}_+$, *one has*

$$\|T_*^{s,t}(\omega_s)\varphi - T_*^{s,t}(\omega_s)\psi\|_1 \le \lambda \|\varphi - \psi\|_1$$

*for at least one* $t \in \mathbb{R}_+$;
*(iv)* *The f.m.p.* $(\mathcal{T}, \omega_0)$ *satisfies the ergodic principle.*

*Proof* The implications (i)$\Rightarrow$ (ii) and (iv)$\Rightarrow$ (i) follow from Theorems 8.6.2 and 8.6.1, respectively. The implication (ii)$\Rightarrow$ (iii) is obvious.

So, we have to consider (iii)$\Rightarrow$ (iv). We first suppose that the q.q.s.p. $\{P^{s,t}\}$ is of type (A). Let us fix states $\varphi, \psi \in S$ and a number $s \in \mathbb{R}_+$. By condition (iii) there is a number $t_1 \in \mathbb{R}_+$ such that

$$\|T_*^{s,t_1}(\omega_s)\varphi - T_*^{s,t_1}(\omega_s)\psi\|_1 \le \lambda \|\varphi - \psi\|_1. \tag{8.66}$$

Again applying condition (iii) to $T_*^{s,t_1}(\omega_s)\varphi, T_*^{s,t_1}(\omega_s)\psi$ and the number $t_1$, we can find a number $t_2 \in \mathbb{R}_+$ such that

$$\|T_*^{t_1,t_1+t_2}(\omega_{t_1})T_*^{s,t_1}(\omega_s)\varphi - T_*^{t_1,t_1+t_2}(\omega_{t_1})T_*^{s,t_1}(\omega_s)\psi\|_1$$

$$\le \lambda \|T_*^{s,t}(\omega_s)\varphi - T_*^{s,t}(\omega_s)\psi\|_1$$

$$\le \lambda^2 \|\varphi - \psi\|_1. \tag{8.67}$$

Here we have used inequality (8.66).

By Proposition 8.6.3, for any state $\varphi \in S$ one has

$$T_*^{\tau,t}(\omega_\tau)T_*^{s,\tau}(\omega_s)\varphi = T^{s,t}(\omega_s)\varphi. \qquad (8.68)$$

Using this equation, from (8.67) we obtain

$$\|T_*^{s,t_1+t_2}(\omega_s)\varphi - T_*^{s,t_1+t_2}(\omega_s)\psi\|_1 \leq \lambda^2\|\varphi - \psi\|_1.$$

By a similar argument as in the proof of Theorem 8.6.2 we find

$$\|T_*^{s,K_m}(\omega_s)\varphi - T_*^{s,K_m}(\omega_s)\psi\|_1 \leq \lambda^m\|\varphi - \psi\|_1.$$

Here, as before, $K_m = \sum_{i=1}^m t_i$. Thus, one has

$$\|T_*^{s,t}(\omega_s)\varphi - T_*^{s,t}(\omega_s)\psi\|_1 \to 0, \qquad t \to \infty. \qquad (8.69)$$

Let $\sigma \in S$ be an arbitrary state. For any $\tau \geq s + 1$, from (8.68) and (8.69) it follows that

$$\|T_*^{s,t}(\sigma)\varphi - T_*^{s,t}(\sigma)\psi\|_1 = \|T_*^{\tau,t}(\omega_\tau)T_*^{s,\tau}(\sigma)\varphi - T_*^{\tau,t}(\omega_\tau)T_*^{s,\tau}(\sigma)\psi\|_1$$
$$= \|T_*^{\tau,t}(\omega_\tau)\tilde{\varphi} - T_*^{\tau,t}(\omega_\tau)\tilde{\psi}\|_1 \to 0, \qquad t \to \infty,$$

where $\tilde{\varphi} = T_*^{s,\tau}\varphi$, $\tilde{\psi} = T_*^{s,\tau}\psi$.

Thus, Theorem 8.6.1 yields the ergodic principle.

When the q.q.s.p. $\{P^{s,t}\}$ is of type (B), by a similar argument as in the proof of Theorem 8.6.2 we get the required assertion.

## 8.7  Conjugate Quantum Quadratic Stochastic Processes

The expansion of q.q.s.p.s enables us to introduce a notion of conjugacy of two q.q.s.p.s defined on a von Neumann algebra.

Let $(\{P^{s,t}\}, \omega_0)$, $(\{Q^{s,t}\}, \omega_0)$ be q.q.s.p.s on a von Neumann algebra $\mathcal{M}$ and let $(\mathcal{T}_P, \omega_0)$, $(\mathcal{T}_Q, \omega_0)$ be their expansions into f.m.p.s. We say that the two q.q.s.p.s $(\{P^{s,t}\}, \omega_0)$ and $(\{Q^{s,t}\}, \omega_0)$ are *conjugate* and write $P^{s,t} \sim Q^{s,t}$ if there is a family of automorphisms $\{\theta^{s,t}(\varphi) : s, t \in \mathbb{R}_+, t - s \geq 1, \varphi \in \mathcal{M}_*\}$ such that for all numbers $s, t \in \mathbb{R}_+$ with $t - s \geq 1$ and any functional $\varphi \in \mathcal{M}_*$ one has

$$T_P^{s,t}(\varphi)\theta^{s,t}(\varphi) = T_Q^{s,t}(\varphi). \qquad (8.70)$$

This is clearly an equivalence relation. Let us give some examples.

*Example 8.7.1* Let $Z$ be an ultraweakly continuous Markov operator on a von Neumann algebra $\mathcal{M}$. Suppose that a state $\omega_0 \in S$ is invariant with respect to $Z$, that is, $\omega_0(Zx) = \omega_0(x)$ for all $x \in \mathscr{M}$. Then

$$P_Z^{k,n}x = \frac{1}{2}(Z^{k,n}x \otimes \mathbb{1} + \mathbb{1} \otimes Z^{k,n}x),$$

where

$$Z^{k,n}x = \frac{1}{2^{n-k-1}}(Z^{n-k}x + (2^{n-k-1} - 1)\omega_0(x)\mathbb{1}), \quad x \in \mathscr{M},$$

$k \geq 0$, $n \in N$, $k < n$, determines a q.q.s.p. with an initial state $\omega_0$.

Let $\theta : \mathscr{M} \to \mathscr{M}$ be an automorphism such that $\omega_0 \circ \theta = \omega_0$ and $\theta Z = Z\theta$. We put $Z_1 = Z\theta$. Clearly, $Z_1$ is a Markov operator with invariant state $\omega_0$. Reasoning as above, we define another q.q.s.p. corresponding to $Z_1$ by

$$P_{Z_1}^{k,n}x = \frac{1}{2}(Z_1^{k,n}x \otimes \mathbb{1} + \mathbb{1} \otimes Z_1^{k,n}x),$$

where

$$Z_1^{k,n}x = \frac{1}{2^{n-k-1}}(Z_1^{n-k}x + (2^{n-k-1} - 1)\omega_0(x)\mathbb{1}), \quad x \in \mathscr{M},$$

$k \geq 0$, $n \in N$, $k < n$. The corresponding f.m.p.s are given by

$$T_Z^{k,n}(\varphi)x = \frac{1}{2}(\varphi(Z^{k,n}x)\mathbb{1} + \varphi(\mathbb{1})Z^{k,n}x),$$

$$T_{Z_1}^{k,n}(\varphi)x = \frac{1}{2}(\varphi(Z_1^{k,n}x)\mathbb{1} + \varphi(\mathbb{1})Z_1^{n,k}x),$$

where $\varphi \in \mathscr{M}_*$, $x \in \mathscr{M}$. Define automorphisms $\theta^{k,n}(\varphi) : \mathscr{M} \to \mathscr{M}$ by

$$\theta^{k,n}(\varphi) = \theta^{n-k}, \quad \forall \varphi \in \mathscr{M}_*.$$

Then we easily see that (8.70) holds. Hence, $P_Z^{k,n}$ and $P_{Z_1}^{k,n}$ are conjugate.

*Example 8.7.2* Now we consider a commutative finite-dimensional von Neumann algebra $\ell_2^\infty$. In this case the q.q.s.p. coincides with the q.s.p. defined by $p_{ij,k}^{s,t}$:

$$p_{ij,k}^{s,t} \geq 0, \ p_{ij,k}^{s,t} = p_{ji,k}^{s,t}, \ \sum_{k=1}^2 p_{ij,k}^{s,t} = 1,$$

where $i, j, k \in \{1, 2\}$.

Let an initial state be $x_0 = (x, 1 - x)$, $x \in [0, 1]$. Define the coefficients of $P^{s,t}$ and $Q^{s,t}$, respectively, by

$$p_{11,1}^{s,t} = \frac{1}{2^{t-s-1}}((2^{t-s-1} - 1)x + 1),$$

$$p_{12,1}^{s,t} = p_{21,1}^{s,t} = \frac{1}{2^{t-s-1}}((2^{t-s-1}-1)x + \frac{1}{2}),$$

$$p_{22,1}^{s,t} = \frac{1}{2^{t-s-1}}(2^{t-s-1}-1)x,$$

$$p_{ij,2}^{s,t} = 1 - p_{ij,1}^{s,t}, \quad i,j \in \{1,2\},$$

$$q_{11,1}^{s,t} = \frac{1}{2^{t-s-1}}(2^{t-s-1}-1)(1-x),$$

$$q_{12,1}^{s,t} = q_{21,1}^{s,t} = \frac{1}{2^{t-s-1}}((2^{t-s-1}-1)(1-x) + \frac{1}{2}),$$

$$q_{22,1}^{s,t} = \frac{1}{2^{t-s-1}}((2^{t-s-1}-1)(1-x) + 1),$$

$$q_{ij,2}^{s,t} = 1 - q_{ij,1}^{s,t}, \quad i,j \in \{1,2\}.$$

Then these processes are conjugate by the automorphisms $\theta^{s,t}(\varphi) : \ell_2^\infty \to \ell_2^\infty$ given by

$$\theta^{s,t}(\varphi)(y_1, y_2) = (y_2, y_1), \quad \forall s, t \in R_+, t - s \geq 1,$$

where $y = (y_1, y_2) \in \ell_2^\infty, \varphi \in \ell_2^1((\ell_2^1)^* = \ell_2^\infty)$.

What can be said about a q.q.s.p. if its conjugate q.q.s.p. satisfies the ergodic principle? The following theorem answers this question.

**Theorem 8.7.1** *Let* $(\{P^{s,t}\}, \omega_0)$ *and* $(\{Q^{s,t}\}, \omega_0)$ *be two conjugate q.q.s.p.s on a von Neumann algebra* $\mathcal{M}$. *Then* $(\{P^{s,t}\}, \omega_0)$ *satisfies the ergodic principle iff* $(\{Q^{s,t}\}, \omega_0)$ *satisfies the ergodic principle.*

*Proof* By symmetry it suffices to prove the "only if" part. Suppose that $(\{P^{s,t}\}, \omega_0)$ satisfies the ergodic principle. Then by Theorem 8.6.1 we have

$$\lim_{t\to\infty} \|T_{*,P}^{s,t}(\sigma)\varphi - T_{*,P}^{s,t}(\sigma)\psi\|_1 = 0,$$

for any states $\varphi, \psi, \sigma \in S$. Using (8.70) we find that

$$T_{*,Q}^{s,t}(\sigma)\varphi(x) = \varphi(T_Q^{s,t}(\sigma)x) = \varphi(T_P^{s,t}(\sigma)\theta^{s,t}(\sigma)x)$$

$$= \theta_*^{s,t}(\sigma)T_{*,P}^{s,t}(\sigma)\varphi(x), \quad \varphi \in S, \; x \in \mathcal{M}.$$

Therefore, one gets

$$\|T_{*,Q}^{s,t}(\sigma)\varphi - T_{*,Q}^{s,t}(\sigma)\psi\|_1 = \|\theta_*^{s,t}(\sigma)(T_{*,P}^{s,t}(\sigma)\varphi - T_{*,P}^{s,t}(\sigma)\psi)\|_1$$

$$\leq \|T_{*,P}^{s,t}(\sigma)\varphi - T_{*,P}^{s,t}(\sigma)\psi\|_1.$$

It follows that

$$\lim_{t \to \infty} \|T_{*,Q}^{s,t}(\sigma)\varphi - T_{*,Q}^{s,t}(\sigma)\psi\|_1 = 0.$$

Again using Theorem 8.6.1 we obtain the desired assertion. This completes the proof.


## 8.8   Quantum Quadratic Stochastic Processes and Related Markov Processes

In this section we are going to consider the relation between q.q.s.p.s and Markov processes.

Let $\mathscr{M}$ be a von Neumann algebra. First recall that a family $\{Q^{s,t} : \mathscr{M} \to \mathscr{M} : s, t \in \mathbb{R}_+, \ t - s \geq 1\}$ of completely positive Markov operators is called a *Markov process* if

$$Q^{s,t} = Q^{s,\tau} Q^{\tau,t} \tag{8.71}$$

holds for any $s, \tau, t \in \mathbb{R}_+$ such that $t - \tau \geq 1$ and $\tau - s \geq 1$.

A Markov process $\{Q^{s,t}\}$ is said to satisfy the *ergodic principle* if for every $\varphi, \psi \in S$ and $s \in \mathbb{R}_+$ one has

$$\lim_{t \to \infty} \|Q_*^{s,t}\varphi - Q_*^{s,t}\psi\|_1 = 0.$$

Let $(\{P^{s,t}\}, \omega_0)$ be a q.q.s.p. and $(\mathscr{T}_P, \omega_0)$ be its expansion into an f.m.p. Now we define a new process $Q_P^{s,t} : \mathscr{M} \to \mathscr{M}$ by

$$Q_P^{s,t}x = T^{s,t}(\omega_s)(x), \quad x \in \mathscr{M}. \tag{8.72}$$

Using (8.47) the process $Q_P^{s,t}$ can be rewritten as follows

$$Q_P^{s,t} = E_{\omega_s} P^{s,t}. \tag{8.73}$$

Then according to Proposition 8.6.3, $\{Q_P^{s,t}\}$ is a Markov process associated with the q.q.s.p. It is evident that the defined process satisfies the ergodic principle if the q.q.s.p. satisfies the ergodic principle. An interesting question is the converse. From Corollary 8.6.4 we get an affirmative response to the last question.

**Theorem 8.8.1** *Let* $(\{P^{s,t}\}, \omega_0)$ *be a q.q.s.p. on a von Neumann algebra* $\mathscr{M}$ *and let* $\{Q_P^{s,t}\}$ *be the corresponding Markov process. Then the following conditions are equivalent:*

*(i)* $(\{P^{s,t}\}, \omega_0)$ *satisfies the ergodic principle;*

*(ii)  $\{Q_P^{s,t}\}$ satisfies the ergodic principle;*

*(iii)  There is a number $\lambda \in [0, 1)$ such that, given any states $\varphi, \psi \in S^2$ and a number $s \in \mathbb{R}_+$ one has*

$$\|Q_{P,*}^{s,t}\varphi - Q_{P,*}^{s,t}\psi\|_1 \leq \lambda\|\varphi - \psi\|_1$$

*for at least one $t \in \mathbb{R}_+$.*

### 8.8.1   Q.q.s.p.s of Type (A)

Let us assume that the q.q.s.p. $(\{P^{s,t}\}, \omega_0)$ has type (A).

Now for each $s$ and $t$ let us define a process $H^{s,t} : \mathcal{M} \otimes \mathcal{M} \to \mathcal{M} \otimes \mathcal{M}$ by

$$H_P^{s,t}\mathbf{x} = P^{s,t}E_{\omega_t}\mathbf{x}, \qquad \mathbf{x} \in \mathcal{M} \otimes \mathcal{M}. \tag{8.74}$$

It is clear that every $H_P^{s,t}$ is a Markov operator. It turns out that $\{H_P^{s,t}\}$ is a Markov process. Indeed, using (ii)$_A$ of Definition 8.1.1 one has

$$\begin{aligned}
H_P^{s,t}\mathbf{x} &= P^{s,t}E_{\omega_t}\mathbf{x} \\
&= P^{s,\tau}E_{\omega_\tau}(P^{\tau,t}E_{\omega_t}\mathbf{x}) \\
&= H_P^{s,\tau}H_P^{\tau,t}\mathbf{x},
\end{aligned}$$

which is the assertion.

One can see that the defined two Markov processes $\{Q_P^{s,t}\}$ and $\{H_P^{s,t}\}$ relate to each other by the following equalities

$$E_{\omega_s}(H_P^{s,t}\mathbf{x}) = E_{\omega_s}(P^{s,t}(E_{\omega_t}(\mathbf{x})) = Q_P^{s,t}(E_{\omega_t}(\mathbf{x}))$$

for every $\mathbf{x} \in \mathcal{M} \otimes \mathcal{M}$. Moreover, $H_P^{s,t}$ has the following properties

$$H_P^{s,t}\mathbf{x} = P^{s,t}(E_{\omega_t}(\mathbf{x})) = P^{s,t}E_{\omega_t}E_{\omega_t}(\mathbf{x}) = H_P^{s,t}(E_{\omega_t}(\mathbf{x}) \otimes \mathbb{1}) \tag{8.75}$$

$$UH_P^{s,t} = H_P^{s,t}, \quad H_P^{s,t}(x \otimes \mathbb{1}) = P^{s,t}x, \quad x \in \mathcal{M}.$$

From (8.75) one gets $H_P^{s,t}(\mathbb{1} \otimes x) = \omega_t(x)\mathbb{1} \otimes \mathbb{1}$. Here we can represent

$$\begin{aligned}
\omega_t(x) &= \omega_0 \otimes \omega_0(P^{0,t}x) = \omega_0(Q_P^{0,t}x), \\
\omega_t(x) &= \omega_0 \otimes \omega_0(H_P^{0,t}(x \otimes \mathbb{1})).
\end{aligned}$$

Now if we suppose that we have two Markov processes with the above indicated properties, can we recover the q.q.s.p.?

Let $\{Q^{s,t} : \mathcal{M} \to \mathcal{M}\}$ and $\{H^{s,t} : \mathcal{M} \otimes \mathcal{M} \to \mathcal{M} \otimes \mathcal{M}\}$ be two Markov processes with an initial state $\omega_0 \in S$. Define

$$\varphi_t(x) = \omega_0(Q^{0,t}x), \quad \psi_t(x) = \omega_0 \otimes \omega_0(H^{0,t}(x \otimes \mathbb{1})).$$

Assume that

(i)   $UH^{s,t} = H^{s,t}$;
(ii)  $E_{\psi_s} H^{s,t}\mathbf{x} = Q^{s,t}E_{\varphi_t}\mathbf{x}$ for all $\mathbf{x} \in \mathcal{M} \otimes \mathcal{M}$;
(iii) $H^{s,t}\mathbf{x} = H^{s,t}(E_{\psi_t}(\mathbf{x}) \otimes \mathbb{1})$.

First note that if we take $\mathbf{x} = \mathbb{1} \otimes x$ in (iii) then we get

$$\begin{aligned}
H^{s,t}(\mathbb{1} \otimes x) &= H^{s,t}(E_{\psi_t}(\mathbb{1} \otimes x) \otimes \mathbb{1}) \\
&= H^{s,t}(\psi_t(x)\mathbb{1} \otimes \mathbb{1}) \\
&= \psi_t(x)\mathbb{1} \otimes \mathbb{1}.
\end{aligned} \tag{8.76}$$

Now from (ii) and (8.76) we have

$$\begin{aligned}
E_{\psi_s} H^{s,t}(\mathbb{1} \otimes x) &= E_{\psi_s}(\psi_t(x)\mathbb{1} \otimes \mathbb{1}) \\
&= \psi_t(x)\mathbb{1} \\
&= Q^{s,t}E_{\varphi_t}(\mathbb{1} \otimes x) \\
&= \varphi_t(x)\mathbb{1}.
\end{aligned} \tag{8.77}$$

This means that $\varphi_t = \psi_t$. Therefore in the sequel we use $\omega_t := \varphi_t = \psi_t$.

**Theorem 8.8.2** *Let $\{Q^{s,t}\}$ and $\{H^{s,t}\}$ be Markov processes with an initial state $\omega_0 \in S$ such that (i)–(iii) are satisfied. Then by the equality $P^{s,t}x = H^{s,t}(x \otimes \mathbb{1})$ one defines a q.q.s.p. $(\{P^{s,t}\}, \omega_0)$ of type (A). Moreover, one has*

(a) $P^{s,t} = H^{s,\tau}P^{\tau,t}$ *for any* $\tau - s \geq 1, t - \tau \geq 1$,
(b) $Q^{s,t} = E_{\omega_s}P^{s,t}$.

*Proof* We have to check only condition (ii)$_A$ of the definition of q.q.s.p.s. Take any $\tau - s \geq 1, t - \tau \geq 1$. Then using the assumption (iii) we have

$$\begin{aligned}
P^{s,\tau}E_{\omega_\tau}(P^{\tau,t}x) &= H^{s,\tau}(E_{\omega_\tau}H^{\tau,t}(x \otimes \mathbb{1}) \otimes \mathbb{1}) \\
&= H^{s,\tau}H^{\tau,t}(x \otimes \mathbb{1}) \\
&= H^{s,t}(x \otimes \mathbb{1}) \\
&= P^{s,t}x, \quad x \in M.
\end{aligned}$$

Now from the Markov property of $H^{s,t}$ we immediately get (a).

If we put $\mathbf{x} = x \otimes \mathbb{1}$ in (iii) then from (ii) and the definition of the expectation one finds

$$E_{\omega_s} P^{s,t} x = E_{\omega_s} H^{s,t}(x \otimes \mathbb{1}) = Q^{s,t} E_{\omega_t}(x \otimes \mathbb{1}) = Q^{s,t} x.$$

This completes the proof.

These two Markov processes $\{Q^{s,t}\}$ and $\{H^{s,t}\}$ are called *marginal Markov processes* associated with the q.q.s.p. $(\{P^{s,t}\}, \omega_0)$. So, according to Theorem 8.8.2, the marginal Markov processes uniquely define a q.q.s.p.

Now we define another process $\{Z^{s,t} : \mathcal{M} \otimes \mathcal{M} \to \mathcal{M} \otimes \mathcal{M}\}$ by

$$Z^{s,t}\mathbf{x} = E_{\omega_s} H^{s,t}(\mathbf{x}) \otimes \mathbb{1}, \quad \mathbf{x} \in \mathcal{M} \otimes \mathcal{M}. \tag{8.78}$$

From (ii) one gets $Z^{s,t}\mathbf{x} = Q^{s,t} E_{\omega_t}\mathbf{x} \otimes \mathbb{1}$. In particular,

$$Z^{s,t}(x \otimes \mathbb{1}) = Q^{s,t}x \otimes \mathbb{1},$$

$$Z^{s,t}(\mathbb{1} \otimes x) = \omega_t(x)\mathbb{1} \otimes \mathbb{1}.$$

**Proposition 8.8.3** *The process $\{Z^{s,t}\}$ is a Markov process.*

*Proof* Take any $\tau - s \geq 1$, $t - \tau \geq 1$. Then using the assumption (iii) and the Markovianity of $H^{s,t}$ we have

$$\begin{aligned}
Z^{s,\tau} Z^{\tau,t}\mathbf{x} &= E_{\omega_s} H^{s,\tau}(E_{\omega_\tau} H^{\tau,t}(\mathbf{x}) \otimes \mathbb{1}) \otimes \mathbb{1} \\
&= E_{\omega_s} H^{s,\tau} H^{\tau,t}(\mathbf{x}) \otimes \mathbb{1} \\
&= E_{\omega_s} H^{s,t}(\mathbf{x}) \otimes \mathbb{1} \\
&= Z^{s,t}\mathbf{x},
\end{aligned}$$

for every $\mathbf{x} \in \mathcal{M} \otimes \mathcal{M}$, which is the assertion.

**Observation 8.8.1** *Consider a q.q.s.p. $(\{P^{s,t}\}, \omega_0)$ of type (A). Let $H^{s,t}$, $Z^{s,t}$ be the associated Markov processes. Take any $\varphi \in S^2$. Then from (8.74), taking into account the form of the expectation (see (5.8)), one concludes that*

$$\varphi(H^{s,t}\mathbf{x}) = P^{s,t}_* \varphi(E_{\omega_t}(\mathbf{x})) = P^{s,t}_* \varphi \otimes \omega_t(\mathbf{x}), \tag{8.79}$$

*for any $\mathbf{x} \in \mathcal{M} \otimes \mathcal{M}$.*

*Similarly, using (8.78), for $Z^{s,t}$ we have*

$$\sigma \otimes \psi(Z^{s,t}\mathbf{x}) = P^{s,t}_*(\sigma \otimes \omega_s) \otimes \omega_t(\mathbf{x}), \tag{8.80}$$

*for every $\sigma, \psi \in S$ and $\mathbf{x} \in \mathcal{M} \otimes \mathcal{M}$.*

From Corollary 8.6.4 and Theorem 8.8.1 we have

**Corollary 8.8.4** *Let $(\{P^{s,t}\}, \omega_0)$ be a q.q.s.p. of type (A) on $\mathcal{M}$, and $\{Q^{s,t}\}$, $\{H^{s,t}\}$ be its marginal processes. Then the following statements are equivalent*

 (i) *$(\{P^{s,t}\}, \omega_0)$ satisfies the ergodic principle;*
 (ii) *$\{Q^{s,t}\}$ satisfies the ergodic principle;*
(iii) *$\{H^{s,t}\}$ satisfies the ergodic principle;*
(iv) *$\{Z^{s,t}\}$ satisfies the ergodic principle.*

*Proof* From (8.79) for every $f, g \in S^2$ one has

$$\|H_*^{s,t}f - H_*^{s,t}g\|_1 = \|P_*^{s,t}f \otimes \omega_t - P_*^{s,t}g \otimes \omega_t\|_1$$
$$= \|P_*^{s,t}(f - g) \otimes \omega_t\|_1$$
$$= \|P_*^{s,t}f - P_*^{s,t}g\|_1. \tag{8.81}$$

Similarly, from (8.80) for every $\sigma, \varphi, \psi \in S$ one has

$$\|Z_*^{s,t}(\sigma \otimes \varphi) - Z_*^{s,t}(\sigma \otimes \psi)\|_1 = \|P_*^{s,t}(\sigma \otimes \varphi) - P_*^{s,t}(\sigma \otimes \psi)\|_1. \tag{8.82}$$

The obtained equalities (8.81) and (8.82) immediately yield the assertions.

### 8.8.2  Q.q.s.p.s of Type (B)

Now suppose that a q.q.s.p. $(\{P^{s,t}\}, \omega_0)$ has type (B).

Similarly as in the previous case (see (8.74)), we define a process $\mathfrak{h}_P^{s,t} : \mathcal{M} \otimes \mathcal{M} \to \mathcal{M} \otimes \mathcal{M}$ by

$$\mathfrak{h}_P^{s,t}\mathbf{x} = P^{s,t}E_{\omega_t}\mathbf{x}, \qquad \mathbf{x} \in \mathcal{M} \otimes \mathcal{M}. \tag{8.83}$$

Note that the defined process $\{\mathfrak{h}^{s,t}\}$ is not Markov, but it satisfies another equation. Namely, using (ii)$_B$ of Definition 8.1.1 and (8.73) we get

$$\mathfrak{h}_P^{s,t}\mathbf{x} = E_{\omega_s}P^{s,\tau} \otimes E_{\omega_s}P^{s,\tau}(P^{\tau,t}E_{\omega_t}\mathbf{x})$$
$$= Q_P^{s,\tau} \otimes Q_P^{s,\tau}(\mathfrak{h}_P^{\tau,t}\mathbf{x}),$$

where $\mathbf{x} \in \mathcal{M} \otimes \mathcal{M}$.

Note that the process $\{\mathfrak{h}_P^{s,t}\}$ has the same properties as $\{H_P^{s,t}\}$.

Similarly to Theorem 8.8.2, we can formulate the following.

**Theorem 8.8.5** *Let $\{Q^{s,t} : \mathcal{M} \to \mathcal{M}\}$ be a Markov process and $\{\mathfrak{h}^{s,t} : \mathcal{M} \otimes \mathcal{M} \to \mathcal{M} \otimes \mathcal{M}\}$ be processes with an initial state $\omega_0 \in S$ such that (i)–(iii) are satisfied,* *and one has*

$$\mathfrak{h}^{s,t} = Q^{s,\tau} \otimes Q^{s,\tau} \circ \mathfrak{h}^{\tau,t} \tag{8.84}$$

*for any $\tau - s \geq 1$, $t - \tau \geq 1$. Then by the equality $P^{s,t}x = \mathfrak{h}^{s,t}(x \otimes \mathbb{1})$ one defines a q.q.s.p. $(\{P^{s,t}\}, \omega_0)$ of type (B). Moreover, one has $Q^{s,t} = E_{\omega_s}P^{s,t}$.*

*Proof* We have to check only condition $(ii)_B$ of Definition 8.1.1. Note that the assumption (ii) implies that

$$E_{\omega_s}\mathfrak{h}^{s,t}(x \otimes \mathbb{1}) = Q^{s,t}E_{\omega_t}(\cdot \otimes \mathbb{1}) = Q^{s,t}x, \quad x \in \mathcal{M}.$$

Using this equality with (8.84) for any $\tau - s \geq 1, t - \tau \geq 1$ we have

$$
\begin{aligned}
E_{\omega_s}P^{s,\tau} \otimes E_{\omega_s}P^{s,\tau}(P^{\tau,t}x) &= E_{\omega_s}\mathfrak{h}^{s,\tau}(\cdot \otimes \mathbb{1}) \otimes E_{\omega_s}\mathfrak{h}^{s,\tau}(\cdot \otimes \mathbb{1})(\mathfrak{h}^{\tau,t}(x \otimes \mathbb{1})) \\
&= Q^{s,\tau} \otimes Q^{s,\tau}(\mathfrak{h}^{\tau,t}(x \otimes \mathbb{1})) \\
&= \mathfrak{h}^{s,t}(x \otimes \mathbb{1}) \\
&= P^{s,t}x
\end{aligned}
$$

for any $x \in \mathcal{M}$. This completes the proof.

These two processes $\{Q^{s,t}\}$ and $\{\mathfrak{h}^{s,t}\}$ are called *marginal processes* associated with the q.q.s.p. $(\{P^{s,t}\}, \omega_0)$.

Now define a process $\{z^{s,t} : \mathcal{M} \otimes \mathcal{M} \rightarrow \mathcal{M} \otimes \mathcal{M}\}$ by

$$z^{s,t}\mathbf{x} = E_{\omega_s}h^{s,t}(\mathbf{x}) \otimes \mathbb{1}, \quad \mathbf{x} \in \mathcal{M} \otimes \mathcal{M}. \tag{8.85}$$

For this process (8.79) also holds.

**Proposition 8.8.6** *The process $z^{s,t}$ is a Markov process.*

*Proof* First from Theorem 8.8.5, (8.47) and (8.65) we conclude that

$$E_{\omega_s}Q^{s,t} = E_{\omega_t}. \tag{8.86}$$

Let us take any $s, \tau, t \in \mathbb{R}_+$ with $\tau - s \geq 1, t - \tau \geq 1$. Then from (8.85) with (8.84) and (8.86) one gets

$$
\begin{aligned}
z^{s,t}\mathbf{x} &= E_{\omega_s}(Q^{s,\tau} \otimes Q^{s,\tau}(h^{\tau,t}(\mathbf{x})) \otimes \mathbb{1} \\
&= Q^{s,\tau}E_{\omega_s}Q^{s,\tau}(h^{\tau,t}(\mathbf{x})) \otimes \mathbb{1} \\
&= Q^{s,\tau}E_{\omega_\tau}(\mathfrak{h}^{\tau,t}(\mathbf{x})) \otimes \mathbb{1}.
\end{aligned}
\tag{8.87}
$$

On the other hand, using conditions (ii) and (iii) we obtain

$$z^{s,\tau} z^{\tau,t} \mathbf{x} = E_{\omega_s} h^{s,\tau} (E_{\omega_\tau} \mathfrak{h}^{\tau,t}(\mathbf{x}) \otimes \mathbb{1}) \otimes \mathbb{1}$$
$$= E_{\omega_s} h^{s,\tau} \mathfrak{h}^{\tau,t}(\mathbf{x}) \otimes \mathbb{1}$$
$$= Q^{s,\tau} E_{\omega_\tau} \mathfrak{h}^{\tau,t}(\mathbf{x}) \otimes \mathbb{1}$$

for every $\mathbf{x} \in \mathcal{M} \otimes \mathcal{M}$. This relation with (8.87) proves the assertion.

**Corollary 8.8.7** *Let* $(\{P^{s,t}\}, \omega_0)$ *be a q.q.s.p. of type (B) on* $\mathcal{M}$, *and* $\{Q^{s,t}\}$, $\{\mathfrak{h}^{s,t}\}$ *be its marginal processes. Then the following statements are equivalent:*

*(i)* $(\{P^{s,t}\}, \omega_0)$ *satisfies the ergodic principle;*
*(ii)* $\{Q^{s,t}\}$ *satisfies the ergodic principle;*
*(iii)* $\{\mathfrak{h}^{s,t}\}$ *satisfies the ergodic principle;*
*(iv)* $\{z^{s,t}\}$ *satisfies the ergodic principle.*


## 8.9   Tensor Products of q.s.p.s and q.q.s.p.s

Let $(\{p_{\alpha\beta,\gamma}^{[s,t]}\}, \lambda^{(0)})$ be a q.s.p. with a state space $E$ (here $E$ could also be countable), and $(\{P^{s,t}\}, \omega_0)$ be a q.s.p. on a von Neumann algebra $\mathcal{M}$. Now we consider a von Neumann algebra

$$\ell^\infty(E; \mathcal{M}) = \{\mathbf{x} = \{x_\alpha\}_{\alpha \in E} : x_\alpha \in \mathcal{M}, \alpha \in E, \sup \|x_\alpha\| < \infty\}.$$

Due to the equality $\ell^\infty(E \times E; \mathcal{M} \otimes \mathcal{M}) = \ell^\infty(E; \mathcal{M}) \otimes \ell^\infty(E; \mathcal{M})$, in what follows we will deal with $\ell^\infty(E \times E; \mathcal{M} \otimes \mathcal{M})$.

For each $s$ and $t$ let us define an operator $\Psi^{s,t} : \ell^\infty(E; \mathcal{M}) \to \ell^\infty(E \times E; \mathcal{M} \otimes \mathcal{M})$ by

$$(\Psi^{s,t}(\mathbf{x}))_{\alpha\beta} = \sum_{\gamma \in A} p_{\alpha\beta,\gamma}^{[s,t]} P^{s,t} x_\gamma, \tag{8.88}$$

where $\mathbf{x} = \{x_\gamma\}_{\gamma \in E} \in \ell^\infty(E; \mathcal{M})$.

Take an initial state on $\ell^\infty(E; \mathcal{M})$ defined by

$$\tilde{\omega}_0(\mathbf{x}) = \sum_{\gamma \in E} \lambda_\gamma^{(0)} \omega_0(x_\gamma). \tag{8.89}$$

**Theorem 8.9.1** *Let* $(\{p_{\alpha\beta,\gamma}^{[s,t]}\}, \lambda^{(0)})$ *be a q.s.p. with a state space $E$ of type (A) (resp. (B)) and* $(\{P^{s,t}\}, \omega_0)$ *be a q.q.s.p. of type (A) (resp. (B)) on a von Neumann algebra* $\mathcal{M}$. *Then the process* $(\{\Psi^{s,t}\}, \tilde{\omega}_0)$ *is a q.q.s.p. of type (A) (resp. (B)) on the von Neumann algebra* $\ell^\infty(E; \mathcal{M})$.

*Proof* The complete positivity of $P^{s,t}$ with [246, Proposition 4.23, p. 218] implies that $\Psi^{s,t}$ is also completely positive. Moreover, $\Psi^{s,t}$ is unital.

From (8.88) one sees that

$$\tilde{\omega}_t(\mathbf{x}) = \tilde{\omega}_0 \otimes \tilde{\omega}_0(\Psi^{0,t}\mathbf{x}) = \sum_{\gamma \in E} \lambda_\gamma^{(t)} \omega_t(x_\gamma). \qquad (8.90)$$

Let us note that for $\{x_{\alpha\beta}\} \in \ell^\infty(E \times E; \mathcal{M} \otimes \mathcal{M})$ from (8.89) one finds that

$$E_{\tilde{\omega}_t}(\{x_{\alpha\beta}\}) = \left\{ \sum_{b \in E} \lambda_\beta^{(t)} E_{\omega_t}(x_{\alpha\beta}) \right\}. \qquad (8.91)$$

Let us check (ii)$_A$ of Definition 8.1.1. Take arbitrary numbers $s, \tau, t \in \mathbb{R}_+$ such that $\tau - s \geq 1$, $t - \tau \geq 1$, then using (8.89) and (8.91) we have

$$(\Psi^{s,t}(\mathbf{x}))_{\alpha\beta} = \sum_\gamma p_{\alpha\beta,\gamma}^{[s,t]} P^{s,t} x_\gamma$$

$$= \sum_\gamma \left( \sum_{u,v} p_{\alpha\beta,u}^{[s,\tau]} p_{uv,\gamma}^{[\tau,t]} \lambda_v^{(\tau)} \right) P^{s,\tau} E_{\omega_\tau}(P^{\tau,t} x_\gamma)$$

$$= \sum_u p_{\alpha\beta,u}^{[s,\tau]} P^{s,t} \left( \sum_v \lambda_v^{(\tau)} E_{\omega_\tau} \left( \sum_\gamma p_{uv,\gamma}^{[\tau,t]} P^{\tau,t} x_\gamma \right) \right)$$

$$= \sum_u p_{\alpha\beta,u}^{[s,\tau]} P^{s,\tau} \left( \sum_v \lambda_v^{(\tau)} E_{\omega_\tau}((\Psi^{\tau,t}(\mathbf{x}))_{uv}) \right)$$

$$= \sum_u p_{\alpha\beta,u}^{[s,\tau]} P^{s,\tau} \left( \sum_v \lambda_v^{(\tau)} E_{\omega_\tau}((\Psi^{\tau,t}(\mathbf{x}))_{uv}) \right)$$

$$= \Psi^{s,\tau}(E_{\tilde{\omega}_\tau} \Psi^{\tau,t}(\mathbf{x}))_{\alpha\beta}, \quad \alpha, \beta \in E,$$

which proves (ii)$_A$. Therefore, $\Psi^{\tau,t}$ is a q.q.s.p. of type (A). In a similar manner one can prove the result for type (B) q.q.s.p.s. This completes the proof.

The defined q.q.s.p. $(\{\Psi^{s,t}\}, \tilde{\omega}_0)$ is the *tensor product* of the q.s.p. $(\{p_{\alpha\beta,\gamma}^{[s,t]}\}, \lambda^{(0)})$ and the q.s.p. $(\{P^{s,t}\}, \omega_0)$. Note that this theorem allows us to construct a lot of non-trivial examples of q.q.s.p.s.

*Remark 8.9.1* We note that the provided construction will still be valid if one replaces a q.s.p. $\{p_{\alpha\beta,\gamma}^{[s,t]}\}$ by a more general q.s.p. $\{P(s, x, y, t, A)\}$ given on a measurable space $(E, \mathcal{F})$.

Now let us find the marginal Markov processes associated with the constructed q.q.s.p. (8.88).

Let us first of all write down the Markov processes associated with the q.s.p.s $(\{p_{\alpha\beta,\gamma}^{[s,t]}\}, \lambda^{(0)})$ and $(\{P^{s,t}\}, \omega_0)$. According to (8.72) one has

$$q_{\alpha,\gamma}^{s,t} = \sum_{\beta} p_{\alpha\beta,\gamma}^{[s,t]} \lambda^{(s)}, \quad Q_P^{s,t} = E_{\omega_s} P^{s,t}. \tag{8.92}$$

Take $\mathbf{x} = \{x_\gamma\} \in \ell^\infty(E; \mathcal{M})$. Then the Markov process associated with $\{\Psi^{s,t}\}$ is defined as follows (see (8.72))

$$\begin{aligned}
(Q_\Psi^{s,t}\mathbf{x})_\alpha &= E_{\tilde{\omega}_s}((\Psi^{s,t}\mathbf{x})_{\alpha\beta}) = E_{\tilde{\omega}_s}\left(\sum_\gamma p_{\alpha\beta,\gamma}^{[s,t]} P^{s,t} x_\gamma\right) \\
&= \sum_\beta \lambda_\beta^{(s)} E_{\omega_s}\left(\sum_\gamma p_{\alpha\beta,\gamma}^{[s,t]} P^{s,t} x_\gamma\right) \\
&= \sum_\gamma \sum_\beta p_{\alpha\beta,\gamma}^{[s,t]} \lambda_\beta^{(s)} E_{\omega_s}(P^{s,t} x_\gamma) \\
&= \sum_\gamma q_{\alpha,\gamma}^{s,t} Q_P^{s,t}(x_\gamma).
\end{aligned} \tag{8.93}$$

Here we have used (8.91) and (8.92).

Note that the matrix $\{q_{\alpha\gamma}^{s,t}\}$ defines an operator $\mathbf{q}^{s,t} : \ell^\infty(E) \to \ell^\infty(E)$ given by

$$\mathbf{q}^{s,t}(\{b_\gamma\}) = \left\{\sum_\gamma q_{\alpha\gamma}^{s,t} b_\gamma\right\}. \tag{8.94}$$

Then from (8.93) one finds that the operator $Q_\Psi^{s,t}$ can be viewed on $\ell^\infty(E) \otimes \mathcal{M}$ as follows[1]

$$Q_\Psi^{s,t} = \mathbf{q}^{s,t} \otimes Q_P^{s,t}. \tag{8.95}$$

From (8.74) one finds the Markov process corresponding to the q.s.p. by

$$h_{(uv),(\alpha\beta)}^{s,t} = p_{uv,\alpha}^{[s,t]} \lambda_\beta^{(t)}. \tag{8.96}$$

By means of (8.74) let us find $H_\Psi^{s,t}$. Take $\{x_{\alpha\beta}\} \in \ell^\infty(E \times E; \mathcal{M} \otimes \mathcal{M})$ then

$$\left(H_\Psi^{s,t}(\{x_{\alpha\beta}\})\right)_{uv} = (\Psi^{s,t} E_{\tilde{\omega}_t}(\{x_{\alpha\beta}\}))_{uv} = \sum_\beta \lambda_\beta^{(t)} \Psi^{s,t}\left(\{E_{\omega_t}(x_{\alpha\beta})\}\right)_{uv}$$

---

[1]Note that here $\ell^\infty(E) \otimes \mathcal{M}$ is identified with $\ell^\infty(E; \mathcal{M})$.

$$= \sum_{\beta} \lambda_{\beta}^{(t)} \sum_{\alpha} p_{uv,\alpha}^{[s,t]} P^{s,t} E_{\omega_t}(x_{\alpha\beta})$$

$$= \sum_{\alpha,\beta} p_{uv,\alpha}^{[s,t]} \lambda_{\beta}^{(t)} P^{s,t} E_{\omega_t}(x_{\alpha\beta})$$

$$= \sum_{\alpha,\beta} h_{(uv),(\alpha\beta)}^{s,t} H_P^{s,t}(x_{\alpha\beta}). \tag{8.97}$$

Now we are interested in the ergodic principle for the constructed q.q.s.p. One has the following.

**Theorem 8.9.2** *Let* $(\{p_{\alpha\beta,\gamma}^{[s,t]}\}, \lambda^{(0)})$ *be a q.s.p. with a state space* $E$ *and* $(\{P^{s,t}\}, \omega_0)$ *be a q.q.s.p. on a von Neumann algebra* $\mathcal{M}$. *Then the q.q.s.p* $(\{\Psi^{s,t}, \tilde{\omega}_0)$ *given by* (8.88) *satisfies the ergodic principle if and only if the given q.s.p. and q.q.s.p. satisfy that principle.*

*Proof* "If" part. Let $(\{\Psi^{s,t}\}, \tilde{\omega}_0)$ satisfy the ergodic principle. By choosing an element $\mathbf{x}_\gamma$ given by

$$\mathbf{x}_\gamma = (\underbrace{0, \dots, 0}_{\gamma}, \mathbb{1}, 0, \dots)$$

we have $(\Psi^{s,t}(\mathbf{x}_\gamma))_{\alpha\beta} = p_{\alpha\beta,\gamma}^{[s,t]}$. Therefore one concludes that $\{p_{\alpha\beta,\gamma}^{[s,t]}\}$ satisfies the ergodic principle. Similarly, by taking an element $\mathbf{x} = (x, \dots, x, \dots)$ one finds $(\Psi^{s,t}(\mathbf{x}))_{\alpha\beta} = P^{s,t}(x)$, which means that $\{P^{s,t}\}$ satisfies the ergodic principle.

"Only if" part. Now assume that the given q.s.p. and q.q.s.p. satisfy the ergodic principle. To prove the assertion, by Corollaries 8.8.4 and 8.8.7 it is enough to show that the associated Markov process $Q_\psi^{s,t}$ satisfies the ergodic principle.

Now let us note that the pre-dual to $\ell^\infty(E; \mathcal{M})$ is

$$\ell^1(E; \mathcal{M}_*) = \{\mathbf{x} = \{x_\alpha\}_{\alpha \in E} : x_\alpha \in \mathcal{M}_*, \ \alpha \in E, \ \sum_{\alpha \in E} \|x_\alpha\|_1 < \infty\},$$

which is isometrically isomorphic to $\ell^1(E) \otimes \mathcal{M}_*$.

Let us consider the following set

$$\mathscr{G}_+ = \left\{ \mathbf{f} \in \ell^1(E)_+ \otimes \mathcal{M}_{*,+} : \mathbf{f} = \sum_{k=1}^n \mu_k \otimes \varphi_k, \ \{\mu_k\} \subset \ell^1(E)_+, \{\varphi_k\} \subset \mathcal{M}_{*,+}. n \in \mathbb{N} \right\}.$$

Then the set $\mathscr{G} := \mathscr{G}_+ - \mathscr{G}_+$ is an $\mathbb{R}$-linear subspace of $\ell^1(E)_{sa} \otimes \mathcal{M}_{*,sa}$ and dense in it (in the norm sense). Then the set $\mathscr{F}_0(\mathscr{G})$ (see (8.23)) is dense in $\mathscr{F}_0(\ell^1(E; \mathcal{M}_*))$. Therefore, to establish the ergodic principle it is enough to prove

$$\|Q_{\Psi,*}^{s,t} \mathbf{f}\|_1 \to 0 \quad \text{as} \quad t \to \infty$$

for every $\mathbf{f} \in \mathscr{F}_0(\mathscr{G})$ (see Theorem 8.3.2).

One can see that the set $\mathscr{F}_0(\mathscr{G})$ coincides with

$$\mathscr{F}_0 = \{\mathbf{f} \in \mathscr{G} : \mathbf{f} = \mathbf{f}_+ - \mathbf{f}_-, \ \mathbf{f}_+, \mathbf{f}_- \in \mathscr{G}_+, \|\mathbf{f}_+\|_1 = \|\mathbf{f}_-\|_1\}. \tag{8.98}$$

Therefore, we will deal with $\mathscr{F}_0$.

So, let us take an arbitrary $\mathbf{f} \in \mathscr{F}_0$. Then one has $\mathbf{f} = \mathbf{f}_+ - \mathbf{f}_-$ with

$$\mathbf{f}_+ = \sum_{i=1}^{N} \mu_i \otimes \varphi_i, \quad \mathbf{f}_- = \sum_{j=1}^{M} \lambda_j \otimes \psi_j \tag{8.99}$$

where $N, M \in \mathbb{N}$.

Due to the positivity of the elements $\mathbf{f}_+, \mathbf{f}_-$ we have

$$\|\mathbf{f}_+\|_1 = \sum_{i=1}^{N} \|\mu_i\|_1 \|\varphi_i\|, \quad \|\mathbf{f}_-\|_1 = \sum_{j=1}^{M} \|\lambda_j\|_1 \|\psi_j\|_1.$$

Since $\|\mathbf{f}_+\|_1 = \|\mathbf{f}_-\|_1$, we define

$$\gamma_0 := \sum_{i=1}^{N} \|\mu_i\|_1 \|\varphi_i\| = \sum_{j=1}^{M} \|\lambda_j\|_1 \|\psi_j\|_1. \tag{8.100}$$

Now put

$$\alpha_i = \frac{\|\mu_i\|_1 \|\varphi_i\|_1}{\gamma_0}, \quad \tilde{\mu}_i = \frac{\mu_i}{\|\mu_i\|_1}, \quad \tilde{\varphi}_i = \frac{\varphi_i}{\|\varphi_i\|_1}, \quad i = 1, 2, \ldots, N \tag{8.101}$$

$$\beta_j = \frac{\|\lambda_j\|_1 \|\psi_j\|_1}{\gamma_0}, \quad \tilde{\lambda}_j = \frac{\lambda_j}{\|\lambda_j\|_1}, \quad \tilde{\psi}_j = \frac{\psi_j}{\|\psi_j\|_1}, \quad j = 1, 2, \ldots, M. \tag{8.102}$$

Then taking into account (8.100) from (8.101) and (8.102) one gets

$$\begin{aligned}
\mathbf{f} &= \gamma_0 \left( \sum_{i=1}^{N} \alpha_i \tilde{\mu}_i \otimes \tilde{\varphi}_i - \sum_{j=1}^{M} \beta_j \tilde{\lambda}_j \otimes \tilde{\psi}_j \right) \\
&= \gamma_0 \left( \sum_{i,j=1}^{N,M} \alpha_i \beta_j \tilde{\mu}_i \otimes \tilde{\varphi}_i - \sum_{i,j=1}^{N,M} \alpha_i \beta_j \tilde{\lambda}_j \otimes \tilde{\psi}_j \right) \\
&= \gamma_0 \sum_{i,j=1}^{N,M} \alpha_i \beta_j \left( \tilde{\mu}_i \otimes \tilde{\varphi}_i - \tilde{\lambda}_j \otimes \tilde{\psi}_j \right). \tag{8.103}
\end{aligned}$$

Note that from (8.94) one finds that the conjugate operator to $\mathbf{q}^{s,t}$ is what we denoted by $\mathbf{q}_*^{s,t}$, which is defined as the transpose of the matrix $(\mathbf{q}_{\alpha\gamma}^{s,t})$ and maps

$\ell^1(E)$ to itself. Then from (8.95) we conclude that the conjugate operator to $Q_\psi^{s,t}$ is defined by

$$Q_{\psi,*}^{s,t}(\mu \otimes \varphi) = \mathbf{q}_*^{s,t}(\mu) \otimes Q_{P,*}^{s,t}\varphi, \quad \mu \otimes \varphi \in \ell^1(E) \otimes \mathscr{M}_*. \tag{8.104}$$

Here, $(Q_{P,*}^{s,t}\varphi)(x) = \varphi(Q_P^{s,t}x), x \in \mathscr{M}$.

According to the condition, the ergodic principle for a q.s.p. and q.q.s.p. means

$$\|\mathbf{q}_*^{s,t}(\tilde{\mu}_i) - \mathbf{q}_*^{s,t}(\tilde{\lambda}_j)\|_1 \to 0, \quad \|Q_{P,*}^{s,t}(\tilde{\varphi}_i) - Q_{P,*}^{s,t}(\tilde{\psi}_j)\|_1 \to 0 \ \text{as } t \to \infty, \tag{8.105}$$

for every $i = 1, \ldots, N, j = 1, \ldots, M$.

Now from (8.103), (8.104) and (8.105) we obtain

$$\|Q_{\psi,*}^{s,t}(\mathbf{f})\|_1 \le \gamma_0 \sum_{i,j=1}^{N,M} \alpha_i \beta_j \|Q_{\psi,*}^{s,t}(\tilde{\mu}_i \otimes \tilde{\varphi}_i - \tilde{\lambda}_j \otimes \tilde{\psi}_j)\|$$

$$\le \gamma_0 \sum_{i,j=1}^{N,M} \alpha_i \beta_j \Big( \|(\mathbf{q}_*^{s,t}(\tilde{\mu}_i) - \mathbf{q}_*^{s,t}(\tilde{\lambda}_j)) \otimes Q_{P,*}^{s,t}\tilde{\varphi}_i\|_1$$

$$+ \|\mathbf{q}_*^{s,t}(\tilde{\lambda}_j) \otimes (Q_{P,*}^{s,t}\tilde{\varphi}_i - Q_{P,*}^{s,t}\tilde{\psi}_j)\|_1 \Big)$$

$$\le \gamma_0 \sum_{i,j=1}^{N,M} \alpha_i \beta_j \Big( \|\mathbf{q}_*^{s,t}(\tilde{\mu}_i) - \mathbf{q}_*^{s,t}(\tilde{\lambda}_j)\|_1$$

$$+ \|Q_{P,*}^{s,t}\tilde{\varphi}_i - Q_{P,*}^{s,t}\tilde{\psi}_j\|_1 \Big) \to 0, \quad t \to \infty.$$

This completes the proof.

*Remark 8.9.2* Note that these results are noncommutative analogues of the results presented in Sects. 3.5 and 3.6 of Chap. 3.


## 8.10  Comments and References

Quadratic stochastic processes describe the classical physical systems defined earlier, but they do not cover the cases at a quantum level. So, it is natural to define a concept of quantum quadratic processes. In [56–58] quantum (noncommutative) quadratic stochastic processes (q.q.s.p.s) were defined on a von Neumann algebra and certain ergodic properties were studied (see Sect. 8.1). The results of Sect. 8.2 were published in [84]. The results related to the ergodic principle have been published in [57, 60, 61, 169]. The regularity of q.q.s.p.s has been investigated in [59, 158, 168]. The results of Sects. 8.4–8.8 are taken from [60, 165, 171, 173]. The content of the last Sect. 8.9 is new. These results are noncommutative analogues of the results presented in Chap. 3.

Ergodic type theorems for q.q.s.p.s have been investigated in [161, 162, 166, 167, 170].

# References

1. Aaronson, J., Lin, M., Weiss, B.: Mixing properties of Markov operators and ergodic transformations, and ergodicity of Cartesian products. Isr. J. Math. **33**(3–4), 198–224 (1979)
2. Accardi, L.: On the noncommutative Markov property. Funct. Anal. Appl. **9**, 1–8 (1975)
3. Accardi, L.: Nonrelativistic quantum mechanics as a noncommutative Markov process. Adv. Math. **20**, 329–366 (1976)
4. Accardi, L., Ceccini, C.: Conditional expectation in von Neumann algebras and a theorem of Takesaki. J. Funct. Anal. **45**, 245–273 (1982)
5. Accardi, L., Chruscinski, D., Kossakowski, A., Matsuoka, T., Ohya, M.: On classical and quantum liftings. Open Sys. Inform. Dyn. **17**, 361–387 (2010)
6. Accardi, L., Fidaleo, F.: Quantum Markov fields. Inf. Dim. Anal. Quantum Probab. Related Topics* **6**, 123–138 (2003)
7. Accardi, L., Fidaleo, F.: Non homogeneous quantum Markov states and quantum Markov fields. J. Funct. Anal. **200**, 324–347 (2003)
8. Accardi, L., Fidaleo, F., Mukhamedov, F.: Markov states and chains on the CAR algebra. Inf. Dim. Analysis, Quantum Probab. Related Topics **10**, 165–183 (2007)
9. Accardi, L., Frigerio, A.: Markovian cocycles. Proc. Royal Irish Acad. **83A**, 251–263 (1983)
10. Accardi, L., Mukhamedov, F., Saburov, M.: On quantum Markov chains on Cayley tree I: Uniqueness of the associated chain with XY-model on the Cayley tree of order two. Inf. Dim. Anal. Quantum Probab. Related Topics **14**, 443–463 (2011)
11. Accardi, L., Mukhamedov, F., Saburov, M.: On quantum Markov chains on Cayley tree II: Phase transitions for the associated chain with XY-model on the Cayley tree of order three. Ann. Henri Poincare **12**, 1109–1144 (2011)
12. Accardi, L., Ohno, H., Mukhamedov, F.: Quantum Markov fields on graphs. Inf. Dim. Analysis, Quantum Probab. Related Topics **13**, 165–189 (2010)
13. Balibrea, F., Guirao, J.L., Lampart, M., Llibre, J.: Dynamics of a Lotka-Volterra map. Fundamenta Math. **191**, 265–279 (2006)
14. Bartoszek, W., Brown, T.: On Frobenius-Perron operators which overlap supports. Bull. Pol. Acad. Sci. Math. **45**, 17–24 (1997)
15. Bartoszek, W., Pulka, M.: On mixing in the class of quadratic stochastic operators. Nonlinear Anal. Theory Methods **86**, 95–113 (2013)
16. Bartoszek, K., Pulka, M.: Asymptotic properties of quadratic stochastic operators acting on the $L^1$-space. Nonlinear Anal. Theory Methods **114**, 26–39 (2015)
17. Bengtsson, I., Zyczkowski, K.: Geometry of Quantum States. Cambridge University Press, Cambridge (2006)

© Springer International Publishing Switzerland 2015

F. Mukhamedov, N. Ganikhodjaev, *Quantum Quadratic Operators and Processes*,
Lecture Notes in Mathematics 2133, DOI 10.1007/978-3-319-22837-2

18. Bernstein, S.N.: The solution of a mathematical problem related to the theory of heredity. Uchn. Zapiski NI Kaf. Ukr. Otd. Mat. (1), 83–115 (1924) (Russian)
19. Bhatt, S.J.: Stinespring representability and Kadison's Schwarz inequality in non-unital Banach star algebras and applications. Proc. Indian Acad. Sci. (Math. Sci.) **108**, 283–303 (1998)
20. Bhatia, R.: Positive Definite Matrices. Princeton University Press, Princeton (2009)
21. Bhatia, R., Davies, C.: More operator versions of the Schwarz inequality. Commun. Math. Phys. **215**, 239–244 (2000)
22. Bhatia, R., Sharma, R.: Some inequalities for positive linear maps. Linear Alg. Appl. **436**, 1562–1571 (2012)
23. Blath, J., Jamilov, U., Scheutzow, M.: $(G, \mu)$-quadratic stochastic operators. J. Differ. Eqs. Appl. **20**, 1258–1267 (2014)
24. Boltzmann, L.: Selected Papers. Nauka, Moscow (1984, Russian)
25. Bratteli, O., Robinson, D.W.: Operator Algebras and Quantum Statistical Mechanics I. Springer, New York, Heidelberg, Berlin (1979)
26. Choi, M.-D.: Completely positive maps on complex matrices. Linear Alg. Appl. **10**, 285–290 (1975)
27. Chruscinski, D.: Quantum-correlation breaking channels, quantum conditional probability and Perron–Frobenius theory. Phys. Lett. A **377**, 606–611 (2013)
28. Chruscinski, D.: A class of symmetric Bell diagonal entanglement witnesses – a geometric perspective. J. Phys. A. Math. Theor. **47**, 424033 (2014)
29. Chruscinski, D., Kossakowski, A.: Geometry of quantum states: New construction of positive maps. Phys. Lett. A **373**, 2301–2305 (2009)
30. Chruscinski, D., Kossakowski, A.: Spectral conditions for positive maps. Commun. Math. Phys. **290**, 1051–1064 (2009)
31. Chruscinski, D., Sarbicki, G.: Exposed positive maps in $M_4(C)$. Open Sys. Inform. Dyn. **19**, 1250017 (2012)
32. Cohn, H.: On a paper by Doeblin on non-homogeneous Markov chains. Adv. Appl. Probab. **13**, 388–401 (1981)
33. Cornfeld, I.P., Fomin, S.V., Sinai, Ya.G.: Ergodic theory. In: Grundlehren Math. Wiss., vol. 245. Springer, Berlin (1982)
34. Dahlberg, C.: Mathematical Methods for Population Genetics. Interscience Publishing, New York (1948)
35. Devaney, R.L.: An Introduction to Chaotic Dynamical System. Westview Press, Boulder (2003)
36. Dobrushin, R.L.: Central limit theorem for nonstationary Markov chains. I, II. Theory Probab. Appl. **1**, 65–80; 329–383 (1956)
37. Dohtani, A.: Occurrence of chaos in higher-dimensional discrete-time systems. SIAM J. Appl. Math. **52**, 1707–1721 (1992)
38. Dzhurabayev, A.M.: Toplogical calssification of fixed and periodic points of quadratic stochastic operators. Uzbek. Math. J. (5–6), 12–21 (2000)
39. Eemel'yanov, E.Ya.: Non-spectral asymptotic analysis of one-parameter operator semigroups, In: Operator Theory: Advances and Applications, vol. 173. Birkhauser Verlag, Basel (2007)
40. Elsgolz, L.E., Norkin, S.B.: Introduction to the Theory of Differential Equations with a Deviating Argument. Nauka, Moscow (1971)
41. Fannes, M., Nachtergaele, B., Werner, R.F.: Finitely correlated states on quantum spin chains. Commun. Math. Phys. **144**, 443–490 (1992)
42. Feller, W.: Diffusion processes in genetics. Proc. Second Berkeley Syrup. Math. Star. Prob. 227–246 (1951)
43. Fisher, M.E., Goh, B.S.: Stability in a class of discrete-time models of interacting populations. J. Math. Biol. **4**, 265–274 (1977)
44. Fisher, R.A.: The Genetical Theory of Natural Selection. Clarendon Press, Oxford (1930)
45. Franz, U., Skalski, A.: On ergodic properties of convolution operators associated with compact quantum groups. Colloq. Math. **113**, 13–23 (2008)

46. Ganikhodzhaev (Ganikhodjaev), N.N.: On stochastic processes generated by quadratic operators. J. Theor. Prob. **4**, 639–653 (1991)
47. Ganikhodjaev, N.N.: An application of the theory of Gibbs distributions to mathematical genetics. Dokl. Math. **61**, 321–323 (2000)
48. Ganikhodjaev, N.N.: The random models of heredity in random environment. Dokl. Acad. Nauk RUz No. (12), 6–8 (2001)
49. Ganikhodjaev, N.N.: Lattice gas and thermodynamics in models of heredity. Inter. J. Mod. Phys. Conf. Ser. **9**, 157–162 (2012)
50. Ganikhodjaev, N.N., Akin, H., Mukhamedov, F.M.: On the ergodic principle for Markov and quadratic stochastic processes and its relations. Linear Alg. Appl. **416**, 730–741 (2006)
51. Ganikhodzhaev, N.N., Azizova, S.R.: On nonhomogeneous quadratic stochastic processes. Dokl. Akad. Nauk. UzSSR (4), 3–5 (1990) (Russian)
52. Ganikhodzhaev, N.N., Ganikhodzhaev, R.N., Jamilov, U.: Quadratic stochastic operators and zero-sum game dynamics. Ergod. Theory Dyn. Syst. **35**, 1443–1473 (2015)
53. Ganikhodjaev, N.N., Hamza, N.Z.A.: On Poisson nonlinear transformations. Sci. World J. **2014** (2014). Article ID 832861
54. Ganikhodjaev, N.N., Jamilov, U., Mukhitdinov, R.: On non-ergodic transformations on $S^3$. J. Phys. Conf. Ser. **435**, 012005 (2013)
55. Ganikhodjaev, N.N., Jamilov, U., Mukhitdinov, R.: Nonergodic quadratic operators for a two-sex population. Ukr. Math. J. **65**, 1282–1291 (2014)
56. Ganikhodjaev, N.N., Mukhamedov, F.M.: On quantum quadratic stochastic processes. Dokl. Akad. Nauk. Rep. of Uzb. (3), 13–16 (1997) (Russian)
57. Ganikhodzhaev, N.N., Mukhamedov, F.M.: On quantum quadratic stochastic processes and ergodic theorems for such processes. Uzbek. Math. J. (3), 8–20 (1997) (Russian)
58. Ganikhodjaev, N.N., Mukhamedov, F.M.: Ergodic properties of quantum quadratic stochastic processes defined on von Neumann algebras. Russian Math. Surv. **53**, 1350–1351 (1998)
59. Ganikhodjaev, N.N., Mukhamedov, F.M.: Regularity conditions for quantum quadratic stochastic processes. Dokl. Math. **59**, 226–228 (1999)
60. Ganikhodzhaev, N.N., Mukhamedov, F.M.: On the ergodic properties of discrete quadratic stochastic processes defined on von Neumann algebras. Izvestiya Math. **64**, 873–890 (2000)
61. Ganikhodjaev, N.N., Mukhamedov, F.M., Rozikov, U.A.: Analytic methods in the theory of quantum quadratic stochastic processes. Uzbek. Math. J. (2), 18–23 (2000)
62. Ganikhodzhaev, N.N., Mukhitdinov, R.T.: On a class of measures corresponding to quadratic operators. Dokl. Akad. Nauk Rep. Uzb. (3), 3–6 (1995) (Russian)
63. Ganikhodjaev, N.N., Mukhitdinov, R.T.: Extreme points of a set of quadratic operators on the simplices $S^1$ and $S^2$. Uzbek. Math. J. (3), 35–43 (1999) (Russian)
64. Ganikhodzhaev, N.N., Mukhitdinov, R.T.: On a class of non-Volterra quadratic operators. Uzbek. Math. J. (3–4), 9–12 (2003) (Russian)
65. Ganikhodjaev, N.N., Saburov, M.: On rare mutation, chaos and Darwin's theory. Revel. Sci. **4**, 37–44 (2014)
66. Ganikhodzhaev, N.N., Saburov, M., Jamilov, U.: Mendelian and non-Mendelian quadratic operators Appl. Math. Infor. Sci. **7**, 1721–1729 (2013)
67. Ganikhodjaev, N.N., Saburov, M., Navi, A.M.: Mutation and chaos in nonlinear models of heredity. Sci. World J. **2014**, (2014). Article ID 835069
68. Ganikhodjaev, N.N., Rozikov, U.A.: On quadratic stochastic operators generated by Gibbs distributions. Regul. Chaotic Dyn. **11**, 467–473 (2006)
69. Ganikhodzhaev, N.N., Zanin, D.V.: On a necessary condition for the ergodicity of quadratic operators defined on a two-dimensional simplex. Russian Math. Surv. **59**, 571–572 (2004)
70. Ganikhodzhaev, R.N.: Solution of quadratic operator equations. Dokl. Akad. Nauk UzSSR (5), 8–10 (1977) (Russian)
71. Ganikhodzhaev, R.N.: Fixed points of quadratic operators. Dokl. Akad. Nauk UzSSR (8), 3–4 (1977) (Russian)
72. Ganikhodzhaev, R.N.: A family of quadratic stochastic operators that act in $S^2$. Dokl. Akad. Nauk UzSSR (1), 3–5 (1989) (Russian)

73. Ganikhodzhaev, R.N.: Ergodic principle and regularity of a class of quadratic stochastic operators acting on finite-dimensional simplex. Uzbek. Math. J. (3), 83–87 (1992) (Russian)

74. Ganikhodzhaev, R.N.: Quadratic stochastic operators, Lyapunov functions and tournaments. Acad. Sci. Sb. Math. **76**(2), 489–506 (1993)

75. Ganikhodzhaev, R.N.: On the definition of quadratic bistochastic operators. Russian Math. Surv. **48**, 244–246 (1993)

76. Ganikhodzhaev, R.N.: A chart of fixed points and Lyapunov functions for a class of discrete dynamical systems. Math. Notes **56**, 1125–1131 (1994)

77. Ganikhodzhaev, R.N., Abdirakhmanova, R.E.: Description of quadratic automorphisms of a finite-dimensional simplex. Uzbek. Math. J. (1), 7–16 (2002) (Russian)

78. Ganikhodzhaev, R.N., Abdirakhmanova, R.E.: Fixed and periodic points of quadratic automorphisms of non-Volterra type. Uzbek. Math. J. (2), 6–13 (2002) (Russian)

79. Ganikhodzhaev, R.N., Dzhurabaev, A.M.: The set of equilibrium states of quadratic stochastic operators of type $Vs\pi$. Uzbek. Math. J. (3), 23–27 (1998) (Russian)

80. Ganikhodzhaev, R.N., Eshmamatova, D.B.: On the structure and properties of charts of fixed points of quadratic stochastic operators of Volterra type. Uzbek. Math. J. (5–6), 7–11 (2000) (Russian)

81. Ganikhodzhaev, R.N., Eshmamatova, D.B.: Quadratic automorphisms of a simplex and the asymptotic behavior of their trajectories. Vladikavkaz. Math. J. **8**(2), 12–28 (2006) (Russian)

82. Ganikhodzhaev, R.N., Eshniyazov, A.I.: Bistochastic quadratic operators. Uzbek. Math. J. (3), 29–34 (2004) (Russian)

83. Ganikhodzhaev, R.N., Karimov, A.Z.: Mappings generated by a cyclic permutation of the components of Volterra quadratic stochastic operators whose coefficients are equal in absolute magnitude. Uzbek. Math. J. (4), 16–21 (2000) (Russian)

84. Ganikhodzhaev, R., Mukhamedov, F., Rozikov, R.: Quadratic stochastic operators and processes: Results and open problems. Infin. Dimen. Anal. Quantmum Probab. Related Topics **14**, 270–335 (2011)

85. Ganikhodzhaev, R.N., Mukhamedov, F.M., Saburov, M.: $G$-decompositions of matrices and related problems I. Linear Alg. Appl. **436**, 1344–1366 (2012)

86. Ganikhodzhaev, R.N., Sarymsakov, A.T.: Nonexpansive quadratic stochastic operators. Dokl. Akad. Nauk UzSSR (8), 6–7 (1988) (Russian)

87. Ganikhodzhaev, R.N., Sarymsakov, A.T.: A simple criterion for regularity of quadratic stochastic operators. Dokl. Akad. Nauk UzSSR. (11), 5–6 (1988) (Russian)

88. Ganikhodzhaev, R.N., Sarymsakov, A.T.: On a generalization of an example of S. Ulam. Dokl. Akad. Nauk UzSSR (3), 5–7 (1989) (Russian)

89. Ganikhodzhaev, R.N., Shahidi, F.: Doubly stochastic quadratic operators and Birkhoff's problem. Linear Alg. Appl. **432**, 24–35 (2010)

90. Ganikhodzhaev, R.N., Saburov, M.: A generalized model of nonlinear Volterra type operators and Lyapunov functions. Zhurn. Sib. Federal Univ. Mat.-Fiz ser. **1**(2), 188–196 (2008)

91. Gerontidis, I.I.: Cyclic strong ergodicity in nonhomogeneous Markov systems. SIAM J. Matrix Anal. Appl. **13**, 550–566 (1992)

92. Groh, U.: The peripheral point spectrum of Schwarz operator on $C^*$-algebras. Math. Z. **176**, 311–318 (1981)

93. Groh, U.: Uniform ergodic theorems for identity preserving Schwarz maps on $W^*$-algebras. J. Operator Theory **11**, 395–404 (1984)

94. Gudder, S.: Quantum Markov chains. J. Math. Phys. **49**, 72105 (2008)

95. Gudder, S.: Document transition effect matrices and quantum Markov chains. Found. Phys. **39**, 573–592 (2009)

96. Ha, K.-C.: Entangled states with strong positive partial transpose. Phys. Rev. A **81**, 064101 (2010)

97. Hajnal, J.: Weak ergodicity in non-homogeneous Markov chains. Proc. Cambridge Phil. Soc. **54**, 233–246 (1958)

98. Herkenrath, U.: On ergodic properties of inhomogeneous Markov processes. Rev. Roumaine Math. Pures Appl. **43**, 375–392 (1998)

99. Hofbauer, J., Hutson, V., Jansen, W.: Coexistence for systems governed by difference equations of Lotka–Volterra type. J. Math. Biol. **25**, 553–570 (1987)
100. Hofbauer, J., Sigmund, K.: The Theory of Evolution and Dynamical Systems. Cambridge University Press, Cambridge (1988)
101. Horodecki, M., Horodecki, P., Horodecki, R.: Separability of mixed states: Necessary and sufficient conditions. Phys. Lett. A **223**, 1–8 (1996)
102. Horodecki, R., Horodecki, P., Horodecki, M., Horodecki, K.: Quantum entanglement. Rev. Mod. Phys. **81**, 865 (2009)
103. Iosifescu, M.: On two recent papers on ergodicity in nonhomogeneous Markov chains. Ann. Math. Stat. **43**, 1732–1736 (1972)
104. Iosifecsu, M.: Finite Markov Processes and Their Applications. Wiley, New York (1980)
105. Ipsen, I.C.F., Salee, T.M.: Ergodicity coefficients defined by vector norms. SIAM J. Matrix Anal. Appl. **32**, 153–200 (2011)
106. Isaacson, D.L., Madsen, R.W.: Markov Chains: Theory and Applications. Wiley, New York (1976)
107. Jajte, R.: Strong linit theorems in non-commutative probability. In: Lecture Notes in Mathematics, vol. 1110. Springer, Berlin-Heidelberg (1984)
108. Jamilov, U.: Quadratic stochastic operators corresponding to graphs. Lobach. J. Math. **34**, 148–151 (2013)
109. Jamilov, U., Ganikhoajaev, N.: On sufficient condition of ergodicity of Volterra quadratic stochastic operators of bisexsual population. Uzbek. Math. J. (2), 35–42 (2014)
110. Jamilov, U., Scheutzow, M., Wilke-Berenguer, M.: On the random dynamics of Volterra quadratic operators. Ergod. Theory Dyn. Syst. doi:10.1017/etds.2015.30
111. Jantzen, J.C.: Lectures on Quantum Groups. AMS, Providence (1995)
112. Jenks, R.D.: Homogeneous multidimensional differential systems for mathematical models. J. Diff. Eqs. **4**, 549–565 (1968)
113. Jenks, R.D.: Irreducible tensors and associated homogeneous nonnegative transformations. J. Diff. Eqs. **4**, 566–572 (1968)
114. Jenks, R.D.: Quadratic differential systems for interactive population models. J. Diff. Eqs. **5**, 497–514 (1969)
115. Junge, M., Xu, Q.: Noncommutative maximal ergodic theorems. J. Am. Math. Soc. **20**, 385–439 (2007)
116. Kemeny, J.G., Snell, J.L., Knapp, A.W.: Denumerable Markov Chains. Springer, New York (1976)
117. Kesten, H.: Quadratic transformations: a model for population growth, I, II. Adv. Appl.Probab. **2**(1), 1–82; **2**(2), 179–228 (1970)
118. King, C., Ruskai, M.B.: Minimal entropy of states emerging from noisy quantum channels. IEEE Trans. Info. Theory **47**, 192–209 (2001)
119. Kirzhner, V.M.: On behavior of trajectories of some class genetical systems. Dokl. Akad. Nauk SSSR **209**, 287–290 (1973) (Russian)
120. Kirzhner, V., Lyubich, Y.I.: General evolution equation and a limit theorem for genetical systems without choice. Dokl. Akad. Nauk SSSR **215**, 776–779 (1974) (Russian)
121. Kolmogorov, A.N.: On analytical methods in probability theory. Uspekhi Mat. Nauk (5), 5–51 (1938)
122. Kolmogorov, A.N., Fomin, S.V.: Introductory Real Analysis. Dover, New York (1970)
123. Kolokoltsov, V.N.: Nonlinear Markov semigroups and interacting Levy type processes. J. Stat. Phys. **126**, 585–642 (2007)
124. Kolokoltsov, V.N.: Nonlinear Markov Processes and Kinetic Equations. Cambridge University Press, New York (2010)
125. Kossakowski, A.: A class of linear positive maps in matrix algebras. Open Sys. Inform. Dyn. **10**, 213–220 (2003)
126. Krapivin, A.A.: Fixed points of quadratic operators with positive coefficients. Teor. Funkcii Funkcional. Anal. i Prilozhen **24**, 62–67 (1975) (Russian)

127. Krapivin, A.A., Ljubich, Y.I.: Estimates of Lipschitz constants for polynomial operators in a simplex. Dokl. Akad. Nauk SSSR **234**, 528–531 (1977) (Russian)

128. Krengel, U.: Ergodic Theorems. Walter de Gruyter, Berlin-New York (1985)

129. Kummerer, B.: Quantum Markov processes and applications in physics. Lecture Notes in Mathematics, Springer-Verlag, vol. 1866, pp. 259–330 (2006)

130. Kurganov, K.A.: On fixed points and behavior of trajectories of a quadratic map of four-dimensional simplex. In: Mathematical Analysis, Algebra and Geometry. Proc. Tashkent. State Univ., Fan, Tashkent, 1983, pp. 41–45. (Russian)

131. Kurganov, K.A.: On behavior of trajectories of a quadratic mapping acting four dimensional simplex, In: Mathematical Analysis and Probability Theory. Proc. Tashkent. State Univ., Fan, Tashkent, 1983, pp. 77–80 (Russian)

132. Kurganov, K.A., Ganikhodzhaev, R.N.: On limiting behavior of trajectory of Volterra type quadratic transformations of $S^4$. Dokl. Akad. Nauk UzSSR (8–9), 6–9 (1992) (Russian)

133. Li, S.-T., Li, D.-M., Qu, G.-K.: On stability and chaos of discrete population model for a single-species with harvesting. J. Harbin Univ. Sci. Tech. **6**, 021 (2006)

134. Lotka, A.J.: Undamped oscillations derived from the law of mass action. J. Am. Chem. Soc. **42**, 1595–1599 (1920)

135. Lu, Z., Wang, W.: Permanence and global attractivity for Lotka–Volterra difference systems. J. Math. Biol. **39**, 269–282 (1999)

136. Lusztig, G.: Introduction to Quantum Groups. Birkhauser, Basel (2010)

137. Lyubich, Yu.I.: Basic concepts and theorems of the evolution genetics of free populations. Russian Math. Surv. **26**, 51–116 (1971)

138. Lyubich, Yu.I.: Iterations of quadratic maps, In: Mathematical Economics and Functional Analysis, pp. 109–138. Nauka, Moscow (1974, Russian)

139. Lyubich, Yu.I.: Mathematical Structures in Population Genetics. Springer, Berlin-New York (1992)

140. Lyubich, Yu.I.: Ultranormal case of the Bernstein problem. Func. Anal. Appl. **31**(1), 60–62 (1997)

141. Madsen, R.W., Conn, P.S.: Ergodic behavior for nonnegative kernels. Ann. Prob. **1**, 995–1013 (1973)

142. Majewski, W.A., Marciniak, M.: On nonlinear Koopman's construction. Rep. Math. Phys. **40**, 501–508 (1997)

143. Majewski, W.A., Marciniak, M.: On a characterization of positive maps. J. Phys. A. Math. Gen. **34**, 5863–5874 (2001)

144. Majewski, W.A., Marciniak, M.: On the structure of positive maps between matrix algebras. In: Noncommutative Harmonic Analysis with Applications to Probability, Banach Center Publ. vol. 78, pp. 249–263. Polish Acad. Sci. Inst. Math., Warsaw (2007)

145. Majewski, W.A., Marciniak, M.: $k$-decomposability of positive maps. In: Quantum Probability and Infinite Dimensional Analysis. QP–PQ: Quantum Probab. White Noise Anal., vol. 18, pp. 362–374. World Science Publishing, Hackensack, NJ (2005)

146. Maksimov, V.M.: Necessary and sufficient conditions for the convergence of the convolution of non-identical distributions on a finite group. Teor. Verojatnost. i Primenen **13**, 295–307 (1968) (Russian)

147. Maksimov, V.M.: Cubic stochastic matrices and their probability interpretations. Theory Probab. Appl. **41**, 55–69 (1996)

148. Maličký, P.: On number of interior periodic points of a Lotka–Volterra map. Acta Univ. M. Belii, Ser. Math. **19**, 21–30 (2011)

149. Maličký, P.: Interior periodic points of a Lotka–Volterra map. J. Diff. Equ. Appl. **18**, 553–567 (2012)

150. Maruyama, T.: Stochastic Problems in Population Genetics. Springer, Berlin (1977)

151. May, R.M.: Biological populations obeying difference equations: stable points, stable cycles and chaos. Theory Biol. **51**, 511–524 (1975)

152. May, R.M.: Simple mathematical models with very complicated dynamics. Nature **261**, 459–467 (1976)

153. May, R.M., Oster, G.F.: Bifurcations and dynamic complexity in simple ecological models. Am. Nat. **110**, 573–599 (1976)
154. Menzel, M.T., Stein, P.R., Ulam, S.M.: Quadratic Transformations. Los Alamos Scientific Laboratory, Los Alamos (1959)
155. Meyliev, Kh.Zh.: Description of surjective quadratic operators and classification of the extreme points of a set of quadratic operators defined on $S^3$. Uzbek. Math. J. (3), 39–48 (1997) (Russian)
156. Meyliev, Kh.Zh., Mukhitdinov, R.T., Rozikov, U.A.: On two classes of quadratic operators that correspond to Potts models and $\lambda$-models. Uzbek. Math. J. (1), 23–28 (2001) (Russian)
157. Moran, P.A.P.: Some remarks on animal population dynamics. Biometrics **6**, 250–258 (1950)
158. Mukhamedov, F.M.: Ergodic properties of conjugate quadratic operators. Uzbek. Math. J. (1), 71–79 (1998) (Russian)
159. Mukhamedov, F.M.: On compactness of some sets of positive maps on von Neumann algebras. Methods Funct. Anal. Topol. **5**, 26–34 (1999)
160. Mukhamedov, F.M.: Weighted ergodic theorems for finite dimensional dynamical systems. Uzbek. Math. J. (2), 48–53 (1999) (Russian)
161. Mukhamedov, F.M.: On uniform ergodic theorem for quadratic processes on $C^*$-algebras. Sbornik: Math. **191**, 1891–1903 (2000)
162. Mukhamedov, F.M.: On the Blum–Hanson theorem for quantum quadratic processes. Math. Notes **67**, 81–86 (2000)
163. Mukhamedov, F.M.: On infinite-dimensional quadratic Volterra operators. Russian Math. Surv. **55**, 1161–1162 (2000)
164. Mukhamedov, F.M.: On the compactness of a set of quadratic operators defined on a von Neumann algebra. Uzbek. Math. J. (3), 21–25 (2000) (Russian)
165. Mukhamedov, F.M.: Expansion of quantum quadratic stochastic prcesses. Dokl. Math. **61**, 195–197 (2000)
166. Mukhamedov, F.M.: On a limit theorem for quantum quadratic processes. Dokl. Natl. Acad. Ukraine, (11), 25–27 (2000) (Russian)
167. Mukhamedov, F.M.: On ergodic properties of discrete quadratic dynamical system on $C^*$-algebras. Methods Funct. Anal. Topol. **7**(1), 63–75 (2001)
168. Mukhamedov, F.M.: On a regularity condition for quantum quadratic stochastic processes. Ukrainian Math. J. **53**, 1657–1672 (2001)
169. Mukhamedov, F.M.: On the ergodic principle for Markov processes associated with quantum quadratic stochastic processes. Russian Math. Surv. **57**, 1236–1237 (2002)
170. Mukhamedov, F.M.: An individual ergodic theorem on time subsequences for quantum quadratic dynamical systems. Uzbek. Math. J. (2), 46–50 (2002) (Russian)
171. Mukhamedov, F.M.: On the decomposition of quantum quadratic stochastic processes into layer-Markov processes defined on von Neumann algebras. Izv. Math. **68**, 1009–1024 (2004)
172. Mukhamedov, F.: Dynamics of quantum quadratic stochastic operators on $M_2(\mathbb{C})$, In: Darus, M., Owa, S. (eds.) Proc. Inter. Symp. New Developments of Geometric Function Theory and its Appl., pp. 425–430, Kuala Lumpur, 10–13 November 2008. National University, Malaysia (2008)
173. Mukhamedov, F.: On marginal markov processes of quantum quadratic stochastic processes. In: Barhoumi, A., Ouerdiane, H. (eds.) Quantum Probability and White Noise Analysis, vol. 25, pp. 203–215. Proceedings of the 29th Conference Hammamet, Tunis 13–18 October 2008. World Scientific, Singapore (2010)
174. Mukhamedov, F.: On $L_1$-Weak ergodicity of nonhomogeneous discrete Markov processes and its applications. Rev. Mat. Comput. **26**, 799–813 (2013)
175. Mukhamedov, F.: Dobrushin ergodicity coefficient and ergodicity of noncommutative Markov chains. J. Math. Anal. Appl. **408**, 364–373 (2013)
176. Mukhamedov, F.: Weak ergodicity of nonhomogeneous Markov chains on noncommutative $L^1$-spaces. Banach J. Math. Anal. **7**, 53–73 (2013)
177. Mukhamedov, F.: On $L_1$-Weak Ergodicity of nonhomogeneous continuous-time Markov processes. Bull. Iran. Math. Soc. **40**, 1227–1242 (2014)

178. Mukhamedov, F.: On pure quasi-quantum quadratic operators of $M_2(\mathbb{C})$ II (submitted)
179. Mukhamedov, F., Abduganiev, A.: On description of bistochastic Kadison–Schwarz operators on $M_2(\mathbb{C})$. Open Sys. Inform. Dyn. **17**, 245–253 (2010)
180. Mukhamedov, F., Abduganiev, A.: On Kadison–Schwarz property of quantum quadratic operators on $M_2(\mathbb{C})$. In: Accardi, L., et al. (eds.) Quantum Bio-Informatics IV, vol. 27. From Quantum Information to Bio-informatics, Tokyo University of Science, Japan 10–13 March 2010. World Scientific, Singapore (2011)
181. Mukhamedov, F., Abduganiev, A.: On Kadison–Schwarz type quantum quadratic operators on $M_2(C)$. Abst. Appl. Anal. **2013**, (2013). Article ID 278606, 9 p
182. Mukhamedov, F., Abduganiev, A.: On pure quasi quantum quadratic operators of $M_2(\mathbb{C})$. Open Sys. Inform. Dyn. **20**, 1350018 (2013)
183. Mukhamedov, F., Abduganiev, A., Mukhamedov, M.: On dynamics of quantum quadratic operators on $M_2(C)$. Proc. Inter. Conf. on Mathematical Applications in Engineering (ICMAE'10), pp. 14–18. Kuala Lumpur, 3-4 August 2010, Inter. Islamic Univ. Malaysia (2010)
184. Mukhamedov, F., Akin, H., Temir, S.: On infinite dimensional quadratic Volterra operators. J. Math. Anal. Appl. **310**, 533–556 (2005)
185. Mukhamedov, F., Akin, H., Temir, S., Abduganiev, A.: On quantum quadratic operators on $M_2(\mathbb{C})$ and their dynamics. J. Math. Anal. Appl. **376**, 641–655 (2011)
186. Mukhamedov, F., Jamal, A.H.M.: On $\xi^s$-quadratic stochastic operators in 2-dimensional simplex. Proc. the 6th IMT-GT Conf. Math., Statistics and its Applications (ICMSA2010). Kuala Lumpur, 3–4 November 2010, pp. 159–172. Universiti Tunku Abdul Rahman, Malaysia (2010)
187. Mukhamedov, F.M., Normatov, I.Kh., Rozikov, U.A.: The evolution equations for one class of finite dimensional quadratic stochastic processes. Uzbek. Math. J. (4), 41–46 (1999) (Russian)
188. Mukhamedov, F., Qaralleh, I.: On derivations of genetic algebras. J. Phys. Conf. Ser. **553**, 012004 (2014)
189. Mukhamedov, F., Qaralleh, I., Rozali, W.N.F.A.W.: On $\xi^{(a)}$-quadratic stochastic operators on 2D simplex. Sains Malaysiana **43**, 1275–1281 (2014)
190. Mukhamedov, F., Saburov, M.: On infinite dimensional Volterra type operators. J. Appl. Funct. Anal. **4**, 580–588 (2009)
191. Mukhamedov, F.M., Saburov, M.: On homotopy of Volterrian quadratic stochastic operators. Appl. Math. Inform. Sci. **4**, 47–62 (2010)
192. Mukhamedov, F., Saburov, M.: On dynamics of Lotka–Volterra type operators. Bull. Malay. Math. Sci. Soc. **37**, 59–64 (2014)
193. Mukhamedov, F., Saburov, M., Jamal, A.H.M.: On dynamics of $\xi^s$-quadratic stochastic operators. Inter. J. Modern Phys. Conf. Ser. **9**, 299–307 (2012)
194. Mukhamedov, F., Saburov, M., Qaralleh, I.: On $\xi^{(s)}$-quadratic stochastic operators on two dimensional simplex and their behavior. Abst. Appl. Anal. **2013**, (2013). Article ID 942038
195. Mukhamedov, F., Saburov, M., Qaralleh, I.: Classification of $\xi^{(s)}$-Quadratic Stochastic Operators on 2D simplex. J. Phys. Conf. Ser. **435**, 012003 (2013)
196. Mukhamedov, F., Supar, N.A.: On marginal processes of quadratic stochastic processes. Bull. Malay. Math. Sci. Soc. **38**, 1281–1296 (2015)
197. Mukhamedov, F., Supar, A., Pah, Ch.H.: On quadratic stochastic processes and related differential equations. J. Phys. Conf. Ser. **435**, 012013 (2013)
198. Mukhamedov, F., Taha, H.M.: On Volterra and orthogonality preserving quadratic stochastic operators. Miskloc Math. Notes (in press). arXiv:1401.3114
199. Nielsen, M.A., Chuang I.L.: Quantum Computation and Quantum Information. Cambridge University Press, Cambridge (2000)
200. Ohya, M., Volovich, I.: Mathematical Foundations of Quantum Information and Computation and its Applications to Nano- and Bio-Systems. Springer, New York (2011)
201. Ozawa, M.: Continuous affine functions on the space of Markov kernels. Theory Probab. Appl. **30**, 516–528 (1986)

202. Parthasarathy, K.R.: An introduction to quantum stochastic calculus. In: Monographs in Mathematics, vol. 85. Birkhhauser Verlag, Basel (1992)
203. Paulsen, V.: Completely Bounded Maps and Operator Algebras. Cambridge University Press, Cambridge (2002)
204. Plank, M., Losert, V.: Hamiltonian structures for the n-dimensional Lotka–Volterra equations. J. Math. Phys. **36**, 3520–3543 (1995)
205. Podles, P., Muller, E.: Introduction to quantum groups. Rev. Math. Phys. **10**, 511–551 (1998)
206. Pulka, M.: On the mixing property and the ergodic principle for nonhomogeneous Markov chains. Linear Alg. Appl. **434**, 1475–1488 (2011)
207. Ratner, V.A.: Mathematical theory of evolution of Mendel populations. Probl. Evolutsii (3), 151–213 (1973) (Russian)
208. Robertson, A.P., Robertson, W.J.: Topological Vector Spaces. Cambridge University Press, Cambridge (1964)
209. Robertson, A.G.: Schwarz inequalities and the decomposition of positive maps on $C^*$-algebras. Math. Proc. Camb. Philos. Soc. **94**, 291–296 (1983)
210. Roy, N.: Extreme points and $\ell_1(\Gamma)$-spaces. Proc. Am. Math. Soc. **86**, 216–218 (1982)
211. Rozikov, U.A., Nasir, S.: Separable quadratic stochastic operators. Lobachevskii J. Math. **31**, 214–220 (2010)
212. Rozikov, U.A., Shamsiddinov, N.B.: On non-Volterra quadratic stochastic operators generated by a product measure. Stochastic Anal. Appl. **27**(2), 353–362 (2009)
213. Rozikov, U.A., Zada, A.: On $\ell$-Volterra quadratic stochastic operators. Dokl. Math. **79**, 32–34 (2009)
214. Rozikov, U.A., Zada, A.: On $\ell$-Volterra quadratic stochastic operators. Inter. J. Biomath. **3**, 143–159 (2010)
215. Rozikov, U.A., Zhamilov, U.U.: On $F$-quadratic stochastic operators. Math. Notes **83**, 554–559 (2008)
216. Rozikov, U.A., Zhamilov, U.U.: On dynamics of strictly non-Volterra quadratic operators defined on the two dimensional simplex. Sbornik: Math. **200**(9), 81–94 (2009)
217. Rozikov, U.A., Zhamilov, U.U.: Volterra quadratic stochastic of a two-sex population. Ukr. Math. J. **63**, 1136–1153 (2011)
218. Ruelle, D.: Historic behavior in smooth dynamical systems. In: Broer, H.W., et al. (eds.) Global Anal. Dynamical Syst. Institute of Physics Publishing, Bristol (2001)
219. Ruskai, M.B., Szarek, S., Werner, E.: An analysis of completely positive trace-preserving maps on $M_2$. Linear Alg. Appl. **347**, 159–187 (2002)
220. Saburov, M.Kh.: On asymptotically behaviors of homogeneous quadratic systems of the differential equations of Volterra type. Uzbek. Math. J. (2), 85–93 (2006) (Russian)
221. Saburov, M.Kh.: On ergodic theorem for quadratic stochastic operators. Dokl. Acad. Nauk Rep. Uzb. (6), 8–11 (2007) (Russian)
222. Saburov, M.Kh.: Some strange properties of quadratic stochastic Volterra operators. World Appl. Sci. J. **21**, 94–97 (2013)
223. Saburov, M.: Quadratic plus linear operators which preserve pure states of quantum Ssstems: Small dimensions. J. Phys. Conf. Ser. **553**, 012003 (2014)
224. Saburov, M., Saburov, Kh.: Mathematical models of nonlinear uniform consensus. Sci. Asia **40**, 306–312 (2014)
225. Saburov, M.Kh., Shahidi, F.A.: On localization of fixed and periodic points of quadratic authomorphisms of the simplex. Uzbek. Math. J. (3), 81–87 (2007) (Russian)
226. Sakai, S.: $C^*$-algebras and $W^*$-algebras. In: Ergeb. Math. Grenzgeb. (2), vol. 60. Springer, Berlin (1971)
227. Sarymsakov, A.T.: On the trajectories of some quadratic transformations of a two-dimensional simplex. Izv. Akad. Nauk UzSSR Ser. Fiz.-Mat. Nauk (1), 34–37 (1981) (Russian)
228. Sarymsakov, A.T.: Quadratic transformations that preserve a simplex. Izv. Akad. Nauk UzSSR Ser. Fiz.-Mat. Nauk (2), 16–19 (1982) (Russian)

229. Sarymsakov, A.T.: On homogeneous second order differential equations on one-dimensional and two-dimensional simplexes. Dokl. Akad. Nauk UzSSR (6), 9–10 (1982) (Russian)

230. Sarymsakov, A.T.: Ergodic principle for quadratic stochastic processes. Izv. Akad. Nauk UzSSR, Ser. Fiz.-Mat. Nauk (3), 39–41 (1990) (Russian)

231. Sarymsakov, A.T., Ganikhodzhaev, R.N.: Asymptotic behavior of trajectories of certain quadratic transformations of a three-dimensional simplex into itself. Dokl. Akad. Nauk UzSSR (5), 7–8 (1985) (Russian)

232. Sarymsakov, T.A., Ganikhodzhaev, R.N.: The ergodic principle for quadratic stochastic operators. Izv. Izv. Akad. Nauk UzSSR, Ser. Fiz.-Mat. Nauk (6), 34–38 (1979) (Russian)

233. Sarymsakov, A.T., Ganikhodzhaev, R.N.: The ergodic principle and regularity for a class of quadratic stochastic operators that act in a finite-dimensional simplex. Uzbek. Mat. Zh. (3–4), 83–87 (1992) (Russian)

234. Sarymsakov, T.A., Ganikhodzhaev N.N.: Analytic methods in the theory of quadratic stochastic operators. Sov. Math. Dokl. **39**, 369–373 (1989)

235. Sarymsakov, T.A., Ganikhodzhaev, N.N.: Analytic methods in the theory of quadratic stochastic processes. J. Theor. Prob. **3**, 51–70 (1990)

236. Sarymsakov, T.A., Ganikhodzhaev, N.N.: On the ergodic principle for quadratic processes. Sov. Math. Dokl. **43**, 279–283 (1991)

237. Sarymsakov, T.A., Ganikhodzhaev, N.N.: Central limit theorem for quadratic chains. Uzbek. Math. J. (1), 57–64 (1991)

238. Sarymsakov, T.A., Ganikhodzhaev, N.N.: On some probabilistic problems in the theory of quadratic operators. In: Nonlinearity with Disorder. Springer Proceeding in Physics, Berlin, vol. 67, pp. 143–149 (1992)

239. Sarymsakov, T.A., Zimakov, N.P.: Ergodic properties of Markov operators in norm ordered spaces with a base. In: Operator Algebras and Functional Spaces, pp. 45–53. Fan, Tashkent (1985)

240. Soltan, P.M.: Quantum $SO(3)$ groups and quantum group actions on $M_2$. J. Noncommut. Geom. **4**, 1–28 (2010)

241. Stormer, E.: Positive linear maps of operator algebras. Acta Math. **110**, 233–278 (1963)

242. Stratila, S., Zsido, L.: Lectures in von Neumann Algebras. Macmillan Education, Australi (1979)

243. Svirezhev, Yu.M., Logofet, D.O.: Stability of Biological Populations. Nauka, Moscow (1978) (Russian)

244. Swirszcz, G.: On a certain map of a triangle. Fund. Math. **155**, 45–57 (1998)

245. Takens, F.: Orbits with historic behavior, or non-existence of averages. Nonlinearity **21**, T33–T36 (2008)

246. Takesaki, M.: Theory of Operator Algebras, I. Springer, Berlin–Heidelberg–New York (1979)

247. Tan, Ch.P.: On the weak ergodicity of nonhomogeneous Markov chains. Stat. Probab. Lett. **26**, 293–295 (1996)

248. Takeuchi, Y.: Global Dynamical Properties of Lotka–Volterra Systems. World Scientific, Singapore (1996)

249. Udwadia, F.E., Raju, N.: Some global properties of a pair of coupled maps: quasi-symmetry, periodicity and syncronicity. Phys. D **111**, 16–26 (1998)

250. Ulam, S.M.: Problems in Modern Mathematics. Wiley, New York (1964)

251. Vallander, S.S.: On the limit behaviour of iteration sequences of certain quadratic transformations. Sov. Math. Dokl. **13**, 123–126 (1972)

252. Volterra, V.: Lois de fluctuation de la population de plusieurs espèces coexistant dans le même milieu. Association Franc. Lyon **1926**, 96–98 (1927)

253. Woronowicz, S.L.: Compact matrix pseudogroups. Comm. Math. Phys. **111**, 613–665 (1987)

254. Zakharevich, M.I.: The behavior of trajectories and the ergodic hypothesis for quadratic mappings of a simplex. Russian Math. Surv. **33**, 207–208 (1978)

255. Zaharopol, R.: Invariant Probabilities of Markov-Feller Operatos and Their Supports. Birkhäuser Verlag, Basel (2005)
256. Zeifman, A.I.: On the weak ergodicity of nonhomogeneous continuous-time Markov chains. J. Math. Sci. **93**, 612–615 (1999)
257. Zimakov, N.P.: Finite-dimensional discrete linear stochastic accelerated-time systems and their application to quadratic stochastic dynamical systems. Math. Notes **59**, 511–517 (1996)

# Index

© Springer International Publishing Switzerland 2015
F. Mukhamedov, N. Ganikhodjaev, *Quantum Quadratic Operators and Processes*,
Lecture Notes in Mathematics 2133, DOI 10.1007/978-3-319-22837-2

# LECTURE NOTES IN MATHEMATICS  Springer

Edited by J.-M. Morel, B. Teissier; P.K. Maini

**Editorial Policy** (for the publication of monographs)

1. Lecture Notes aim to report new developments in all areas of mathematics and their applications - quickly, informally and at a high level. Mathematical texts analysing new developments in modelling and numerical simulation are welcome.

   Monograph manuscripts should be reasonably self-contained and rounded off. Thus they may, and often will, present not only results of the author but also related work by other people. They may be based on specialised lecture courses. Furthermore, the manuscripts should provide sufficient motivation, examples and applications. This clearly distinguishes Lecture Notes from journal articles or technical reports which normally are very concise. Articles intended for a journal but too long to be accepted by most journals, usually do not have this "lecture notes" character. For similar reasons it is unusual for doctoral theses to be accepted for the Lecture Notes series, though habilitation theses may be appropriate.

2. Manuscripts should be submitted either online at www.editorialmanager.com/lnm to Springer's mathematics editorial in Heidelberg, or to one of the series editors. In general, manuscripts will be sent out to 2 external referees for evaluation. If a decision cannot yet be reached on the basis of the first 2 reports, further referees may be contacted: The author will be informed of this. A final decision to publish can be made only on the basis of the complete manuscript, however a refereeing process leading to a preliminary decision can be based on a pre-final or incomplete manuscript. The strict minimum amount of material that will be considered should include a detailed outline describing the planned contents of each chapter, a bibliography and several sample chapters.

   Authors should be aware that incomplete or insufficiently close to final manuscripts almost always result in longer refereeing times and nevertheless unclear referees' recommendations, making further refereeing of a final draft necessary.

   Authors should also be aware that parallel submission of their manuscript to another publisher while under consideration for LNM will in general lead to immediate rejection.

3. Manuscripts should in general be submitted in English. Final manuscripts should contain at least 100 pages of mathematical text and should always include

   – a table of contents;
   – an informative introduction, with adequate motivation and perhaps some historical remarks: it should be accessible to a reader not intimately familiar with the topic treated;
   – a subject index: as a rule this is genuinely helpful for the reader.

   For evaluation purposes, manuscripts may be submitted in print or electronic form (print form is still preferred by most referees), in the latter case preferably as pdf- or zipped ps-files. Lecture Notes volumes are, as a rule, printed digitally from the authors' files. To ensure best results, authors are asked to use the LaTeX2e style files available from Springer's web-server at:

   ftp://ftp.springer.de/pub/tex/latex/svmonot1/ (for monographs) and
   ftp://ftp.springer.de/pub/tex/latex/svmultt1/ (for summer schools/tutorials).

Additional technical instructions, if necessary, are available on request from lnm@springer.com.

4. Careful preparation of the manuscripts will help keep production time short besides ensuring satisfactory appearance of the finished book in print and online. After acceptance of the manuscript authors will be asked to prepare the final LaTeX source files and also the corresponding dvi-, pdf- or zipped ps-file. The LaTeX source files are essential for producing the full-text online version of the book (see http://www.springerlink.com/openurl.asp?genre=journal&issn=0075-8434 for the existing online volumes of LNM). The actual production of a Lecture Notes volume takes approximately 12 weeks.

5. Authors receive a total of 50 free copies of their volume, but no royalties. They are entitled to a discount of 33.3 % on the price of Springer books purchased for their personal use, if ordering directly from Springer.

6. Commitment to publish is made by letter of intent rather than by signing a formal contract. Springer-Verlag secures the copyright for each volume. Authors are free to reuse material contained in their LNM volumes in later publications: a brief written (or e-mail) request for formal permission is sufficient.

**Addresses:**
Professor J.-M. Morel, CMLA,
École Normale Supérieure de Cachan,
61 Avenue du Président Wilson, 94235 Cachan Cedex, France
E-mail: morel@cmla.ens-cachan.fr

Professor B. Teissier, Institut Mathématique de Jussieu,
UMR 7586 du CNRS, Équipe "Géométrie et Dynamique",
175 rue du Chevaleret
75013 Paris, France
E-mail: teissier@math.jussieu.fr

*For the "Mathematical Biosciences Subseries" of LNM:*

Professor P. K. Maini, Center for Mathematical Biology,
Mathematical Institute, 24-29 St Giles,
Oxford OX1 3LP, UK
E-mail: maini@maths.ox.ac.uk

Springer, Mathematics Editorial, Tiergartenstr. 17,
69121 Heidelberg, Germany,
Tel.: +49 (6221) 4876-8259

Fax: +49 (6221) 4876-8259
E-mail: lnm@springer.com

Printed in the United States
By Bookmasters